GW00382185

THE PURE SOCIETY

THE PURE SOCIETY

From Darwin to Hitler

ANDRÉ PICHOT

Translated by David Fernbach

London • New York

This work was published with the help of the French Ministry of Culture – Centre National du Livre

This work is supported by the French Ministry for Foreign Affairs, as part of the Burgess programme headed for the French Embassy in London by the Institut Français du Royaume-Uni.

First published as *La Société pure de Darwin à Hitler*

This edition published by Verso 2009

1 3 5 7 9 10 8 6 4 2

Verso
UK: 6 Meard Street, London W1F 0EG
US: 20 Jay Street, Suite 1010, Brooklyn, NY 11201
www.versobooks.com

Verso is the imprint of New Left Books

ISBN-13: 978-1-84467-244-8

British Library Cataloguing in Publication Data
A catalogue record for this book is available from the British Library

Library of Congress Cataloging-in-Publication Data
A catalog record for this book is available from the Library of Congress

Typeset by Hewer Text UK Ltd, Edinburgh
Printed in the US by Maple Vail

CONTENTS

INTRODUCTION: A Hard Subject to Tackle

This book takes me somewhat away from my areas of specialization – epistemology and the conceptual history of biology – and tackles questions that I have not always been very familiar with, though I did make a previous incursion into the sociology of science with a small book on eugenics.[1] An exercise of this kind is not without risk, and my reason for embarking on it is that there is a definite gap between the constantly proclaimed necessity to study the relationship between biology and society, and a certain inability of sociologists and specialists in social and political history to take into account things that are well known to epistemologists and historians of science – an inability shared by lawyers and politicians (not to speak of the media). I see evidence of this in the difficulty they have in taking a position in relation to recent advances in biology and the declarations of certain biologists.

It is not that the subject matter is particularly arduous, but rather that it amounts to walking through a minefield. On the one hand, that is because of the constant need to interweave data deriving from heterogeneous disciplines, and to distinguish between genuinely scientific approaches and more or less crude analogies. On the other hand, it is because the question inevitably refers to the doctrines of Nazism, and because the relationship between those doctrines and contemporaneous biology is a taboo subject that historians avoid tackling, while the media only caricature it (with the approval of biologists, who prefer that it not be examined too closely). The social application of biology is a field governed by what remains unsaid, and even today it is best not to contravene this rule.

1 A. Pichot, *L'Eugénisme, ou les généticiens saisis par la philanthropie* (Paris: Hatier, 1995).

A good example of this is provided by public reaction to a lecture, or a journalist's comment in an interview, when it is asserted that in Nazi Germany eugenics was responsible for more deaths than anti-Semitism. The accusation of revisionism is immediately raised. More specifically, we saw this kind of reaction when, by way of reply to the public demand for the gritty details of the eugenics of Alexis Carrel, I explained that he had practically no role in this field, despite being a Pétainist. (Indeed, this leads to one's being treated as a Vichyite or a Le Pen supporter.) Yet a glance at the specialist literature shows that, far from being a marginal phenomenon, a mere side effect of the Judaeocide, eugenics was in fact internationally widespread well before Nazism, and that in Germany, where it took a particularly virulent form and was taken to its logical conclusion, it claimed a particularly large number of victims.[2]

The same literature clearly shows how a few familiar lines in Carrel's 1935 book *L'Homme, cet inconnu*, taken as proof of the biologist's involvement in eugenics,[3] were simply a commonplace of the science of the time, ideas shared by the great majority of biologists and doctors. It would be easy to produce dozens of quotations of this kind from a wide range of well-reputed authors. Carrel simply shared the dominant opinion without being himself an active player in this field, on the one hand because he was a surgeon concerned with tissue culture and not a geneticist, on the other hand because in France – even under Pétain – there was no eugenics in the strict sense of the term, but simply a particular public-health policy. The only Vichy legislation that could be termed eugenic was the law of 16 December 1942, which provided for an obligatory premarital medical examination and was retained after the war, and it is far from certain that Carrel had any particular responsibility for this.[4] In the absence of proof to the contrary, there is no known instance of Carrel's involvement in eugenic measures, and he does not even seem to

2 J. Sutter, *L'Eugénique, problème, méthodes, résultats*, Cahier no. 11 of the Institut National d'Études Démographiques (Paris: PUF, 1950); D. J. Kevles, *In the Name of Eugenics* (London: Harvard University Press, 1995); M. B. Adams (ed.), *The Wellborn Science. Eugenics in Germany, France, Brazil and Russia* (Oxford: OUP, 1990); P. Weindling and B. Massin, *L'Hygiène de la race, hygiène raciale et eugénisme médical en Allemagne*, vol. 1 (Paris: La Découverte, 1998).

3 A. Carrel, *L'Homme, cet inconnu* (Paris: Plon, 1941), pp. 359, 363 and passim.

4 A. Carol, *Histoire de l'eugénisme en France* (Paris: Seuil, 1995), p. 330.

have held extreme views on this subject. (Readers should note that I am far indeed from sharing Carrel's political ideas, but as against those who launched this discussion and have kept it going, I see no need to add to the ill repute of this individual by retailing unsupported allegations. Historical truth may well be a difficult and distant ideal, but this does not mean it should be completely neglected.)

On the question of eugenics at that time, it would be far more appropriate to quote a geneticist such as Julian Huxley, who in 1941 – when, with the full knowledge of the world, the Nazis were gassing the mentally ill – could write that eugenics would 'inevitably become part of the religion of the future'.[5] Or again Hermann J. Muller, who tried in the 1930s to convince Stalin to adopt a biological policy with a eugenic component (a positive rather than a negative eugenics, such as was then applied in the United States, Germany and the Scandinavian countries).[6] But it is inconvenient to bring up geneticists like these: Huxley was a humanistic social democrat and was appointed director of UNESCO in 1946, while Muller, a Communist, received the Nobel Prize that same year (and was a German Jew by origin, into the bargain). To the contemporary mindset, this does not fit well with the idea of eugenics going hand in hand with Hitler. It thus becomes far simpler to focus on Carrel (even if there is little to reproach him with in this field), as though his notorious support for Pétain can explain everything.

Carrel's political opinions were never in any way secret (their 'rediscovery' was media hype),[7] but until recently they were not counterposed to his scientific work. This is most likely because his Pétainism was well known, and did not present a strong enough argument create a scandal (scientists often share in the ideology of their time, however detestable), so that the accusation of eugenics was added without much concern for historical truth. Curious as it might appear, Carrel's *L'Homme, cet inconnu* is actually quite moderate, and in no way shocking when placed in a context that

5 J. Huxley, *The Uniqueness of Man* (London: Chatto & Windus, 1941), p. 34. This chapter was originally published in *The Eugenics Review*. Huxley's talk of a 'religion of the future' is taken from Galton, the inventor of eugenics.

6 H. J. Muller, *Out of the Night: A Biologist's View of the Future* (London: Gollancz, 1936).

7 See for example G. W. Corner's article on Carrel in C. C. Gillespie (ed.), *Dictionary of Scientific Biography* (New York: Scribner's, 1981), vol. 3, pp. 90–2; or R. Soupault, *Alexis Carrel* (Paris: Plon, 1952), passim.

is full of the most extreme texts. The book in fact brought Carrel the reputation of a great humanitarian – humanism in 1935 was in no way what it is today.

As a pendant to Carrel, I could have mentioned several other geneticists besides Huxley and Muller, but it was not by accident that I chose these two. Quite aside from their political opinions, which show how eugenic ideas were common to scientists of all persuasions, it is interesting to note that the horrors of Nazism did not lead them to any reassessment of those ideas. In the 1960s, Muller appealed to Huxley as a donor when he sought to create in the United States, with the backing of the multimillionaire Robert K. Graham, a 'Repository for Germinal Choice', i.e., a sperm bank of Nobel Prize winners and similar geniuses (one of the projects he had tried in vain to 'sell' to Stalin in the 1930s, when he was a Communist working in the Soviet Union). This sperm bank was indeed established in 1971, four years after Muller's death, and it is quite likely that Huxley contributed to it, at the age of eighty-four.[8]

The virtuous indignation that has been so profuse in reaction to Carrel's eugenic views, therefore, is somewhat misplaced – which in no way means approving his support for Pétain. Its main effect is to falsify the historical reality of eugenics in the first half of the twentieth century (and even a little later), by enclosing it in the narrow ambit of Nazism, Pétainism and a few ideologically driven biologists.

We may even suspect that this is not merely its main effect but also its major function. Compare, for example, the insistence with which the French media raked over the Carrel affair and the superficial treatment they accorded eugenics in this connection, or again the way they handled the 'discovery' of the eugenic policies of Swedish social democracy that persisted long after the Second World War. The press, generally so greedy for horror and massacre, refrained from conducting any serious discussion, clearly not seeking a real understanding of the premises of the question or its logical conclusion. It was rather comic to see the media discover that eugenic legislation still existed in Sweden in the 1970s (after all, this was not so long ago, and journalists are supposed to be well informed), declare themselves scandalized, and hasten to bury the affair.

The history of eugenics is indeed a delicate subject to

8 D. J. Kevles, *In the Name of Eugenics*, pp. 262–3.

tackle, and it is not easy to know in what way to approach it, given the number and diversity of pitfalls surrounding it. It is also, above all, a very embarrassing question from more than one point of view. Silence is the rule, and it is inadvisable to transgress the taboo.

Psychoanalysis claims that censorship endows things with significance, and this is an idea worth pursuing here. Not that there has been censorship on this subject by any political authority, but rather a kind of collective repression. The study of eugenics and its history clearly reveals the state of Western society in the first half of the last century and its relationship with Nazism. But this revelation is hardly 'politically correct'. The Carrel affair was so useful because it comforted established prejudice, and commentators were careful not to speak of what was truly significant in the matter (even what was truly significant in the so-called 'Carrel affair' itself, in which a number of elements, very interesting but rather embarrassing for certain glorious institutions still at work, were systematically obscured, as we shall see in Part Two).

In the case of racism – another major aspect of the relationship between biology and society – the terrain is somewhat less of a minefield, since contrary to what happened with eugenics, biologists were not so severely compromised and managed to find an escape route. But the relationship between racism and biology has nonetheless been a source of misunderstanding. This could be seen in October 1996, when, following Jean-Marie Le Pen's speeches on the inequality of human races, some eminent biologists believed it worthwhile to solemnly gather at the Musée de l'Homme in Paris and declare that human races did not exist. As if the issue involved in racism was the existence of different human races, and the problem could be resolved by proclaiming their non-existence.[9]

This declaration more or less amounted to a denial of the role of biology and biologists in racism, and it is reminiscent of an earlier one made in 1935 (the year that the Nuremberg laws were promulgated in Germany) by Julian Huxley in *We Europeans*: 'This scientific work . . . tends to demonstrate that "races"

9 See A. Pichot, 'Des biologists et des races', *La Recherche* 295, February 1997, pp. 9–10. The participants in the meeting at the Musée de l'Homme were Luca Cavalli-Sforza, Jean Dausset, François Jacob, Axel Kahn, André Langaney, Alberto Piazza and Jacques Ruffié.

– a term that our age has over-used, and that serves to justify political ambitions, economic ends, social resentments and class prejudices – have no biological existence.'[10] Huxley proposed that instead of 'race' one should speak of 'ethnic group' or 'people',[11] just as modern geneticists prefer to use the term 'population' and current fashion favours 'visually impaired' over 'blind'. The general lines of his assertion are completely comparable with those proposed today, even if they of course made no reference to genome analysis.

Huxley's purpose in this book was certainly to combat racism. But the ineffectiveness of this method has been often since been demonstrated. For example,

> What essentially fuels racism is not the assertion of the existence of races. It is certainly hard to define a race: even if the old categories that used to be taught in the textbooks – black, yellow and white races – certainly exist. To deny the existence of races is a procedure that most commonly leads those anti-racists who use it against racist arguments into confusion . . . This was why M. L. C. Dunn, as UNESCO rapporteur, wrote in June 1951: 'The anthropologist, just like the man in the street, knows perfectly well that races exist; the former because he is able to classify the varieties of the human species, the latter because he cannot doubt the evidence of his senses.'[12]

To deny the existence of races or to replace the word 'race' by a synonym, hoping to produce some effect on the question of racism, displays only stupidity and bad faith.

It is permissible to wonder, therefore, whether Huxley was not seeking above all in 1935 to acquit biologists of any responsibility for the Nazis making practical application of racial theories that were widely current in the science of the time. Huxley could himself write, in *The Uniqueness of Man* (the book in which he praised eugenics):

> The existence of marked genetic differences in physical characters (as between yellow, black, white and brown) make it *prima facie* that differences in intelligence and temperament exist also. For instance, I regard

10 J. Huxley, A. D. Haddon and A. M. Carr-Saunders, *Nous Européens* (Paris: Minuit, 1947), from the blurb of this French edition.

11 J. Huxley et al., *We Europeans* (London: Jonathan Cape, 1935).

12 F. de Fontette, *Le Racisme* (Paris: PUF, 1981), pp. 7–8.

> it as wholly probable that true negroes have a slightly lower than average intelligence than the whites or yellows.[13]

This rather contradicts what he wrote in *We Europeans* about the non-existence of human races. His critics may well note that, as its title indicates, this book is about Europeans, in particular the indigenous white population. Huxley begins, however, with a little game in which the reader is asked to attribute a nationality to each person in a series of sixteen photographs – the difficulty of this being taken as confirming the non-existence of races. It goes without saying that all the individuals in the photographs are white Europeans. One chapter (authored by A. M. Carr-Saunders) does focus on 'Europe overseas', in other words the colonial population, but the indigenous peoples of this 'overseas' are scarcely mentioned, and in a very 'politically incorrect' fashion:

> Many of the territories into which the Europeans overflowed were so sparsely peopled by their native inhabitants as to have been, for all practical purposes, empty. There may have been about a million Indians in America north of the Rio Grande, about 150,000 aborigines in Australia, and about 60,000 Maoris in New Zealand when Europeans first settled in those countries. These primitive peoples were easily pushed aside, and with the exception of the Maoris have greatly diminished in numbers. On account of the fact that, again with the partial exception of the Maoris, they have played no part in the building up of the communities which now inhabit these areas, they can be disregarded in what follows. The case is otherwise in regard to South America and in South Africa. There the native population was more dense and could not be swept aside.[14]

We can dispense with the euphemism. The North American Indians and the Australian aborigines were easy to 'sweep aside' (in other words, exterminate) and have 'greatly diminished in numbers', whereas the Indians of South America and the African blacks were too numerous to be 'swept aside' in this way. Of course, the question at issue in *We Europeans* was not the fate of indigenous peoples but rather that of racism among whites, as had been established (or rather, institutionalized) by the German anti-Semitic legislation of 1935. As for the situation of black, Indian, yellow, etc. peoples on the American continent and in the colonies

13 J. Huxley, *The Uniqueness of Man*, p. 53.
14 A. M. Carr-Saunders, in J. Huxley et al., *We Europeans*, pp. 241–2.

of Africa and Asia, this was not an issue, and Huxley could happily declare them intellectually inferior in *The Uniqueness of Man*.

Huxley was not especially racist, indeed far less so than most other geneticists. He was certainly an honest man, but he shared the prejudices of his time and sought to justify them in terms of genetics. We should recall that, in 1937, applying the eugenic (but not yet racist) legislation of July 1933, the Nazis began to sterilize mixed-race children who had been left in Germany by the French colonial forces that occupied the Rhineland and Ruhr after the First World War – to avoid the 'degeneration of the white race'. Huxley's comment about the intelligence of 'true negroes' was thus at least as politically untimely as his praise of eugenics at the very time that the Nazis were gassing the mentally ill. Most disturbing is the fact that it is uncertain whether he disapproved of this sterilization in 1937, since, according to G. Lemaine and B. Matalon (though they give no reference), he wrote an article in 1936 that 'still maintained that blacks are inferior to whites in intelligence, and that recourse to segregation and sterilization would certainly be needed to deal with those who proved unable to succeed in the improved environment to which he looked forward.'[15]

The Uniqueness of Man – the title is a nod to the commercial success of Carrel's book *L'Homme, cet inconnu* – was translated into French in 1947. Huxley was at that time director of UNESCO; one can imagine the reaction if the present holder of this post were to publish a text of this kind. The year 1947 also saw the French translation of *We Europeans*, with its denial of the existence of races. Clearly, neither Huxley nor his readers found the contradiction problematic, just as comparable contradictions do not faze today's geneticists, who are capable in the same breath of praising racial mixing, declaring that certain hereditary diseases are particularly found in certain races, and then proclaiming that races do not exist as soon as a dubious politician tries to make use of the notion.[16]

15 G. Lemaine and B. Matalon, *Hommes supérieurs, hommes inférieurs? La controverse sur l'hérédité de l'intelligence* (Paris: Armand Colin, 1985), p. 44.

16 It should be noted that one of the co-authors of *We Europeans*, Alfred Cort Haddon, professor of ethnology at Cambridge, had published a book before the First World War in which, while acknowledging the difficulty of definition, he classified the different human races in a geographical rather than a hierarchical sense – certainly the least bad way of tackling the question (A. C. Haddon, *The Races of Man and their Distribution* [Halifax: Milner, 1912]).

This kind of 'anti-racist' discourse on the part of geneticists has essentially a decorative purpose, with no effect on scientific practice as such. In 1936, for example, the prestigious periodical *Nature*, for which Huxley often wrote, gave a favourable response to his view of the biological non-existence of human races, and also criticized Nazi racism – the Nuremberg laws having just been adopted. This did not, however, prevent *Nature* from publishing the following item without any reproving comment: 'An Institute of Racial Biology is to be erected at Copenhagen by funding from the Rockefeller Foundation and the Danish Government.'[17]

The situation is rife with contradiction. Asserting the biological non-existence of races did not make the idea of racial biology meaningless, and it is far from irrelevant that this institute was financed by official and respectable institutions at the same time as the Nazis were in power in Germany, and the use they were making of the notion of race was perfectly well known – in addition, there were racist laws throughout the world, especially in the United States and South Africa.

I shall return to this question in Part Three of this book. All these geneticists, past and present, show a great confusion of thought. Their reasoning mingles categories that are completely heterogeneous, those of taxonomy and those of politics.

Biologists have long recognized the limited validity of taxonomic categories (including that of species and, *a fortiori*, that of race, which is far less strict). They know that these contain an arbitrary and conventional element, and refer – to a greater or lesser degree – to a supposedly 'natural order' of living forms. Contrary to what the geneticists who met at the Musée de l'Homme seem to believe, they did not need to await molecular biology to know all this, as the limited validity of taxonomic categories is a basic commonplace inherent in their very definition. (In this field, arguments deriving from genome analysis are simply verbiage designed to give a scientific appearance.)

However, to consider the limited validity of taxonomic categories (of which race is one of the weakest) as an argument relevant to questions of social organization is to mix two domains that have nothing to do with each other – biology and politics. It is to place anti-racism on the same ground as racism, and commit a similar error as the latter. Whereas racism argues from the notion of race to justify a certain social order, this supposed anti-racism argues from

17 *Nature*, 1936, vol. 137, no. 3475, p. 942.

the difficulty of defining this notion to reject the social order based on it; but both arguments agree in referring political questions to biology. This so-called anti-racism assumes therefore that it would be legitimate to apply the notion of race to social organization if it could be more strictly defined, and that it is only because it resists such definition that the social order need pay no attention to it. This is a confusion of thought that bears witness to the imperialism of genetics and its worrying propensity to interfere with everything, especially things that do not concern it.

In biology, the category of race has the value that taxonomy gives it, and molecular geneticists are not the last to make use of it, despite being unable to define it in terms of their discipline. In politics, it has no value at all, not because it has no value in taxonomy, but because the categories of politics have nothing to do with those of biology – need we recall that to base politics on biology is characteristic of Nazism?

But here, too, just as on the question of eugenics, the media that shape public opinion are unwilling to get into such subtleties. Their rule is simplicity, and it suits them perfectly to believe that proclaiming the non-existence of races is a remedy for racism, and that to maintain the contrary can only be a sign of impenitent racism. This is the message they hammer home (most often in quite inappropriate terms, and with crude approximations),[18] giving support in this way to the idea that politics should refer to categories of biology, and that the role of geneticists is to lay down the law.

Proclaiming the truth here requires more than simply refuting a stupid argument. And by seeking to base anti-racism on the (erroneous) assertion that races do not exist in biology (whether or not they are 'natural entities'), one risks finding oneself compelled to accept racism as soon as the existence of such races is demonstrated. The media thus maintain a confusion that is very convenient in several respects, but at the same time dangerous.

18 One can often read, for example, that there is only one human race, whereas given that a race is a subdivision of a species it cannot be unique; either there are two or more races or there is no race at all. It is the human species that is actually unique in its genus (today the genus *Homo* contains only the species *Homo sapiens*). The expression 'human *genus*' would be more acceptable than 'human *species*', given that species is the fundamental taxonomic category; a species is not defined as a subdivision of a genus, but a genus as a collection of species (though in this case containing only one). Besides, it is traditionally the name of the genus rather than that of the species that is used in everyday language (thus 'man' for *Homo*).

Certainly they do not bear the whole responsibility, but one may well ask why journalists, often so critical in other fields, behave as vulgar mouthpieces for certain 'mandarins' (always the same ones) whenever science is concerned.

To complete this introduction, I shall now turn to the way in which specialists in social and political history introduce the role of biology and the sciences in general into their studies – or, rather, fail to do so.

In the case of eugenics, there is an almost total absence. Though eugenics was massively widespread in the first half of the twentieth century, it is practically never taken into consideration by historians, at least in general works (I am not referring here to the few books specifically devoted to this question). Joseph Rovan, for example, in his *Histoire de l'Allemagne*, has not a word to say about eugenics under Nazism, and devotes only two lines to the extermination of the mentally ill (after they had already been sterilized). Even then, he does this not for its own sake, but only to illustrate opposition to Hitler from the Catholic Church.[19] Marc Ferro's *Chronologie universelle du monde contemporain*, which lists ten thousand events of the nineteenth and twentieth centuries, does not mention a single eugenic law – neither American, nor German, nor any other; neither the sterilization of 'Rhineland bastards' in 1937 nor Hitler's decree to exterminate the mentally ill, let alone the consequences of these laws and decrees. (It would appear that texts that led to the sterilization of hundreds of thousands of people, as well as the deaths of many of them, do not belong to the ten thousand most important events of the contemporary world – and not for lack of space, since for the year 1907, for example, the year of the first eugenic legislation in the US, the book records the launch of the Ford Model T for $850.)[20]

On the rare occasions that eugenics and the elimination of the mentally ill and handicapped are recalled, this is always in a minor key, as a vague consequence of the Nazi extermination of the Jews. Even studies devoted to the latter question often ignore or skim over it. Thus Raul Hilberg, in his book *The Destruction of*

19 J. Rovan, *Histoire de l'Allemagne des origines à nos jours* (Paris: Points-Seuil, 1998), p. 691.

20 M. Ferro (ed.), *Chronologie universelle du monde contemporain* (Paris: Nathan, 1993).

the European Jews, mentions only very briefly the extermination of the mentally ill and handicapped (one and a half pages in a book of over a thousand), and gives only a very poor (and partial) numerical estimate of this.[21] In the same work, Hilberg writes that the extermination of the Jews was the first massacre to be systematically planned and quasi-industrial, and that 'never before in history had people been killed on an assembly-line basis'.[22] Hilberg was certainly well aware that the systematic extermination of the mentally ill and handicapped began with the start of the war (thus well before that of the Jews), that it, too, was systematically planned and massive, and that the 'technicians' who constructed the gas chambers at Auschwitz were precisely those who had built the gas chambers designed for the mentally ill in 1939 (which indicates very well a continuity between the two processes, at least on the technical level). But it is as if this extermination does not count. Hilberg at least mentions it, but how many historians pass over it in complete silence, because it fails to fit the *a priori* framework they have decided to adopt?

It is not the object here to place the victims of Nazism in a kind of competitive order, with points for those groups with the most killed, those killed first, etc. However abominable it might seem, there was a certain logic to the procedure: the extermination of Jews followed that of the mentally ill not simply in a chronological sense but also in a logical sense, and it is impossible to ignore the one if one wants to understand the other. Yet it is as if historians did not grant any human value to the mentally ill and, as a result, considered their sterilization and subsequent extermination a mere 'detail', a non-event without interest for the understanding of the genesis of Nazi crimes, a kind of epiphenomenon. Victims of this kind, and other '*Untermenschen*', are then wiped out of history as radically as the Nazis wiped them out of society.

The same could be said of the sterilization of the 'Rhineland bastards' in 1937, in the name of the eugenic legislation of 1933, a sterilization that is scarcely mentioned, even though it obviously forms a major element in the 'practical' connection between eugenics and racism, just as the gassing of mental patients plays the same role in the relationship between eugenics (sterilization) and racial extermination (genocide).

---

21 R. Hilberg, *The Destruction of the European Jews* (New Haven: Yale University Press, 2003), p. 930–1.

22 Ibid., p. 922.

Moreover, the silence of historians on these questions sometimes leads to somewhat curious situations. One of the great leitmotifs in this story, at least in its vulgarized versions, consists in asking who was aware – and when – of the existence of the gas chambers. (This was prominently heard at the time of the Maurice Papon trial for complicity in crimes against humanity, and in a more general sense is what underlines the indignation at the silence of Pius XII, deemed to have been *au courant* from the start but having said nothing.) The question is especially glaring in light of the fact that public protests by Germans themselves against the extermination of mental patients, the chronically infirm, and others led to the closure of the gassing centres in August 1941, after which the elimination continued in other forms. Even if this continuation shows the inadequacy of such protests, it in no way invalidates them. 'Public protest' about the existence of gas chambers means 'public knowledge' of them, by August 1941 at the latest, and thus even before the gassing of the Jews. Yet here again, it seems as if these gas chambers, and the knowledge of their existence, do not count, and that no link can be made between them and the gas chambers of Auschwitz.

This curious amnesia, moreover, is not confined to the Nazi exterminations. The above remarks can also be applied to other fields. In François Jacob's *The Logic of Life*, for example, everything related to population genetics (the major support of eugenics in the first half of the twentieth century) has been smuggled out. The statistical dimension of evolutionism and genetics is derived here from contemporary theories of mechanics and statistical dynamics,[23] whereas in reality its source lay in Galton's biometrics. It is certainly less embarrassing, and more 'trendy', to refer to physicists such as Maxwell, Gibbs and Boltzmann than to biometricians and geneticists such as Galton, Pearson, Fisher, and so on, all of whom were more or less compromised by eugenics, racism and even Nazism, and whom François Jacob therefore expels from the history of genetics.

Sociology shows a comparable amnesia. In 1959, to celebrate the centenary of *The Origin of Species*, there was a conference devoted to the influence of Darwinism on the study of society. In the published proceedings, *Darwinism and the Study of Society: A Centenary Symposium*, eugenics is dispatched in five lines, followed by a note stating that in the wake of the Nazi horrors, the reaction against ideas stressing the role of the environment

23 F. Jacob, *The Logic of Living Systems* (London: Allen Lane, 1974).

as opposed to heredity possibly went too far.[24] To be sure, it is admitted here, more or less in passing, that the ideas of social Darwinism did indeed play a certain role in what took place in Germany between 1933 and 1945, but this is only in very general terms, without emphasis and in a 'well-mannered' way – given that the eminent biologists who would have had to be put in the dock were still alive at this time.

A further example is the way that writings on Lysenko mention only rarely the presence of Hermann Muller in the Soviet Union in the 1930s, or else omit to indicate Muller's campaign in support of eugenics.[25] Yet the antagonism between the two men is well known, and it is more or less certain that Muller's sudden departure for Spain in April 1937 was due to pressure from Lysenko. It is therefore inappropriate to neglect one member of the pair in a study of the other.

In the 1930s, Muller was one of the world's leading geneticists. He took part in the work for which Thomas Morgan received the Nobel Prize in 1933 (thus playing a major role in the cutting-edge genetics of the time, the study of *Drosophila*), and when he parted company with Morgan this was to develop his own research, which in turn brought him the Nobel Prize in 1946. His presence in the Soviet Union in the 1930s was thus equivalent to what the presence of Jacques Monod or François Jacob would have been in the 1960s. This shows how strange it is that books on Lysenko treat the matter of Muller so lightly and totally neglect his programme of political biology (which he published in 1935, after having already been in the USSR for two years – see note 6 above).

Generally, texts on the Lysenko affair maintain a deep silence about eugenics, the support it received from Western genetics (in opposition to Lysenko's genetics), and Muller's activity in the USSR. In 1949, Julian Huxley's study of Soviet genetics briefly indicated that Muller's book had been poorly received in high places in the Soviet Union, and that this led to his fall from grace. But as a eugenicist himself, Huxley praised Muller's book as 'one of the most interesting books on eugenics I know', and went on to offer a new rejection of both racism and the way that Lysenko's followers argued from this rejection to reject Western genetics

24 M. Ginsberg, 'Social Evolution', in M. Banton (ed.), *Darwinism and the Study of Society, a Centenary Symposium* (London: Tavistock, 1961), p. 105.

25 For example, Z. Medvedev; *The Rise and Fall of T. D. Lysenko* (New York: Columbia University Press, 1969); D. Lecourt, *Proletarian Science? The Case of Lysenko* (London: NLB, 1977).

as well.[26] When Huxley made this assertion, the results of Nazi eugenic practices were already well known – the doctors' trial at Nuremberg in 1946 having revealed all the details. It is evident that Huxley was incapable not only of analysing this but also of imagining what the Stalinism he criticized would have been like had it been coupled with a eugenic policy, whether championed by Muller or by anyone else. The process of concealment, if not repression, had already begun.

Specialists in political and social history display comparable, though less flagrant, gaps in their presentation of racism. All the studies mention Arthur de Gobineau, Georges Vacher de Lapouge and others of the same stamp. Gobineau, however, was not a biologist but a littérateur, and his nineteenth-century work on the inequality of human races does not in fact contain what its title might be thought to imply.[27] (See Part Three for more detail on this.) As for Vacher de Lapouge, he was no more than a disputed and marginal figure who gained little acceptance from the academic world and whose audience remained limited to those with a pre-existing ideological commitment.[28]

---

26 J. Huxley, *Soviet Genetics and World Science* (London: Chatto & Windus, 1949), p. 184.

27 J. A. de Gobineau, *Essai sur l'inégalité des races humaines* [1853–7] (Paris: Belfond, 1966).

28 Vacher de Lapouge (1854–1936) studied law but did not practise the profession. He became deputy librarian at the University of Montpellier, then librarian at Rennes and subsequently Poitiers, where he died in general indifference. In parallel with his library work, he acquired a wide range of knowledge which enabled him to give 'free lectures'; these made up the greater part of his books. The main characteristic of his thought was an obsession with cranial form; he studied 'dolichocephalic' and 'brachycephalic' skulls in all possible populations and conditions. A typical example, though one of the most comic, was the relation between skull shape and cycling, as estimated from the tax paid by different populations. 'Tax on bicycles. – The cycle is an instrument of sport for snobs and of transport for serious people. Taking both into consideration, ten million dolichocephalics pay the petty sum of 643,000 francs in tax, while brachycephalics pay 386,000. The twenty dolichocephalic departments pay 868,000 francs, and the twenty brachycephalic departments 283,000. The development of cycling is thus in proportion to cephalic index, with the long skulls showing a passion for the new invention, and the short skulls being resistant to this advance just as to others. The department of Seine pays 360,000 francs in tax on cycles. In this activity, Paris is on a par with Britain and America, both dolichocephalic countries!' G. Vacher de Lapouge, *Race et milieu social* (Paris: Rivière, 1909), p. 148. Dozens of similar examples could be given; you can find anything you want in this author. He was a highly heterogeneous writer, at times very lucid and prophetic, at other times completely ridiculous. The best that can be said about him is that his books supply a pretty good caricature of the biologico-social ideas of his era.

Studies on racism, on the other hand, rarely mention the name of Ernst Haeckel, who was a leading light in biology in the late nineteenth and early twentieth centuries, as well as the main modern writer (also one of the first – in 1868, scarcely nine years after Darwin's *Origin of Species*) to have proposed a hierarchical classification of human races within an evolutionary framework which moved from blacks, supposedly close to the ape, up to what he considered the most developed races, the 'Indo-Germanians' (i.e., Germans, Anglo-Saxons and Scandinavians).[29]

Huxley's book *We Europeans* does not betray any familiarity with Haeckel. He attributes to Gobineau the idea of an innate inferiority of certain races and immediately goes on to mention Vacher de Lapouge.[30] Nor have I found any trace of Haeckel in the texts on racism published by UNESCO after the Second World War (ten different authors, all specialists in the question); these texts were republished under the title *The Race Question in Modern Science* but have nothing to say about the classification of races that this science invented.[31] François de Fontette's book on racism, in the popular 'Que sais-je?' series, also fails to mention Haeckel and his classification of races.[32] As a final and recent example, the index to *La Force du préjugé, essai sur le racisme et ses doubles* by Pierre-André Taguieff, published by the left-wing publisher La Découverte, includes just two references for Haeckel (one in the notes), whereas there are thirty-two for Vacher de Lapouge and twenty-five for Gobineau (only Kant has more, with thirty-three).[33]

And yet Haeckel was an eminent scientist, a member of more than ninety academies and learned societies across the world.[34] He was the universal popularizer of Darwin, even to the point that it was often his own theories, rather than Darwin's, that the educated public came to know as 'Darwinism'. His books were published in enormous print runs and translated into every language. The book in which he presented his classification of human races, *The History of Creation*, is one of his most famous, even if, according

29 E. Haeckel, *The History of Creation* (London: H. S. King, 1876), pp. 410–46.

30 J. Huxley et al., *We Europeans*, p. 66.

31 UNESCO, *The Race Question and Modern Science* (London: Allen & Unwin, 1975).

32 F. de Fontette, *Le Racisme*.

33 Pierre-André Taguieff, *La Force du préjugé, essai sur le racisme et ses doubles* (Paris: La Découverte, 1987).

34 G. Uschmann, *'Haeckel', in Dictionary of Scientific Biography*, vol. 6, pp. 6–11.

to Jacques Roger, its French translation tones down a number of passages that would have been badly received in France.[35]

Even today, certain biologists[36] refer to Haeckel with praise – less for his actual scientific work, which today is rather forgotten, than for the 'monist' philosophy that he developed. What these biologists generally celebrate is Haeckel's materialism and anti-religious attitude, but they silently overlook his racism, pan-Germanism and the rather esoteric fantasies he developed towards the end of his life. Haeckel belongs to the pantheon of modern biology, along with Darwin, Weismann and a few others. His weight in the early twentieth century was incomparably greater than someone like Gobineau, who, despite the celebrity he eventually gained (belatedly, and in Germany above all), was nothing more than a romantic whose opinions had no scientific value. (Besides, he died in 1882, whereas Haeckel lived until 1919.) *A fortiori*, Haeckel's audience and authority as an eminent representative of international science far outranked that of the librarian Vacher de Lapouge (who, by the way, translated some of Haeckel's books, perhaps for the sake of income but presumably also because of an affinity for the ideas).[37]

It is legitimate to inquire, therefore, why on the question of racism historians have systematically highlighted a misled fabulist, attached to a flimsy philosophy, and a marginal figure who never had any place in the academy, whereas they have failed to cite one of the most glorious 'mandarins' of the biology of the time. Given that racism appealed to biological theory, why not explain what established science claimed? The only reason can be that a form of racism underpinned by the ideas of a romantic littérateur or an ideologically driven marginal figure is far less embarrassing than a racism supported by the theories of eminent biologists and relayed by all kinds of institutions, both public and private. This is easy to understand but, again, a curious way to write history.

In a general sense, everything to do with science seems to pose

35 J. Roger, 'L'eugénisme, 1850–1950', in *Pour une histoire des sciences à part entière* (Paris: Albin Michel, 1995), p. 412.

36 For example, J.-P. Changeux, *L'Homme neuronal* (Paris: Fayard, 1983), p. 342 and *passim*.

37 Vacher de Lapouge translated Haeckel's book on his 'monist' philosophy, which appeared in French as *Le Monisme, lien entre la religion et la science* (Paris: Schleicher, 1897). On the links between Vacher de Lapouge and Haeckel, see the chapter 'The Monism of Georges Vacher de Lapouge and Gustave Le Bon', in D. Gasman, *Haeckel's Monism and the Birth of Fascist Ideology* (New York: Lang, 1998), pp. 135–64.

severe problems for specialists in political and social history. Yet science occupies, at least in the modern age, an important place in human activity, and it is hard for history to overlook this. How can the seventeenth century be understood without reference to the mechanics of Galileo and Descartes, the eighteenth – the Age of Enlightenment – without Newtonian physics, the nineteenth without the determinism of Laplace, the twentieth without Darwinism?[38]

It is trivial to say that the virtual absence of science from historical studies is caused by historians being generally unable to read scientific texts or assess their importance – leading them to refer to the texts of such ideologues as Gobineau, Vacher de Lapouge and the like, who have only a vague relation to science, no matter their claims to be scientific. A second reason is that the role of science in political and social history has primarily been envisaged only in terms of technology. Historians thus take into consideration, above all, the material effects of technology on social development. As for historians of science, their preoccupation is with the action of society on science, which they sometimes consider an 'ideological contamination' – at least when they do not try to reduce science to a simple social production.

Practically no one, however, is interested in the action of science on society through ideology rather than through technology, or else this action is envisaged only in a very general sense, in terms of scientism, i.e., an 'ideology of science' that characterizes this or that society at a particular epoch. Science as such is not deemed to be a producer of ideologies; in its immaculate splendour, it can only be contaminated by them.

It is clear, however, that the sciences have marked social and political history in ways other than by technology, and in a more particular way than by a generalized scientism. In the case of biology, which is our concern here, it is hard to believe that the work of Pasteur had an impact on society only by way of techniques such as vaccination, pasteurization and asepsis – or again that Darwinism, though without leading to practical applications such as these, played no role in the way society was understood.

38 It is often considered that the twentieth century is the century of physics (relativity and quantum mechanics), but if it is true that this discipline has played a major role in society, this is above all by way of its applications (nuclear weapons, for example). In actual fact, modern physics, doubtless because of a complexity that is intelligible only with difficulty to the uninitiated, moved straight from the laboratory into science fiction, without stopping in the realm of ideology. It was Darwinism, and not physics, that really characterized twentieth-century society from this ideological point of view.

I. SOCIOLOGY AND BIOLOGY

1

The Naturalization of Society

The second half of the nineteenth century was a golden age for biology. Three names stand out above all – Claude Bernard, Louis Pasteur and Charles Darwin, who respectively represent the following disciplines:

- modern physiology and its application to experimental medicine;
- microbiology and its medical, public health, industrial and agricultural applications;
- evolutionism, with no immediate technological applications.[1]

This triumphant biology could not but have an impact on society, but its various currents did not always do so in the same fashion. Their impact also varied from country to country, according to the recognition that these authors received.

The case of Claude Bernard is the simplest, inasmuch as his physiology had no immediate practical outcome. It certainly did inaugurate experimental medicine, but it was only later that this led to a significant improvement in medical techniques. This does not, however, mean that Bernard had no ideological influence. One aspect of this was undoubtedly that his physiology decisively introduced life into the sciences of physics and chemistry. Life became naturalized, no longer needing recourse to a vital force – even if Bernard himself remained a vitalist to some degree. The old definition of life as the persistence of being was freed from the supernatural and reduced to homeostasis, i.e., maintenance of the composition of an internal state through regulation. This was a move towards a materialist conception of life, and even if Bernard

1 The term 'evolutionism' was commonly used, without prejudice, in the nineteenth century; today the word is sometimes counterposed to 'creationism' by the religious opponents of evolutionary thought

himself rejected this, it followed very clearly from the general lines of his physiology.[2]

A second possible influence of his physiology derived from the way it strengthened the old analogies between society and the organism – society as organism or, reciprocally, the organism as society. This notion existed long before Bernard's time. It had even been extended, by Lamarck and especially by H. Milne-Edwards, through the claim that an organism was more perfect if composed of organs specialized for different tasks. This idea was undoubtedly borrowed from Adam Smith's sociological theories on the division of labour, and suggested an analogy between social and biological organization.[3]

Claude Bernard's physiology, however, went further than this, showing how the organs of the body were not only specialized but also linked to one another by regulatory mechanisms in such a way as to ensure the survival of the organism. Survival was thus their final purpose, their raison d'être as well as that which permitted them to exist. The organs maintain the constancy of the internal environment, which in return enables them to exist and function; each is involved with the others in a common goal, necessitating close interdependence. The biological metaphor of society as organism, or vice versa, could only be reinforced by this approach, thus in a sense inclining to totalitarianism, the primacy of the group over the individual, following the model of the primacy of the organism over the organ or the cell – the latter in particular being an easily replaceable cog.

Bernard's influence on social ideologies did not go beyond these points, which predated him and which he simply consolidated. The influence of Pasteur was undoubtedly stronger, first of all because his microbiology had immediate applications, but not for this reason alone.

Public health legislation based on Pasteur's principles, such as compulsory vaccination, the declaration of contagious diseases, quarantine, etc., necessarily had results that were not simply practical but also ideological – affecting the very manner in which society was understood. Although such legislation was framed in terms of medical techniques, it inevitably translated into a

2 A. *Pichot, Histoire de la notion de vie* (Paris: Gallimard, 1993).

3 C. Bouglé, *La Démocratie devant la science* (Paris: Alcan, 1923), pp. 22ff., 137ff. Adam Smith's views on the division of labour were particularly developed in the first three chapters of Book I of *The Wealth of Nations*.

certain 'biologization' of politics. On the one hand, it highlighted the biological dimension of society. On the other, it gave this dimension a particular importance, to the extent that it was far more controllable than the majority of other social dimensions – politics, law, the economy and so on.

Pasteurianism was not simply a biomedical technology but also a social one. It opened up the possibility of acting on society in a different way than by politics, law and economics in the narrow sense. It involved a biologization of society, that is to say, a 'naturalization' that made possible the 'scientization' of politics – a kind of ideal in a scientistic epoch.

Pasteur's techniques were applied first of all to flocks of sheep suffering from anthrax, to silkworms affected by pébrine, and the like; subsequently they were applied to human populations suffering from tuberculosis, cholera, syphilis, rabies and other infectious diseases that were widespread in Europe at the time. These human populations reacted to Pasteurian methods (hygiene, vaccination, quarantine) exactly as did sheep and other domestic animals. The art of managing humans was thus brought closer to that of managing a herd. The naturalization of society brought politics closer to biological techniques.

This then opened the way for the further influence of biology on society, in the form of evolutionism and genetics. Both genetics and evolutionary biology were emerging disciplines which also had a somewhat 'herding' vision of society but could not yet give rise to such successful techniques as pasteurization, and which without their example would hardly have been able to develop their sociobiologizing claims. Pasteurianism and Darwinism were exactly contemporary, but while the latter remained purely speculative, the former, right from the start, was accompanied by practical applications whose importance and prestige grew along with its development – from the study of fermentation to the practice of vaccination against rabies. Such applications immediately brought Pasteurianism an immense international reputation.[4]

Yet it was evolutionism and genetics, far more than Pasteurianism, that led to the most pronounced biological conceptions of society. Paradoxically, this was precisely because of their lack of technical success. Whilst Pasteurian hygiene brought

4 Cf. L. Pasteur, *Écrits scientifiques et médicaux*, selected and edited by A. Pichot (Paris: Flammarion, 1994).

to light the biological dimension of human society, its scientific principles were applied in the service of society as it then existed, since vaccination and asepsis work under any political regime. It undoubtedly tended to reinforce the existing social order by the constraints it imposed (in the form of legislation or principles of conduct), but these constraints were essentially hygienic, and only secondarily political. The techniques were effective in themselves, and relatively neutral from the political point of view – even if it is indeed the case, for instance, that vaccination makes sense only in societies or among social classes that present certain conditions of life. The same rule is even more important for the application of prophylactic public health measures.

Evolution and genetics, on the other hand, did not (and still do not) offer comparable techniques; Darwinism was inspired by, but is not the basis of, the empirical selection methods of plant and animal breeders. Their social applications, accordingly, do not consist in pure techniques that add to a given political order without greatly modifying it. They directly touch on this political order, seeking to correct it so as to bring it in line with a supposed natural order, a biological order that conforms to their principles. These disciplines thus claim to substitute a new social order for an existing one: a supposedly natural (biological) order based on science, whereas the old order rested on tradition, religion and the whole of the 'obscurantism' that scientism and progressivism battled with in the late nineteenth century. ('Pasteurianism as the triumph of science over obscurantism' was the theme annexed by these ideologies, even though Pasteur himself was typically 'priest-ridden'.)

Public health, as the application of Pasteurian microbiology, was essentially technical in both its aim and its effect. Its political aspect was secondary and generally neglected – even if not completely negligible. The applications of genetics and of evolutionary biology, on the other hand, were essentially political, even when they claimed to be based on science and technology. The eugenics and racism already mentioned are prime examples.

The cases of Pasteurianism and Darwinism are related to the extent that, in precisely the same era, they both contributed to a biologization of society. But they differ in the way that this biologization was effected. Pasteurianism, by virtue of its rather technical character and the relative dissociation of its political dimension, had an international influence, which could be also said of the physiology and experimental medicine of Claude Bernard.

Darwinism, by the same logic, was less universally accepted, both on the scientific level and in terms of its 'social applications'.

Pasteurianism, sometimes complemented by a Lamarckian vision of heredity,[5] underpinned social hygienics as it developed in France. Darwinism, complemented by the genetics of Weismann and Mendel, lay at the root of the eugenics that developed in the Anglo-Saxon countries, Germany and Scandinavia. We can also note here a particular form of racism, inherent in the evolutionary hierarchy of races imagined by Haeckel.

This is not simply an *a posteriori* historical reconstruction. From the end of the nineteenth century, these biologizations of society, and especially that inherent in Darwinism, posed a problem and were the object of a broad philosophical and sociological literature that is somewhat forgotten today – suffering from the same repression as the subject it addresses.

Here, for example, is a text from 1873 in which Léon Dumont, in a book presenting Haeckel's theories, explained to French conservative politicians – whom he found rather lukewarm in their response to Darwinism – why they had nothing to fear from this doctrine, which according to him was in perfect agreement with their own political ideas and which alone could 'provide their scientific justification' – that is, he proposed a resort to biology to justify a political and social thesis.

The quotation from Dumont is rather long, but it is highly enlightening, as indeed is the author's entire introduction. To grasp all its nuances, we should remember that its date of publication was just after the Franco-Prussian War of 1870–71 and the Paris Commune of 1871; not only was Haeckel a German, but certain supporters of the Commune had sought to appeal to Darwinism:

> What though is the moral doctrine that is rigorously implied by transformism? It is quite simply political economy, which has never to our knowledge passed as an enemy of the social order. Darwin himself declared, on many occasions, that all he did was extend to the origin of species the theories of the economists, and that his views had been particularly suggested to him by the reading of Malthus. If the conservative party were rather less blind, it would recognize that the theory of evolution bears within it the

5 'Pseudo-Lamarckian' would be more correct, as Lamarck never explicitly offered a theory of heredity. What has become known as 'Lamarckian heredity' is essentially a legend born in the late nineteenth century with the opposition between Weismann and the neo-Lamarckians.

very philosophy of conservative doctrine, and that it alone can provide a scientific justification of this.

What there is rather more reason to fear is that Darwinism, when its practical consequences are more generally understood, might provoke on the contrary an exaggerated reaction against democratic and egalitarian tendencies, and lead us to the political principles of Hobbes, Machiavelli and J. de Maistre. Socialist utopias naturally and necessarily derive from idealism, and this is why they are a particular danger for the Latin nations, these being the most idealist in the world. From the moment that an absolute value is attached to individuals, that they are considered as made up of both a body and a soul that is a substance in itself, from the moment that each man has the beginning of his existence and the free basis of all his actions in himself, it must be found right and proper that all men start life under similar conditions of existence, and this leads necessarily to communism: from this point on, no social inequality is tolerable, and even physiological inequalities come to be seen as an injustice of nature that society has the mission of attenuating and correcting as far as possible.

It is true that Darwinism does not rule out idealism, but it certainly has a lesser affinity with it than with any other metaphysical system, and tends on the contrary to consider the individual as an ensemble or series of facts that is no more than the result and continuation, by way of heredity, of an indefinite number of other earlier series. No two embryos are alike, as all are produced by accumulations of forces that have been subject to the most varied circumstances in their long development. Men therefore do not start out from the same point: faculties, instincts, organs, everything about them is due to heredity, and it is because heredity preserves and perpetuates differences and inequalities that it is a principle of selection and progress. The same is true of the goods of the individual, which are an outward extension of his personality and arise from accumulations of material forces capitalized by his ancestors. Justice for society consists not in re-establishing equality among individuals, but rather in protecting in an equal fashion the continuous development of all these hereditary series. These consequences of Darwinism completely rule out socialism, all forms of which have a common hostility to heredity. If we see Darwinists such as Büchner, Naquet, etc. profess opinions that lead to social utopias, this can only be explained by the fact that they had long been members of certain political parties before their conversion to the ideas of the struggle for existence and natural selection.[6]

6 L. A. Dumont, *Haeckel et la théorie de l'évolution en Allemagne* (Paris: Germer-Baillière, 1873), pp. 6–9.

Dumont makes it all quite clear. This quotation brings up the problems that Darwinism might pose to the 'idealist' conservative bourgeoisie, but also its perfect compatibility with this same bourgeoisie's political and economic principles, for which it offers a scientific justification. The risks of totalitarianism that the transposition of Darwinism to politics might entail are already described. And finally, this author indicates the mistake made by those who seek to recruit Darwin in support of socialist theses – the mistake made by the left-wing biologists who latched onto Haeckel while forgetting his racism and pan-Germanism.[7] In 1873 Dumont could only glimpse the movement of thought that was beginning to form. A few years later, the extension of Darwinism to man and society was achieved in the broadest possible fashion, leading to reactions that were often virulent.

In 1928 Pitrim A. Sorokin proposed the following classification of biologically inspired sociological theories:

7 This is how Vacher de Lapouge, in 1899, made fun of those leftists who criticized his ideas and those of like-minded thinkers (i.e., the precursors of Nazism): 'I experience a malicious but lively pleasure in mentioning the embarrassment that shows through a number of recent articles, published in socialist, anarchist or supposedly democratic periodicals. These saw above all in Darwinism, or in a more general fashion in the scientific doctrines on the origin of species and the world, an argument to oppose to religion, and in our country to the Catholic church, which is creationist and bound to the book of Genesis. They failed to realize that Darwinism applied to man and social life excludes in future any non-scientific element of social explanation, i.e., the admission of supernatural causes outside the general causality of the universe. This is the case with free-thinkers, or those who see themselves as such, as right from the start the churches saw the consequences of the new theories, and took the opportunity to denigrate them. This exclusion of the supernatural is thus all the more fatal to doctrines that have freed themselves from religious dogma, but remain enslaved to the rest, in that they lack the supreme recourse that Christians have ... Their only choice is between a pure and simple return to the theological doctrines from which they started, or pure and simple adherence to the scientific [i.e., Darwinian] explanation of social phenomena, and the abandonment of all the philosophical principles that make up their political doctrine . . . Liberals, socialists and anarchists are already treating the Darwinists as barbarians. So be it! The barbarians are coming, the besiegers are under siege, and their last hope of resistance is to barricade themselves in the citadel they were attacking. The near future will show our sons this curious spectacle, the theorists of false modern democracy compelled to lock themselves up in the citadel of clericalism . . . In the face of the new dogmas, the alliance between men of the Church and men of the Revolution will be tomorrow's agenda, but this alliance will be unable to postpone the hour of destiny, nor will any genius be able to push humanity back into ignorance. We are running towards the unknown, but the past will never return, never!' (L'Aryen, *son rôle social* [Paris: Fontemoing, 1899], pp. 513–4).

> The extraordinary progress of biology during the last seventy years has given an additional impetus to biological interpretations in sociology. Hence, the contemporary biological theories in social science. These are numerous and vary in their concrete forms, but nevertheless, it is possible to group them in a relatively few fundamental classes. The principal concepts of the post-Darwinian biology are: organism, heredity, selection, variation, adaptation, struggle for existence, and the inherited drives (reflexes, instincts, unconditioned responses) of an organism. Correspondingly we have: 1. *The Bio-Organismic Interpretation of Social Phenomena;* 2. *The Anthropo-Racial School*, which interprets social phenomena in terms of heredity, selection, and variation through selection; 3. *The Darwinian School of the Struggle for Existence*, which emphasizes the role of this factor; and 4. *The Instinctivist School*, which views social behaviour and social processes as a manifestation of various inherited or instinctive drives. Besides these, there are many 'mixed' theories, which in their analysis of social facts, combine biological factors with the non-biological ones.[8]

The second and third categories in this classification are specifically Darwinian, while the fourth is at least broadly so, to the extent that the psychological characteristics invoked are often ascribed to hereditary biological characteristics.

André Béjin sees the chronological development of these various social Darwinian theories classified by Sorokin as follows:

> It is possible to distinguish, by and large, three phases in the development of these theories. From 1853 to 1883, we move steadily from an early and fairly liberal Social Darwinism to a second form that is more socialistic and statist. This second form is dominant from 1884 to 1904, often with a eugenicist, racist or imperialist cast. Finally, from 1905 to 1935, this current of thought ceases to produce any major innovations, but inspires political applications that are heavy with consequences.[9]

A classification of this kind, with its developmental movement, is only broadly valid. On the one hand, this is because these are not really well-constructed and unitary theories, but rather a set of Darwinian biological themes, located, whether complementarily or contradictorily, within theories that are

8 P. A. Sorokin, *Contemporary Sociological Theories* (New York: Harper, 1928), p. 194.

9 A. Béjin, 'Théories socio-politiques de la lutte pour la vie', in P. Ory (ed.), *Nouvelle Histoire des idées politiques* (Paris: Hachette-Pluriel, 1996), p. 407.

themselves fairly heterogeneous, rarely being sociobiological in any strict sense. On the other hand, almost all possible variants can be found somewhere in these theories, and this is indeed their principal characteristic: they took all possible forms and could justify everything. The plastic character of Darwinian principles transplanted into sociology lent itself to this only too well.

Here is an almost archetypal example, taken from a quite moderate English writer, Walter Bagehot (1826–77). If the object is to explain progress (no matter what kind) in this or that society, then social Darwinism offers an adequate explanation; struggle, competition and selection are there to ensure the triumph of the superior form:

> First. In every particular state of the world, those nations which are strongest tend to prevail over the others; and in certain marked peculiarities the strongest tend to be the best.
>
> Secondly. Within every particular nation the type or types of character then and there most attractive tend to prevail; and the most attractive, though with exceptions, is what we call the best character . . .
>
> These are the sort of doctrines with which, under the name of 'natural selection' in physical science, we have become familiar; and as every great scientific conception tends to advance its boundaries and to be of use in solving problems not thought of when it was started, so here, what was put forward for mere animal history may, with a change of form, but an identical essence, be applied to human history.[10]

Conversely, if the object is to explain the stagnation of a civilization, the absence of progress, then the same explanation can be employed. All that is needed is to modify the result of the struggle: in the case of progress, the most efficient innovations triumph; in the case of stagnation, these innovations all lose out to the one already in place, and we shall see why:

> The reason is, that only those nations can progress which preserve and use the fundamental peculiarity which was given by nature to man's organism as to all other organisms. By a law of which we know no reason, but which is among the first by which Providence guides and governs the world, there is a tendency in descendants to be like their progenitors, and yet a tendency also in descendants to *differ* from their progenitors . . . In certain respects each born generation is not like the last born; and in certain other respects it is like

10 W. Bagehot, *Physics and Politics* (London: H. S. King, 1872) p. 43.

> the last. But the peculiarity of arrested civilisation is to kill out varieties at birth almost; that is, in early childhood, and before they can develop.[11]

The claim is unanswerable. The only problem is that it includes a good deal of rhetoric, but very little science. With principles such as these, anything can be explained, and so can its contrary. Writers of this time, and those who followed them, would have been lost without them. They used the principles of biology, and especially those of Darwinism, in every way possible.

After Bagehot, a moderate writer, we can take the case of an extremist, or one with this reputation, Vacher de Lapouge. Here, in his view, is the role of selection in the life of nations:

> It is selection, which, by constantly modifying the composition of nations, brings about the emergence of new strata, preparing within the masses themselves the determining phenomena in the life and death of nations. This proposition is the fundamental proposition of Darwinian sociology, the credo of the selectionist school.[12]

This is how Lapouge applies his thesis to a case where 'intelligent and bold' conquerors triumph over an inferior and declining race, for the greater good of civilization:

> Let us suppose that in a barbarian country or one occupied by a declining people of inferior race, a handful of conquerors from an intelligent and bold race establish themselves. By the end of a certain time, conquerors and conquered will form one single people, among whom some command and others work. A kind of fertilization takes place, in which the conqueror plays the role of the male element. If nothing happens to disturb the normal development, it is clear that this people will soon reach a high degree of prosperity. The conquerors, by their superior intelligence, their boldness and courage, and the subjects by their many limbs working together for the development of civilization, this association of muscular force giving results that these elements could not have produced in isolation . . .
>
> It is clear that the destiny of a people stands in close correlation with the worse or better quality of the elements that compose and lead it. If it is rich in energetic and intelligent elements, then even the most disastrous events will have no more than a temporary and limited effect on it. The

---

11 Ibid., pp. 53–4.

12 G. Vacher de Lapouge, *Les Sélections sociales* (Paris: Fontemoing, 1896), p. 61.

same circumstances may produce a rapid halt to development, or even a terminal collapse, if intelligence is lacking, action is paralysed by indecision, or discouragement reigns.[13]

All this is certainly somewhat tautological, but on balance it leads us to believe that all is for the best in the best of all possible worlds. With the success of superior beings assured, humanity can march forward to the superman. But this would be a mistake, Lapouge cautions. It turns out that the best are not always the best; often it is the worst that win out, or even the blacks and brachycephalics – with the terrible menace of a rise in the cephalic index![14]

The paradoxical result is that the inferior element gradually reasserts itself, and each step towards purity marks a return to barbarism. If this appears to contradict Darwin's law, in fact it is a rigorous application of it. Superior individuals are relatively inferior when they have less chance of success or posterity owing to the social milieu in which they struggle for life . . . Economists say that bad money drives out good; in the conflict of classes and races, the inferior drives out the superior. The French West Indies, where the white element has practically disappeared, and Haiti where even the mulattos have succumbed, giving way to African barbarism, are well-known examples. We are less familiar with what is happening in our midst, where brachycephalics have almost eliminated European blood. From century to century, the cephalic index has been rising in Europe, ever since the start of the modern age. The race with servile qualities has almost destroyed the indigenous population, and the British Isles are almost alone in showing us the physical type and strongly tempered character of the first inhabitants of Europe . . .

These examples all teach us to see more than an isolated fact in this apparently strange phenomenon of the destruction of the best by the worst among men living in society. The worst, in this case, are better adapted to the milieu – like the parasites, microbes and insects that destroyed the larger, better armed and more intelligent species of the palaeontological worlds. There is in fact no reason why the larger, finer, better armed and more intelligent should triumph in the struggle for existence.[15]

13 Ibid., pp. 64–5, 68.

14 The cephalic index is calculated as follows. Multiply the maximum width of the head by 100, then divide by its maximum height. The higher the index, the wider the head (brachycephalic); the lower the index, the longer the head (dolichocephalic). The convention of writers of this period was to classify as dolichocephalic those individuals with a cephalic index lower than 80, and as brachycephalic those with an index above this value.

15 G. Vacher de Lapouge, *Les Sélections sociales*, pp. 66–7, 457.

All that is very sad, and it is hard to see the need to be larger, finer, more this and that, if it is no use in the face of selection and the brachycephalics thus carry the day. We should note that Vacher de Lapouge was a pessimist, convinced that the social selection that in human society had replaced natural selection was leading humanity to its doom by inverting the natural order (i.e., the pre-eminence of the strongest and best: the blonde dolichocephalics). This was no doubt a question of character, since Otto Ammon, Lapouge's German counterpart (1842–1916), drew exactly the opposite conclusion from the same arguments. This shows very well how everything and anything could be proved with these kinds of theories – as was already the case with Darwin himself.

The following is what Vacher de Lapouge wrote about the 'utopia of progress'. We can compare him here with another pessimist, Gobineau, who also described the end of the world (see the quotation on pp. 240–1 below). Lapouge even imitates Gobineau's style:

> Analysis of social selections decisively leads to the most pessimistic of all possible conclusions. The future will not be for the best, but at most for the mediocre. To the extent that civilization develops, the benefits of natural selection change into bitter plagues for humanity. All apparent progress is paid for out of the capital of strength and energy, will and intelligence, and this capital dries up. That is why so many great peoples have disappeared from the face of the earth, leaving behind them no more than ruins, a page of history, and residues that can no longer be used even to constitute new peoples. The great Eastern civilizations that lasted five or six thousand years came to an end. No education could make a people out of the human dust that remains. The powerful movement of Islam, which erupted in the midst of these worn-out hordes, was unable to lead them out of their eternal torpor. The Graeco-Roman world came to an end, with Christianity only hastening its ruin. Constantly on the move, civilization reached west and north-west Europe, and now we can sense that the life of Europe is ceasing, and the days of our world are numbered. No race can possibly resist this inevitable degradation . . . England, America, countries of European race that are today the centre of civilization, will themselves be unable to escape this process. This is the final stage, and will persist only as long as the active elements of the race persist, coming to an end when only the passive elements remain, blonde and dolichocephalic as they may be. Humanity is in its latter days, on the eve of a long period of convulsions beyond which we can glimpse the beginning of decline. The sum of knowledge and material power will continue increasing in a secular accumulation, but man

> will cease to gain in value, and the grandiose futures that utopians dream of for distant generations will be peopled only by mediocrities, and the fathers of still inferior mediocrities.[16]

We can now turn to the other side of the coin, in the guise of Otto Ammon. Starting from the same principles, he reaches a quite different conclusion. For Ammon, social and natural selection intermingle and harmoniously complement each other. He uses an example that we shall find very often, that of egoistic and altruistic behaviour, and notes that a society in which everyone is egoistic would not be viable, any more than one in which everyone is altruistic. Only the middle way is viable, and this is the way that is effectively followed, since according to Ammon, the effect of selection is to suppress whatever is excessive in one direction or the other (he even imagines two selections operating in opposing directions):

> The admirable balance between egoism and altruism, or, to put it more accurately, between individualist and social instincts in humanity, is the result of the play of natural selection . . . Natural selection always suppresses the excessive. In a society that abandons itself too far to the exclusive pursuit of gain without paying heed to the common good, it casts its weight on the scale on the side of the latter; and where men tend to become angels completely full of disinterest, selection acts to produce a generation more vigorous in action . . . This is a remarkable example of the double significance of natural selection. Selection which is progressive in relation to the altruistic instincts is regressive in relation to the egoistic instincts, and vice versa . . . A proportion of valuable men die prematurely without leaving posterity, and considering only this fact one might believe that humanity must continually degenerate. On the other hand, the evident elimination of the valueless could lead us to an optimistic faith in the rapidity of progress. In reality, there are always two selections working in parallel, a favourable and an unfavourable one, as I have developed in my pamphlet *Der Darwinismus gegen die Sozialdemokratie* [Hamburg, 1891] ... The whole question lies in knowing which of the two selections will get the upper hand. If it is the favourable one, a nation develops on an ascending line; in the other case, it gets bogged down and disappears.[17]

---

16 Ibid., pp. 443–5.

17 O. Ammon, *L'Ordre social et ses bases naturelles, esquisse d'une anthroposociologie* [1894] (Paris: Fontemoing, 1900), pp. 36–7.

As soon as this question is raised, the optimistic Ammon thinks that all is for the best in the best of Bismarckian worlds (though he does accept the need for certain reforms). For him, social selection operates very well in harness with natural selection, and if his social Darwinism does not produce a perfect society, it nevertheless steadily approaches it. Above all, man should not disturb this precise social mechanism by ill-considered intervention, no matter how well intentioned:

> This social organization has steadily continued its development, in correlation with the modifications that have taken place in conditions of life, and countless generations of our ancestors contributed to this with their reason and experience, in successive stages. So many interests are involved here, and often contradictory ones, that it would be hard for us to find the right proportions in creating anything new. What flexibility, what marvellous adaptation in these mechanisms developed over the centuries, when we compare them with anything that the human mind can imagine! The social order that results from all these interested forces may be imperfect in some respects, but it represents nonetheless the closest approximation to an ideal state that it is possible to achieve. The procedures by which the social order realizes natural selection in the measure needed, thanks to the separation of social classes, whilst leaving room for the rise or fall of isolated individuals, taking account of factors that would seem irreconcilable, often provoke our admiration.[18]

Ammon is just as anti-democratic, anti-egalitarian, anti-socialist, etc., as Vacher de Lapouge. He professes the same 'anthropo-sociology' and has the same taste for dolichocephalic blondes. But he is more moderate in character, seeing life through rose-coloured spectacles (Gemany is still imperial and Bismarckian, even after Bismarck), where Vacher de Lapouge sees things in gloomier shades (living as he does in the democracy of the Third Republic). No more than that is needed, so that, with the same theory based on the same principles, the political opinions of Ammon seem no more than an extreme right-wing conservatism, whereas Vacher de Lapouge, with his 'revolutionary' excess, has the air of a forerunner of Nazism.

Despite these divergences, there is a common thread in these various social Darwinist theories: that of a 'naturalization' of society and a 'scientization' of sociology. The two tendencies are

18 Ibid., pp. 227–8.

linked, as the scientization of sociology is most commonly (though not exclusively) effected by a naturalization that places society under laws as determinate and inflexible as those governing nature. These supposedly sociological laws are generally matched to the laws of Darwinian biology, with the result that sociology becomes a kind of 'natural science' articulated to biology, its extension on the supra-individual level.

The application of these principles to politics is immediate. Once sociology places society under the equivalent of natural laws, politics has to take this into account in the development of legislation, if only for reasons of effectiveness. Political legislation must conform to nature, its laws and its strict determinism. Politics becomes in this way a genuine science, in tune with the natural motions of the world. This was the age when 'iron laws' of nature (generally Darwinian) were invoked on all occasions, especially when the object was to justify the pitiless character of certain political and social laws.

The sociologist Ludwig Gumplowicz (1838–1909) gave this question its most clear and pitiless formulation. Sorokin classifies Gumplowicz's sociology in his 'mixed' category, as one mingling biological elements with purely sociological ones. Yet Gumplowicz falls completely in our sphere of interest here, as he was among those who insisted most on the naturalization of society, a naturalization largely understood as biologization.[19] He was also one of the principal theorists of the struggle of the races, a struggle that he placed at the root of all social processes.[20] This was his particular way of naturalizing society, an aspect we shall return to in Part Three.

The second book of Gumplowicz's *Sociology and Politics* is explicitly titled 'History as a Natural Process', and the third book, 'Politics as Applied Science'. He makes it very clear here that politics, if it is to be scientific, must be an application of sociology, itself supposedly made scientific by the naturalization of society and history:

> As a study of the regular tendencies and movements of social groups and collectivities, the value of sociology for practical politics is that it teaches familiarity with the march of social development that is in conformity with natural laws, and is as a result necessary and inevitable, in this way showing

19 Cf. a book I was unable to consult here: E. Brix (ed.), *Ludwig Gumplowicz oder die Gesellschaft als Natur* (Vienna: Hermann Bohlaus, 1986).

20 L. Gumplowicz, *La Lutte des races* [1883] (Paris: Guillaumin, 1893).

> the statesman and politician, as well as each individual, the path they have to follow, if they do not want to come into collision with natural tendencies . . .
>
> As we shall see, only by basing itself on sociology can politics become a positive science.[21]

To illustrate the general spirit of this naturalistic and biologizing sociology, as well as the consequences of its principles, I shall present a few extracts here from Gumplowicz's slightly earlier work, *Outlines of Sociology*. In order to properly appreciate these texts, it is necessary to understand that Gumplowicz was no marginal figure, but a recognized and even famous sociologist – even if his extremism was challenged by some. He taught political science at the University of Graz, during the reign of Emperor Franz Joseph, and the book cited here was a treatise designed for students. It is worth comparing the propositions taught here with the legend that Darwinism had a hard struggle to impose itself against an all-powerful religious creationism.

First of all, for Gumplowicz, it must be recognized that society is governed by blind and inflexible rules comparable with laws of nature, despite the claim of certain sociologists who interpose the phantom of human freedom between natural and social processes:

> Modern natural science has successfully demonstrated that even the 'human mind' is subject to physical laws; that the phenomena of the individual mind are emanations from matter. But in the domain of social phenomena unvarying natural laws have not been completely demonstrated. Between 'mental' phenomena subject to the laws of matter and the social world, strode the conception of human freedom to distract and confuse. It seemed to order and control social relations according to its own choice.[22]

In a sociology seen in the image of natural science, there is no more room for this 'conception of freedom'. In the realm of pure biology this goes without saying, since living beings are explained here in terms of strictly determinate physico-chemical principles (cf. the work of Claude Bernard mentioned above). In the field of psychology, the eviction of the phantom of freedom is permitted by the reduction of thought to an emanation of matter (via the

21 L. Gumplowicz, *Sociologie et politique* [1893] (Paris: Giard, 1898), pp. 196, 203.

22 L. Gumplowicz, *Outlines of Sociology* [1885] (New York: Arno Press, 1975), p. 150.

reduction of psychological processes to biological processes, themselves reduced to the physico-chemical). And for Gumplowicz, this determination of mind by matter is completed by a societal determination, which in his view, moreover, ends up completely coinciding with the physical determination. The text here is quite extraordinary, and may remind the reader of more recent theories:

> The great error of individualistic psychology is the supposition that man thinks. It leads to continual search for the source of thought in the individual and for reasons why the individual thinks so and not otherwise. It prompts naïve theologians and philosophers to consider and even to advise how man ought to think. A chain of errors; for it is not man himself who thinks but his social community. The source of his thoughts is in the social medium in which he lives, the social atmosphere which he breathes, and he cannot think anything else other than what the influences of his social environment concentrating upon his brain necessitate.[23]

Not even philosophy and science, those high points of the free spirit, are spared in this hunt for phantoms. Gumplowicz applies Darwinism even to the production of philosophical and scientific theories:

> Probably human freedom seems under less restraint in the sphere of scientific and philosophical thought than anywhere else. 'Thoughts are free' . . . But the object of intellectual labor is the discovery of truth or knowledge. What has been the result of these 'free' efforts for thousands of years? It is the old story of the bottle and the stoppers. After thousands of failures somebody makes a lucky grab and seizes the right stopper for the philosophical hole. But is that the work of a free mind or of meritorious intellectual labor? The necessity immanent in things and conditions was simply fulfilled. Groping in the dark we hit upon a truth.
>
> Scientific and philosophical investigation, that noblest occupation of 'free minds', is a pure game of chance. Philosophical and scientific truths are like rare prizes among thousands of blanks in a wheel of fortune revolving about us. We 'free thinkers', so proud of our 'intellectual labors', grab awkwardly like innocent children and lo! Among a million blanks someone draws a prize.[24]

The consequences for justice and morality logically follow (and we are far indeed here from the creationist religiosity supposed to have governed the academic minds of this time): justice serves to maintain social inequalities, and morality to support them.

23 Ibid., p. 240.
24 Ibid., p. 275.

This is simply because these inequalities are indispensable to the 'natural' operation of society – an aspect that we shall return to in Part Three:

> Rights are always due to the contact of unlike social elements and every right bears evidence of such an origin. There is not one which does not express inequality, for each is the mediation between unlike social elements, the reconciliation of conflicting interests which was originally enforced by compulsion but has through usage and familiarity acquired the sanction of a new custom . . . In short, every right arises from an inequality and aims to maintain and establish it through the sovereignty of the stronger over the weaker . . .
>
> For morals are nothing but the conviction implanted by the social group in the minds of its members of the propriety (Statthaftigkeit) and manner of life imposed by it on them.[25]

The final conclusion is a logical necessity: social institutions should not overestimate human life. Naturalization means that society must follow the model of nature, and this shows scant concern indeed for individuals (we may compare this with the underlying anti-humanism of the Dumont quotation given on pp. 7–8 above):

> But on the other hand nature can indulge in lighter play with human life – millions of children see the light daily and nature has shrewdly provided for this phenomenon not to cease.
>
> Considering these natural conditions is there any sense of justification in overestimating the worth of an individual life as civilized nations do? How much misfortune and evil men might be spared if all the social, political and juridical institutions which follow from such an exaggerated estimation of human life should fall away.[26]

Whatever we might think of this kind of theory (drawn from a manual designed for Austrian students in political science, and translated into several languages, as was also his book on the struggle of races), it must be acknowledged that Gumplowicz, who took his own life in 1909, had the merit of bringing his ideas to their logical conclusion, and – unlike the majority of his colleagues – not dressing them up hypocritically in great humanist sentiment. He was, after all, a contemporary of Nietzsche.

---

25 Ibid., pp. 262, 251.
26 Ibid., p. 279.

2

The Critique of Social Darwinism

As an example of reaction to the multiform use of Darwinism in sociology, and its consequences in terms of morality and politics, we can cite Jacques Novicow (1849–1912), a particularly ferocious critic of social Darwinism: 'Social Darwinism may be defined as a doctrine that views collective homicide as the cause of the human race's progress.'[1] And yet Novicow was far from being on the left; he even puts Marxism, with its class struggle ('banditry from below'), in the same sack as social Darwinism ('banditry from above'). For him, however, by proclaiming that competition, the struggle for life, and selection were the motors of evolution, Darwinism – let alone its extrapolation to other sciences such as sociology – justified the war of all against all, and did away with the ideas of the Enlightenment and liberal humanism.

As a basis for his opinion, Novicow cited four authors at the beginning of his critique: an Englishman, an American, a German and a Frenchman: Herbert Spencer (1820–1903), Lester F. Ward (1841–1913), Gustav Ratzenhofer (1842–1904) and Ernest Renan (1823–92), all of whom interpreted Darwinism in a more or less warlike sense (see the boxed extracts) and placed competition, struggle and selection at the root of almost all social facts. Novicow countered this bellicose aspect of social Darwinism by systematically advancing (sometimes in a somewhat naive fashion) the values of solidarity, cooperation and mutual aid. Sorokin classifies Novicow in his category of 'organicist' sociologists, meaning that, even if he does not deny the importance of struggle, he privileges organized relationships of co-operation among individuals, after the model of the co-operation of organs in a biological organism. These two biological conceptions of society are often counterposed, generally with the assertion that the

1 J. Novicow, *La Critique du darwinisme social* (Paris: Alcan, 1910), p. 3.

Darwinian model should replace the organicist one – in economic terms, free competition against the organizing state.[2]

Novicow clearly chose the texts of these particular authors as a function of what he sought to demonstrate; note that the harshest text is from a German. There are differences of style between them, but also – and this is not just a matter of Novicow's choice – a certain unity of thinking and a common philosophy that strikes us as prescient a century later.

Social Darwinism and War

1) Texts cited by Jacques Novicow in *La Critique du darwinisme social* (Paris: Alcan, 1910)

Herbert Spencer, *Principles of Sociology* [1876] (London: Macmillan, 1969), pp. 177–8, cited from the 1883 French edition in Novicow, *Critique*, pp. 3–4.

We must recognize the truth that the struggles for existence between societies have been instrumental to their evolution. Neither the consolidation and reconsolidation of small groups into large ones; nor the organization of such compound and doubly compound groups; not the concomitant developments of those aids to a higher life which civilization has brought; would have been possible without inter-tribal and inter-national conflicts. Social co-operation is initiated by joint defence and offence; and from the co-operation thus initiated, all kinds of co-operations have arisen. Inconceivable as have been the horrors caused by this universal antagonism which, beginning with the chronic hostilities of small hordes tens of thousands of years ago, has ended in the occasional vast battles of immense nations, we must nevertheless admit that without it the world would still have been inhabited only by men of feeble types, sheltering in caves and living on wild food . . .

[The] inter-social struggle for existence which has been indispensable in evolving societies . . . [R]ecognizing our indebtedness to war for forming great communities and developing their structures.

*

Lester F. Ward, 'Evolution of Social Structures', *American Journal of Sociology*, March 1905, pp. 594–5, cited in Novicow, *Critique*, pp. 4–5.

Just as organic evolution began with the metazoic stage, so social evolution began with the metasocial stage. So, too, as the metazoic stage was

2 P. A. Sorokin, *Contemporary Sociological Theories*, pp. 194–218.

brought about through the union of several or many unicellular organisms into a multicellular organism, so the metasocial stage was brought about by the union of two or more simple hordes or clans into a compound group of amalgamated hordes or clans . . . Two groups thus brought into proximity may be, and usually are, unknown to each other. The mutual encroachment is certain to produce hostility. War is the result, and one of the two groups is almost certain to prove the superior warrior and to conquer the other. The first step in the whole process is the conquest of one race by another . . . The greater part of the conquered race is enslaved . . . The slaves are compelled to work, and labour in the economic sense begins here. The enslavement of the producers and the compelling them to work was the only way in which mankind could have been taught to labor, and therefore the whole industrial system of society begins here.

*

Gustav Ratzenhofer, *Die sociologische Erkenntnis* (Leipzig: Brockhaus, 1898), pp. 233–4, cited in Novicow, *Critique*, pp. 5-6.

The formation of the state does not result from the play of free interests, as does the formation of the horde, the tribe, and various kinds of parties and associations; no, it is the outcome of antagonistic interests, and as a result is a coercive organization . . . Every evolution is the resultant of competition, but in the case of the state, violence is already the agent that creates it. The criterion by which one may judge whether the state fulfils its mission in social life is whether this violence follows the path of social necessity, and genuinely acts in the perspective of natural interests. Whenever one departs from this fundamental conception of the state, whenever the claim is made that the state can arise simply from the effect of civilization, from a peaceful agreement or any other device of this kind, we come into contradiction with the teachings of sociology and end up with political experiments that terminate in the most lamentable fashion.

*

Ernest Renan, *La Réforme intellectuelle et morale* (Paris: Lévy, 1871), p. 111, cited in Novicow, *Critique*, p. 6.

If the stupidity, negligence, idleness and lack of foresight of states did not have the result of making them fight each other, it is hard to tell what degree of abasement the human species might descend to. War is in this way one of the conditions of progress, the lash that prevents a country from going to sleep, by forcing a self-satisfied mediocrity to emerge from its apathy. Man is sustained only by effort and struggle . . . The day that humanity becomes a great pacified Roman Empire with no further external enemies will be the day morality and intelligence run the greatest dangers.

2) Texts from Walter Bagehot in *Physics and Politics* (London: H.S. King, 1872), pp. 49–50.

The progress of the military art is the most conspicuous, I was about to say the most *showy*, fact in human history.

Each nation tried constantly to be the stronger, and so made or copied the best weapons . . . Conquest improved mankind by the intermixture of strengths; the armed truce, which was then called peace, improved them by the competition of training and the consequent creation of new power. Since the long-headed men first drove the short-headed men out of the best land of Europe, all European history has been the history of the superposition of the more military races over the less military – of the efforts, sometimes successful, sometimes unsuccessful, of each race to get more military; and so the art of war has constantly improved.

[O]n the whole, the energy of civilization grows by the coalescence of strengths and by the competition of strengths.

3) Texts from Georges Vacher de Lapouge, *Les Sélections sociales* (Paris: Fontemoing, 1896), pp. 211, 212, 199.

The most ancient songs of the poets and the oldest historic documents show the constant war of all against all. Homer's blonde warriors, like the Gallic and Scandinavian heroes later on, most often died a violent death. Each tribe was at war with its neighbours, and the main occupation seems to have been that of destroying those nearby. Rereading the heroic songs of the early Aryans, we may ask ourselves in amazement how female fertility was sufficient to make up for the ceaseless destruction of the adult population. There was nothing but massacre and butchery . . . Whether the poet was Greek, Germanic or Hindu, the tale scarcely varies, and it ends only when battle ceases for lack of combatants.

Travellers today tell horrendous stories of the centre of Africa. The Negroes of the interior closely reproduce the social state of our own populations of the Bronze Age and early Iron Age. Their life is a perpetual see-saw between bloody expeditions that bring back slaves, ivory and booty, and sudden catastrophes, night-time surprise attacks, the victor of which will be the vanquished of tomorrow . . . All the fertility of the Negresses is required to compensate the effects of this unbelievable consumption of human lives, but this fertility is so great that Africa would become a terrible threat to civilization if we were to impose on it our peaceful life and our simplest arts. This is the source of the barbarians of tomorrow – not barbarians of a noble race like the Germanic, but the possible destroyers of our civilization.

In the fully civilized countries, man has no more enemies to fear, dangerous animals have been destroyed, he has only to busy himself with seeking the necessities of life, which merchants are on hand to supply. The struggle for existence, and its counterpart that homo homini lupus, is no more. It is exercised only through social acts. But this change in its mode and name does not mean it is less bitter and murderous.

Novicow's book, and his choice of quotations, have often been seen as greatly exaggerated and fanatically anti-Darwinian, tainted as well with a revanchist 'anti-*boche*' spirit, as it is Germany above all that he reproaches for bringing forth this kind of theory. He was not, however, the only one to think this way, and it should also be recognized that in a certain sense he was vindicated by events. War, and war between races in particular, was indeed one of the leitmotifs of the application of Darwinism to sociology.

The role of war in society was also theorized in a Darwinian perspective by a range of authors. One chapter in Bagehot's *Physics and Politics* is significantly titled 'Struggle and Progress'. Bagehot, more moderate than the writers cited by Novicow, was far from apologizing for war, but in his case as well we can find sufficiently 'militarist' assertions (see examples in boxed text above).

Bagehot at least was a thoughtful and serious intellectual, which cannot be said of some of these writers. All the warmongers of the time – and they were numerous enough – were carried away by Darwinism and its social applications. According to Léon Poliakov:

> Already in 1889, Max Nordau noted that Darwin was in the course of becoming a supreme authority for the militarists of all European countries: 'Ever since the theory of evolution has been proclaimed, they can cover their natural barbarism with the name of Darwin and give free rein to their bloody instincts, calling this the last word of science.'[3]

Otto Ammon, for example, was very fond of war, at least on the level of theory: it was the ideal instrument of selection, for both

3 L. Poliakov, *Le Mythe aryen* (Paris: Pocket, 1994), p. 390.

nations and individuals, and thus the ideal instrument of the social Darwinism that he championed:

> War is not just an instrument of natural selection in the sense that it assures the nations superior in vigour and intelligence the supremacy they deserve; it can also act by selection in a beneficial sense on isolated individuals . . . The army might be compared with an enormous sponge, which, at the moment of mobilization, absorbs every man in a condition to bear arms, and which, after a war of short duration, restores a selection of the most adroit, the most vigorous and the most resistant.[4]

Vacher de Lapouge, who cannot be suspected of hostility to Darwinism and its social applications, likewise sees war on all sides. The quotation in the box above gives some examples concerning the peoples of antiquity, primitive civilizations and the society of his time. Here is a passage of his on the subject of future war, i.e., war in the twentieth century:

> I am convinced that in the next century, millions will be slaughtered for one or two degrees more or less in the cephalic index; this is the sign that is replacing the biblical shibboleth and the linguistic affinities by which people recognize one another . . . The last sentimental individuals will witness copious exterminations of peoples.[5]

Vacher de Lapouge was prophetic in his way. This is no doubt why he is so frequently cited by historians (a temptation hard to resist, given the number of 'gems' there are in his books), sometimes to the detriment of authors who enjoyed much greater official recognition in their own day. This prophetic character is due to the fact that he concentrated in his texts everything that gravitated around the Darwinian notions of man and society. He certainly did not invent a great deal in this domain. His successive posts as librarian enabled him to read a good deal and keep up with everything happening in his chosen field, all the more so as he could read several languages. He was perfectly able to grasp the spirit of what he was reading and summarize it in his books. What he subsequently produced is a kind of involuntary caricature that reveals the inner nature of the subject studied. It is perhaps due

4 O. Ammon, *L'Ordre social et ses bases naturelles*, pp. 317–18.
5 Cited, without reference, by F. de Fontette, *Le Racisme* (Paris: PUF, 1981), p. 72.

to this aspect of caricature that the success he enjoyed with the public was not matched by an academic career. For our present purpose, however, there should be no mistake: the caricaturist is not totally responsible for what he caricatures, and he should not be treated as a scapegoat for the benefit of individuals better placed in society, more polite in their expression, but who profess the same theories with far greater authority.

Compared with a number of his social Darwinist contemporaries, moreover, Vacher de Lapouge did not really have a taste for war. Here are his conclusions on the question, which could hardly be more humanist:

> If the forty million men who are killed each century were buried together, separated by one foot, they would reach round the earth. Placed alongside one another and holding hands, they would make a line fifteen thousand leagues in length, or a thousand leagues in fifteen ranks. The wars of a century spill six hundred million litres of blood, or three million barrels of two hectolitres. The annual total is six hundred cubic metres, the daily total sixteen thousand litres. This is a flow of 680 litres per hour that war draws from the veins of humanity! If the direct victims of war from the beginning of historical time were put together, tightly packed like sheaves, their mass would cover the whole of France, the whole of Germany, and a large portion of the rest of Europe.[6]

To understand this obsession with war in Darwinian sociology, we need to bear in mind the degree to which Darwinism marked the biology of the late nineteenth century, and the particular fashion in which it did so. Struggle, competition and selection were made into universal explanations in every field. These principles were applied not just to relationships between the living being and its external environment, but also within the individual, its different parts seen as struggling against each other.

This obsession is summed up by the title of a book by Félix Le Dantec, a prominent biologist of the time, who, although a Lamarckian, adopted the Darwinian principle: *La Lutte universelle*. As a complement to this already explicit title, the book's cover bore the proud motto 'To be is to struggle, to live is to conquer.'[7] An entire programme, and broadly applied to the most varied fields.

6 G. Vacher de Lapouge, *Les Sélections sociales*, p. 223.
7 F. Le Dantec, *La Lutte universelle* (Paris: Flammarion, 1906).

This principle of universal struggle evidently acted as a phagocyte on its immediate competitor: Pasteurianism. Pathology, especially the pathology of infection, as well as immunology and vaccination were now interpreted in terms of struggle. All were thus brought within the range of Darwinism: the struggle against the pathogenic microbe, a struggle in which the individual, if he triumphs, emerges strengthened (immunized) and ready for new victories, all of this on the basis of competition for life and natural selection. This is all very innocent and not wrong, mere rhetoric, but we shall later come to discuss far more fearsome Darwinizations of Pasteurianism (see Part Two).

Bichat's famous definition that 'life is the set of functions that resist death' was here expanded and twisted into a more aggressive formula. No doubt by virtue of the principle of competition – at least in so far as this was not an application of the saying that attack is the best defence – life was not simply the resistance imagined by Bichat but also 'the invasion of the environment by the living being',[8] a notion that perhaps owes rather more to Lamarck than to Darwin. On this basis, struggle becomes the universal mode for understanding the relationships between individual and environment. Everything is war, competition and selection.

This struggle was even imported into the interior of the living being. The first to proceed to this internalization was Francis Galton, Darwin's cousin, who saw the genetic elements provided by the two gametes at the point of fertilization as competing so that one of them would be eliminated (somewhat equivalent to 'chromatic reduction' or meiosis, but a meiosis seen as taking place after fertilization).[9] Later the idea would be taken up in a similar way by August Weismann in his theory of the germ-plasm, in which this supposed competition also had the function of eliminating surplus genetic elements; only the most 'competitive' would contribute to the formation of the individual arising from the fertilized ovum (the others being 'vanquished' and serving as nourishment for the victors).[10] In the meantime, moreover, Wilhelm Roux had generalized this method to explain embryogenesis and

8 Ibid., p. 8.

9 F. Galton, 'A theory of heredity', *Contemporary Review*, 1875, 27, pp. 80–95.

10 A. Weismann, *The Germ-Plasm, a Theory of Heredity* [1892], translated by W. Newton Parker and H. Rönnfeldt (London: Walter Scott, 1893), pp. 49, 264–6, 280ff., 295–6.

cellular differentiation in terms of a struggle among different parts of the organism.[11]

These ideas underwent other developments, both in terms of heredity (with theories such as 'gamomachia' and 'gamophagia', designed to explain the greater or lesser resemblance of the child to one or the other parent by imagining a struggle within the embryo between the paternal and maternal hereditary contributions) and metabolism (with a mutual struggle between diastases, enzymes that were frequently considered at the time as the characteristic molecules of life, or even as living molecules).

Struggle, competition and selection were omnipresent in the biology of the late nineteenth and early twentieth centuries, ideas that could be used for all purposes, to explain everything and anything. They played the exact role that genetic programming plays today in this respect, this also being invoked for every occasion. Almost everything today is 'genetically programmed'. In this earlier period, almost everything was 'war and selection', whether this was expressed in a crude fashion or dressed up in a vague and pseudo-Heraclitean philosophy ('conflict is the father of all things').

Why was Darwinism so insistent on this sort of idea? Quite simply, because there was not yet at this point any scientific explanation of heredity (genetics was not properly established until 1900–15), nor even a theory of variation (the theory of mutation dates from 1901–3, and it took a certain time before it was developed into a usable form). During the whole period from the publication of Darwin's *The Origin of Species* in 1859 until nearly 1915 (which saw the publication of *The Mechanism of Mendelian Heredity* by Thomas H. Morgan, the work that founded modern genetics), the only characteristic notions of Darwinism were struggle, competition and natural selection. It was therefore these that served as its banners and spilled over into other disciplines, in particular the human and social sciences (just as today genetic programming is not confined to biology, but extended to psychological and even sociological facts).

In biology, the advent of genetics (between 1900 and 1915) went some way to correcting this pre-eminence of struggle and selection, focusing the interest of biologists towards the study of

11 W. Roux, *Der Kampf der Theile im Organismus* [1895], (Leipzig: Engelmann, 1895), vol. 1, pp. 135–437.

heredity and variation. In those human and social sciences already contaminated, however, struggle and competition continued their career. This is certainly the reason, as A. Béjin notes (see quotation on p. 10 above), that after 1905 there were no more innovations in social Darwinist theories (biology, with its new focus, no longer supplied them with anything new, and the theorists were unable to imagine these by themselves), and why the next step was to move on to practice, that is, to apply to politics what they had previously just imagined. The first eugenic legislation, for instance, dates from 1907, almost thirty years after the initial formulation of the theories behind it.

Following are a few cases of these Darwinian overspills into the human and social sciences.

In psychology, the theory of a struggle of ideas or images arose very rapidly. It can already be found in Hippolyte Taine's book *De l'intelligence* of 1870. Dumont, who is always very clear, wrote three years later:

> There arises, within each of us, a kind of vital competition between ideas, a combat for admission and preservation; and when the regime of our mind is liberal, when no authoritarian direction comes to intervene, the strongest and most lively thoughts, that is the true ones, always end up by suppressing and driving out the weaker ones, those most contrary to truth. The human mind constructs truth for itself; this truth is the final result of the thinking of generations exercised about the reality of things; the good is the point to which all the moral tendencies of individuals, balancing one another in their mutual actions and reactions, spontaneously lead.[12]

We can call this a caricatured application of Anglo-Saxon economic liberalism to psychology.

The extension of Darwinism to sociology was still easier; it went without saying, since society is the place where competition and selection are exercised. Yet it was still necessary to explain why there was struggle and competition over practically everything. For Darwin, the selection of the fittest took place essentially in competition for food or for females, between individuals of the same species occupying the same territory. This process was very rapidly expanded into a struggle of the living organism against the external environment in all its forms: the physical environment (for example, climatic conditions) as well as the living environment made up of

12 L. A. Dumont, *Haeckel et la théorie de l'évolution en Allemagne*, p. 15.

individuals both of the same species and of others (possible predators or prey). Competition was thus not just for food or females but for absolutely everything, among other things for territory and living space, a fact that sociology had necessarily to register. In his book quoted above, Le Dantec offers an example in the form of a potato plant that shoots up, spreads and gains ground, interpreting this as a veritable territorial conquest.[13] This may be completely anodyne, but we know how questions of territory and living space were interpreted twenty years later, after being transposed to human society.

We clearly find ourselves in a realm of bad analogies. Novicow already stressed this in 1910, and, as against what one might believe at first sight, his critique is in no way excessive. To appreciate this, we need only two quotations – the first by Novicow, from 1910:

> Let us start with the false comparisons. Two plants dispute a field; therefore, say the Darwinians, struggle is a natural law, and therefore citizens of civilized states should massacre one another to the end of time. It would be hard to find a more arbitrary 'therefore', given how enormous the difference between plants in a field and citizens of civilized states. Relationships between plants do not resemble in any way those between people. I have already criticized these superficial comparisons from the standpoint of biology. Here I simply want to show how unsustainable they are from a purely logical point of view. If there is a rule that no reflecting mind should ever forget, it is that comparison should be between comparable facts. Purely external analogies are not sufficient for building a positive science.[14]

Criticism, however, did not prevent such theories from following their course. They can be found in the politics of the 1930s and 1940s, with the notion of *Lebensraum*. Disciplines such as sociobiology and ethology resurrected them long after the war, with all kinds of political extrapolations.

The second telling quotation, from 1961, is by Robert Ardrey, from whom we learn that the human territorial instinct is more imperative than sexuality, all this being clearly tied up with war (and even, indirectly, via the appeal to Arthur Keith, to war between races):

> The territorial drive, as one ancient, animal foundation for that form of human misconduct known as war, is so obvious as to demand small

13 F. Le Dantec, *La Lutte universelle*, pp. 28–9.
14 J. Novicow, *La Critique du darwinisme social*, p. 164.

> attention. When Sir Arthur Keith found himself too old for any active contribution to the Second World War, his broodings produced the marvellous volume, *Essays on Human Evolution*, and the conclusion: 'We have to recognize that the conditions that give rise to war – the separation of animals into social groups, the "right" of each group to its own area, and the evolution of an enmity complex to defend such areas – were on earth long before man made his appearance.' Such an observation of a human instinct probably more compulsive than sex throws into pale context the more wistful conclusions of the romantic fallacy: that wars are a product of munitions makers, or of struggles for markets, or of the class struggle; or that human hostility arises in unhappy family relationships, or in the metaphysical reaches of some organic death throes.[15]

By dint of explaining anything and everything by a Darwinian struggle, it was inevitable that war itself came to be interpreted in Darwinian terms – either to reject it, arguing that it eliminated the young and strong but kept alive the old and sick (a point made by Novicow himself),[16] or on the contrary to praise it, seeing war as the origin of all things (a point made in the boxed texts above that were cited by Novicow).

In a certain sense, the English naturalist Peter Chalmers Mitchell accepted Novicow's criticisms of the warlike character of social Darwinist theories when he waxed indignant in 1916 at the recuperation of Darwinism by the Germans, and especially the use they made of it to claim a racial basis to their bellicose nationalism. This is what Émile Boutroux wrote in his preface to the French translation of Chalmers Mitchell's book, and a good summary of his thesis:

> Now it seems to me that, among the very numerous philosophical theories current in Germany, and which exert such a real and deep influence on German behaviour, one of the most important is the idea of a brutal and ineluctable necessity, inherent in the nature of things, before which all the protestations of human conscience are no more than the vain grizzling of an ill-behaved child. And among these iron laws, one of those that the Germans are most eager to invoke is the 'struggle for existence' so powerfully highlighted by Darwin. In the name of Darwinism, they believe they can

15 R. Ardrey, *African Genesis* (London: Collins, 1961), pp. 189–90. Sir Arthur Keith put forward a theory that more or less assimilated nations to races, and thus interpreted war in a framework of Darwinian biology (see quotation on p. 298 below).

16 Novicow, *La Critique du darwinisme social*, p. 184.

> maintain that science itself, the most modern and solid science, condemns all the nations on earth to be either assimilated or destroyed by the German nation, as the species best equipped for the struggle for existence.[17]

This text dates from 1916, not from the Nazi era, when such ideas were systematized. Their origin, at all events, is made very clear here (whereas in the Nazi era it was suppressed). If Chalmers Mitchell criticizes this German use of Darwinism and sees it as an abuse, this does not erase the affiliation. But the First World War naturally played a part in his protest against the Germans, since on the one hand, similar theories had been proposed by authors of many nationalities (see boxed texts on pp. 22–25), while on the other hand, the racial classifications of Darwinian anthropology had long since grouped Germans and English together in the same superior race, without either side ever protesting this proximity, let alone the supposed superiority. Below is what Haeckel wrote on the subject, and we should note the particular character of the arguments on which he based this superiority: his adherence to Darwinian evolutionism and his own monism. In other words, anyone who criticized these theories only signalled their own membership in an inferior race:

> [A]t present the same [i.e., leading] position is occupied by the Germanic. Its chief representatives are the English and Germans, who are in the present age laying the foundation for a new period of higher mental development, in the recognition and completion of the theory of descent. The recognition of the theory of development and the monistic philosophy based upon it, forms the best criterion for the degree of man's mental development.[18]

The superiority of the Indo-Europeans, as we tend to forget today, was not just that of the blonde Germanics but also of the Anglo-Saxons. There was even a regular fashion in Anglo-Saxon superiority at that time, to which we shall return in Part Three.

Irrespective of this particular point, we see in any case how greatly Darwinism – struggle and selection – marked the mind of this era, and what orientation it gave to sociology. We can understand how Novicow and many others were disturbed by a

---

17 Preface to P. Chalmers-Mitchell, *Le Darwinisme et la Guerre* (Paris: Alcan, 1916), pp. viii–ix.

18 E. Haeckel, *The History of Creation* (London: H. S. King, 1876), vol. 2, p. 332.

state of mind that was warlike to say the least (even warmongering), and supported (if not induced) by scientific authority.

Novicow's work deals with social Darwinism in the strict sense of the term, and criticizes above all its militaristic character. But Darwinism also led to other kinds of sociological theory, which, if less overtly warlike, have scarcely greater value and are generally almost as terrifying for their crass stupidity. Examples of these can be found in almost all countries. It is hard even to take the measure of this phenomenon, for the very same reasons that we can scarcely imagine the degree to which eugenic doctrines were so rampant in the first half of the century: because, quite simply, these are theories and facts which, whether deliberately or not, have been all but forgotten by history, and it is very hard at this point to give them their rightful place. Besides, after the Second World War, Darwinism was modified to make it more presentable, to the point that biologists can pretend not to recognize it in this earlier form.

The idea of applying Darwinism to society and politics was immediate. What Vacher de Lapouge says of this is perfectly clear and correct:

> Ever since the publication of *The Origin of Species*, far-sighted minds have understood that ideas about history and social evolution, the very foundations of morality and politics, could no longer be what they had previously been. Clémence Royer was, I believe, the first to state in the preface to her translation of the book [in 1862] that Darwin's discovery would be still more important from the social point of view than from that of biology . . .
>
> By formulating the principle of the struggle for existence and selection, Darwin not only revolutionized biology and natural philosophy; he transformed political science. Possession of this principle allows us to grasp the laws of the life and death of nations, which escaped the speculation of philosophers.[19]

These social Darwinist theories are of absolutely no interest in themselves, and this is perhaps the reason they are so neglected today, even by historians. They are simplistic and superficial in the extreme, and indeed scarcely even coherent. They can be made to say whatever one wants; all that is needed is a bit of rhetorical skill. They have indeed been used to justify anything and everything,

19 G. Vacher de Lapouge, *Les Sélections sociales*, pp. v–vi, 1–2.

and we find them in infinite variety. It is not the theories that are interesting, but rather the reasons for their success.

We might spontaneously believe that it was the success of Darwinism in biology that spurred the social sciences to borrow its ideas. The problem is that this success is itself rather hard to understand. The Darwinian explanation of the evolution of species was for a long time very shaky. It was only in the 1910s that Darwinism acquired the general shape that it has today, and began to be somewhat more convincing from a scientific point of view – essentially thanks to genetics, which appeared at this time.

For a full half-century, from the publication of *The Origin of Species* in 1859 to the early 1910s, ideas about evolution were extremely confused and could scarcely be taken as a model in the human and social sciences. This period was marked by a muddle of different theories (Galton, Haeckel, Weismann, De Vries, etc.), both overlapping and opposed, but all appealing to Darwin and all more or less agreeing with him on the question of natural selection. This was indeed their only point of agreement, and the point that became central to the rather hazy constellation of biological doctrines gathered under the name of Darwinism. (Even here, the agreement was never complete, such as in the case of De Vries.) Moreover, this was the only point in the theory that was exported to other disciplines (psychology, sociology, and so on).

The incontestable success of Darwinism in biology over these fifty years is thus as much a problem as its success in the human and social sciences, and it was probably due to the same reasons, which have nothing scientific about them. In 1903 Yves Delage, who was not a Darwinian but nonetheless an extremely respected zoologist and biologist, could still write:

> I am . . . absolutely convinced that the reason some people believe in transformism and others do not is not for reasons drawn from natural history, but by virtue of their philosophical opinions. If there was an alternative scientific hypothesis to [Darwinian] descent that explained the origin of species, a number of transformists would abandon their present opinion as insufficiently demonstrated.[20]

In 1859 the idea of evolution was already quite commonplace, even if not yet accepted by everyone. All that was lacking, in

20 Y. Delage, *L'Hérédité et les grands problèmes de la biologie générale* (Paris: Reinwald-Schleicher, 1903), p. 204.

fact, was a convincing explanation. (That given by Lamarck was already out of favour.) This explanation was offered by Darwin, in the form of natural selection, and it is this, rather than the idea of evolution as such, that is the nub of Darwinism.

The idea of explaining the origin of species by the selection of variant forms, moreover, was itself not totally new. It was proposed by Maupertuis in the eighteenth century,[21] taken up by Patrick Matthew in 1831,[22] and it is well known that Alfred Russel Wallace developed it at the same time as Darwin.[23] By the mid-nineteenth century it was very much in the air (not only Matthew, Wallace and Darwin but no doubt many others who have been forgotten). But it only became successful with Darwin's publications, even though Darwin was far from presenting a well-constructed and convincing theory.

If we go back to Novicow and his criticisms, it is notable that he does not refer to the success of Darwinism in biology in order to explain its success in the human and social sciences, but actually challenges the Darwinian explanation of evolution – though certainly not evolution as a fact. He is clearly hard pressed to explain the domination of the intellectual field by Darwinian theories, and offers a number of different reasons.

The first of these is certainly true, but incomplete. According to Novicow, Darwin's ideas appealed both to enlightened minds and to brutal and retrograde ones. Like Dumont's text cited above, this quotation is somewhat long, but it is highly enlightening for the way in which Darwinism was perceived by its contemporaries – not a reconstitution, therefore, but an attestation:

> The doctrine of Darwin, put forward half a century ago, has rapidly spread across the whole world. Right from its first appearance, it has been used to explain almost all natural phenomena, from the formation of celestial nebulae to variations in literary styles. In these fifty years, every science, starting with astronomy and ending with sociology, has been deeply impregnated by Darwinism.

---

21 P. L. Maupertuis, *Vénus physique* [1752] (Paris: Aubier-Montagne, 1980), pp. 133–45; *Système de la nature* [1756], (Paris: Vrin, 1984), p. 164.

22 P. Matthew, *Naval Timber and Arboriculture* (Edinburgh: Black, 1831).

23 To settle the question of priority, their two theses were initially published together: C. Darwin and A. R. Wallace, 'On the tendency of species to form varieties; and On the perpetuation of varieties and species by natural means of selection', Journal of the Proceedings of the Linnaean Society (Zoology), 1858, 3, pp. 45–62.

What is responsible for the immense success of this doctrine? It is the fact that it responds both to the noblest and the basest aspirations of the human soul. It satisfies all sides – conservatives keen on brute force, liberals keen on the idea of justice, positivist and monist freethinkers, and idealist and dualist believers ...

The triumph of Darwinism marks the release of the human mind from the bonds of theology. This makes it one of the most important events in the history of our time. The supernatural, successively expelled from each of the physical sciences, had found its last refuge in the fields of biology and psychology, where it was believed to be beyond challenge. The phenomena of life and mind, it was claimed, could only be explained by the existence in nature of a conscious principle tending towards a determined end. Life and thought proved the existence of God and the truth of dualist philosophy. With Darwin's theory, this whole edifice, patiently constructed over centuries, crumbled away. Since the transformation of species took place by natural means, there was no need for any miracle, either in the field of biology or that of psychology. The whole of nature appeared as a grandiose and magnificent unity. Immutable order in the universe replaced an arbitrary divinity. Man could raise his head, and feel himself the master of the world: he saw unbounded horizons open before his eyes, with no authority now able to stop him in his conquests. With Darwinism bringing the definitive liberation of the human spirit, we can understand the enthusiasm with which it was welcomed by thinkers of a genuine scientific cast – all who had completely freed themselves from outdated traditions, old routines and the ignorance of a barbaric past ...

One other circumstance also aroused a great enthusiasm in favour of Darwinism among all enlightened minds. It propagated the idea of the survival of the fittest, the triumph of the best. This was a declaration that nature practised an incorruptible justice, that the idea of justice was already present in the field of biology ...

These were the elements that assured Darwinism the favour of the most enlightened and liberal minds of our era.

But let us now move to the opposite side.

Darwinism anticipated the archaic instincts of brutality so deeply anchored in the minds of traditionalists, time-servers and the ignorant – who unfortunately form the great majority of the human race. When the theories of Darwin first became fashionable, Marshal von Moltke was able to write with apparent scientific support that war (in other words, collective homicide) 'conformed to the order of things established by God', since this 'order established by God' perfectly corresponded to the expression of the 'laws of nature' to which the positivists and Darwinians appealed.

> All these brutal and violent spirits seized upon Darwinism with enthusiasm. It enabled them to raise the lowest instincts of banditry to the height of a universal law of nature. Since the weakest were necessarily doomed to perish in the struggle for existence, this being the immutable principle of the living world, vae victis was the most rational and legitimate motto imaginable.[24]

A number of elements in this text are worth noting. First of all, the extreme ambiguity of Darwinism, which led to its being taken up by a great range of philosophies and politics that were completely different and even opposed (an ambiguity that we shall often find in the social uses of Darwinism). Then, just as important, the considerable and very rapid success that Darwin's theory experienced, and its export into various disciplines. Novicow describes this as a phenomenon of invasion, proliferating like a cancer or gangrene. We are very far here from the legend that claims Darwinism had to wage a fierce battle to impose itself against a supposedly omnipresent creationism. Novicow is clearly right: at the time of publication of *The Origin of Species* the idea of evolution was already very current, whereas creationism essentially belonged to the natural theology of the eighteenth century, and very little to the biology of the nineteenth – a fact we shall return to below.

The massive export of Darwinism into other disciplines – which attests to the doctrine's success – is also confirmed by a book published at about the same time as Novicow's: *Darwin and the Humanities*, in which the American author James Mark Baldwin (1861–1934) reviewed in turn Darwinian psychology, Darwinian sociology, Darwinian ethics and Darwinian logic.[25] There was also Darwinian linguistics, an example of which is given below (pp. 308ff.), though this was not treated by Baldwin.[26] In practically all the disciplines that form the social sciences and humanities today, Darwinian theories were to be found.

The ambiguity of Darwinism, and its ability to fit in with a great range of points of view – all that is certainly true, but it is not sufficient to explain its omnipresence, its universal and rapid success, in these greatly varied disciplines.

24 J. Novicow, *La Critique du darwinisme social*, pp. 9–11.

25 James Mark Baldwin, *Darwin and the Humanities* (Baltimore: Review, 1909).

26 The main author here is August Schleicher, *Darwinism Tested by the Science of Language* [1863], and *De l'importance du langage pour l'histoire naturelle de l'homme* [1865] (Paris: Vrin, 1980).

Novicow offers a further reason: the simplicity of the theory, making it seductive and accessible to all, and the kind of intellectual terrorism it very rapidly came to exercise in the name of science and modernity – as opposed to the reactionary and religious theories of a supposed creationism:

> Darwinism, like a stream, has carried everything along with it. It has rapidly invaded a wide range of sciences, from astronomy to psychology and sociology. On all of these it has imposed its laws, always as a despotic master; everywhere it has silenced the voices of serious specialists who, unwilling to submit to the fashion of the day, declared that each science had its particular phenomena, which had to be studied directly and separately rather than confusing them with phenomena from different sciences. Universal struggle is all very well, dynamism is even better; but this struggle is expressed in different ways when we move from one field of nature to another. People were unwilling to accept such reservations. Hasty and superficial generalizations, pronounced with the most trenchant assurance, were given out as the definitive result of science. Those who would not go along with this unconsidered tendency were treated with contempt, called time-servers and retrogrades ...
>
> Darwinism, with its simplistic and superficial doctrines, has also turned social science away from the patient and meticulous study of the facts. It has been a poison for professionals as well as for the general public. It has significantly impeded the progress of sociology.[27]

Here again, Novicow is right as far as he goes (intellectual terrorism is still current, and in more or less the same form), but something is still missing from his explanation: What made possible these methods by which Darwinism imposed itself?

To understand this multidisciplinary success, we need to return to the quotation from Dumont given above (pp. 7–8), and compare this with the legendary claim that Darwinism prevailed in biology against an omnipresent creationism. If this struggle against creationism is very largely imaginary, what then does the legend correspond to? What does it mean, and what purpose does it serve?

In the first place, it totally effaces the scientific aspect of the controversy in biology. It confuses Darwinism and evolutionism, by transforming the 'fixism' of the Cuvier school, which was certainly real but on the decline since the master's death in 1832,

27 J. Novicow, *La Critique du darwinisme social*, pp. 384, 390.

with religious creationism, forgetting that the arguments for 'fixism' were scientific – based on Cuvier's principles of anatomy and taxonomy. In place of these biological arguments, to which Darwinism initially had no response, the legend puts forward arguments of a sociological nature – opposing religion (supposedly the basis of 'fixism') to science (Darwinism), and the retrograde spirit to modernism and progress.

Since this Darwinian legend refers to sociological rather than biological arguments, it bears on the role of Darwinism in the social conceptions of the time, and not on its role in biology (where creationism was marginal, and 'fixism' on the decline). It is in the society of this era, and not in biology itself, that the origin of Darwinism's success must be sought, both its biological and its social success. The explanation is not at all hard to find, but it is masked by the entanglement of a number of different doctrines.

The second half of the nineteenth century (which for the present purpose runs from *The Origin of Species* in 1859 to Morgan's *The Mechanism of Mendelian Heredity* of 1915) saw the completion and triumph of the Industrial Revolution. This was the time at which, in different modalities and at different speeds, the bourgeoisie completely and definitively replaced the aristocracy as the dominant class in all European countries. If the aristocracy could justify its social rank, power and wealth by appealing to a right of birth and blood – which had, if not quite divine backing, then at least the blessing of the church – the bourgeoisie could make its stand only on its merits and its work: the social rank, power and wealth it had won for itself. Natural merit replaced 'blue blood', and the aristocracy of divine right was succeeded by a 'meritocracy' founded on nature.

Darwinism itself was a meritocracy, to the extent that it assured the success of the fittest in the competition for life (Novicow sees this as a kind of incarnation of justice in biology – see quotation on pp. 36–8). In this respect, it was a typically bourgeois doctrine. Dumont was completely correct to stress this, even if certain French conservatives, still marked by their Catholicism, bridled somewhat. Besides, the epoch was one of science, and as Dumont again correctly notes, only Darwinism could supply the bourgeoisie with a scientific justification of its rule, in opposition to the divine right of the declining aristocracy.

Social Darwinism simply translates this bourgeois opposition to the aristocracy into crude and direct terms. By transposing Darwinian principles into society, it substitutes for the divine right of the aristocratic hierarchy (creationism that fixes the order of beings) a kind of natural right of the most meritorious, or even the strongest (selection), with a concomitant naturalization of society. And it gives the whole construction the backing of science.

We might therefore be tempted to say that the importation of Darwinian principles into sociology derives from the fact that they gave the bourgeoisie the ideology it needed. This is true as far as it goes, but it still explains only the success of social Darwinism, not that of biological Darwinism. The question is rather more complicated.

Contrary to appearances, social Darwinism is not simply the importation of a biological doctrine into sociology. It was Darwinian biology that had previously imported into biology a sociological doctrine (that of bourgeois political economy, as Dumont explains). Social Darwinism, for its part, was hardly more than the return of this doctrine (naturalized by its transit through biology) to its field of origin, where it could not fail to succeed. Darwin himself recognized that his theory was at least in part the transposition into biology of principles drawn from sociology and economics; he refers explicitly to Malthus. His contemporaries were equally clear on this point, and far more assertive. Marx, for instance, wrote to Engels on the subject in 1862, and Engels to his Russian correspondent Lavrov in 1875:

> It is remarkable how Darwin recognizes among beasts and plants his English society with its division of labour, competition, opening up of new markets, 'inventions', and the Malthusian 'struggle for existence'. It is Hobbes's bellum omnium contra omnes, and one is reminded of Hegel's Phenomenology, where civil society is described as a 'spiritual animal kingdom', while in Darwin the animal kingdom figures as civil society ...
>
> The whole Darwinian teaching of the struggle for existence is simply a transference from society to living nature of Hobbes's doctrine of bellum omnium contra omnes and of the bourgeois-economic doctrine of competition together with Malthus's theory of population. When this conjurer's trick has been performed ... the same theories are transferred back again from organic nature into history and it is now claimed that their validity as eternal laws of human society has been proved. The puerility of this procedure is so obvious that not a word need be said about it.[28]

28 Marx to Engels, 18 June 1862; Engels to Lavrov, 12–17 November 1875. Marx/Engels, *Selected Correpondence* (London: Lawrence & Wishart, n.d.), pp. 156–7, 368.

This is all perfectly clear: the 'naturalist' sociology that appeals to Darwin simply takes back ideas that were initially sociological, but were naturalized by their passage through biology.

Darwinism continued on this course after Darwin's own time, for the entire statistical dimension of population genetics – and thus of the Darwinism whose principal support this was in the first half of the twentieth century – derived, via Galton's biometrics, from the methods employed by Quételet in his social statistics and statistical anthropology.[29] Here again, what we have is a borrowing by biology from the social sciences, one that returns to them by way of the methods of biological sociology (see below).

This back-and-forth between society and nature is found again in the case of heredity, at least at the level of etymology. Dumont, in the quotation given on pp. 7–8, justifies the inheritance of property by presenting this as an extension of biological heredity. In other words, he uses the biological notion of heredity to naturalize something that arises in the social and economic order.

It turns out, however, that the origin of the term 'heredity' (and the Latin *hereditat*, from which it derives) lies in the vocabulary of economics and law. It denotes the goods left by a person on his death, as well as the process of their transmission to his heirs (and the right of the heirs to inherit from him). As of about 1820, the term was applied by analogy to the biological characteristics and aptitudes supposedly transmitted from parents to children, and also to this transmission itself (heredity is the set of characteristics and aptitudes inherited, as well as the process by which this inheritance takes place).

Biological heredity was certainly known well before 1820, and ever since the dawn of time it had been noted that children more or less resemble their parents. The word 'hereditary' had even been used in the eighteenth century to describe certain illnesses, but the word 'heredity' itself was not used in biology, as the corresponding concept was not yet sufficiently precise. (That children resemble their parents was clear, but this quality was not easy to define, and the concept of heredity was still very vague.) It is likely that 'heredity' was given a precise sense in biology under the influence of heredity in the economic sense. It follows from this that Dumont, in using biological heredity to justify the inheritance of property, was simply following in the opposite direction the

29 And not from mechanics and statistical thermodynamics, as F. Jacob suggests (see note 23 on p. xix above).

very road that this notion of biological heredity had taken in being shaped by a naturalization of the inheritance of property.

Thus, in so-called naturalist sociologies, i.e., those that are essentially Darwinian, the image of nature is fashioned after the social model of the industrial revolution (particularly in the English-speaking countries). The ostensibly naturalist conception of society (economics, morality, even logic and religion), a conception which claimed to be a 'scientization' of sociology (as well as economics, morality, and so on), is a mere illusion. It is simply a rather poor justification of a particular social and economic order, by appeal to a nature that is itself conceived after the model of the society to be explained.

It is not therefore the success of Darwinism in biology that led to a proliferation of social Darwinian theories when it was exported into sociology. What explains their proliferation and success is that these social Darwinian theories perfectly corresponded to the spirit of their time. And it is this success of social Darwinism that made possible that of biological Darwinism, despite all its defects – for a period of some fifty years, until it was consolidated by genetics (very largely constructed in this context for this purpose).

Moreover, it was not sociologists who invented social Darwinism on the model of biological Darwinism; it was rather the biologists themselves who dreamed it up, and thereafter continually referred to it in order to justify their biological Darwinism. These biologists thus constructed the ideological support that they needed in order to sustain a vacillating biological theory. Unable to prove the capacity of natural selection to explain biological evolution (*The Origin of Species* does not contain a single example of an evolution explained in this way), they illustrated this by a social metaphor that was all the more effective in that it conformed to the dominant ideology. Sociologists did no more than follow this movement and take advantage of the scientific backing that biology gave to the ideas it had borrowed from them (competition, selection, etc.). When, in the early twentieth century, the advent of genetics shifted biologists' centre of interest, and somewhat tainted their source, sociologists were not capable of taking up the baton, and the exhausted biological sociologies dispersed into practical applications (see the quotation from A. Béjin on p. 10).

This pattern, rather too simple as it stands, is complicated by a number of intervening factors, particularly concerning the question of heredity and the nature of the social order.

To take first of all the question of heredity: The aristocracy was hereditary, resulting from a special kind of blood ('blue blood'), but this heredity was not a matter of genetics in the strict sense. Biological descent certainly played a part but was only one element integrated into a much broader social dimension. The decisive fact was not the transmission of genes ('aristocratic genes'), but that of a name and a title, and the perpetuation of a social order in which each person, whether aristocrat or commoner, had a place from which they were not to move. The hereditary aristocracy, moreover, was so little a matter of genes that by way of consanguine marriage it ended up accumulating hereditary defects. The case of haemophilia in the English and Russian royal families was well known in the era we address here.

Social Darwinism, on the contrary, introduced into bourgeois meritocracy a heredity that was genuinely biological and not at all symbolic. Meritocracy saw the triumph of the fittest, and social Darwinism explained this triumph by a hereditary biological superiority, recognition of which was the key to progress (in precisely the same way as biological Darwinism saw the survival of the fittest in terms of a hereditary biological superiority that ensured the evolution of species). The naturalization of the social order was thus translated into a biological conception of social classes. The hereditary superiority of the bourgeois had its counterpart in the hereditary inferiority of the worker, this relationship then being extended to the relationship between the colonizer and the 'inferior' colonized races (this was the great era of colonization). The heredity of the social hierarchy was now justified by resort to genetics, whereas under the aristocratic system it had been justified by tradition and the vague support of a divine will.[30]

Social Darwinism thus re-established a kind of blood right, one that was no longer a vague symbolism, like the 'blue blood' of the aristocracy, but rather a biological inheritance in the full sense of the term. Meritocracy was the hallmark of the bourgeoisie in its

30 So long as the inheritance of acquired characteristics was still creditable (i.e., until the 1890s, and even somewhat later), this notion of biological/social castes was also able to persist. Through bourgeoisification, the bourgeois integrated his status into his biological inheritance, and his children became still more bourgeois. The same applied to the worker, possibly with his own specialism (the watchmaker's son received a watchmaker's heredity from his father, which he increased by his own work as a watchmaker, transmitting this in turn to his son . . .). See C. Bouglé, *La Démocratie devant la science*, pp. 37–100.

progressive aspect; the biologization of this meritocracy by social Darwinism, making class membership 'hereditary', represents the conservative aspect of this same bourgeoisie once it had established itself (classically, evolution is progress and inheritance is conservation).

The difficulty arises from the fact that these analysts often confused aristocratic inheritance, which was not a matter of genetics in the strict sense (given that biological descent was only one aspect of this), with the hereditary membership of a social class, this indeed being biological and specific to social Darwinism. They saw aristocratic descent as the model, if not the origin, of the hereditary character of social class – and yet the one, the aristocratic form, was vaguely theological and rather reactionary from a political point of view, whereas the other, the bourgeois form, was naturalistic and sometimes progressive, sometimes conservative.

To sum up: in Darwinian sociology, a biological meritocracy (with competition and selection as its touchstones), in other words a pseudo-democracy, replaced an aristocracy of blood. A natural right of the meritorious replaced divine right. To each according to their merit, rather than according to their birth (let alone their need). So much the worse for 'inferior' individuals and races; they get only what they deserve, and occupy the place that falls to them in the natural order. And this is indeed the condition for progress, as Darwinian science attests.

And yet, the place that each individual occupies in this natural (and social) order perhaps does not quite rule out divine providence. A second confusion intervenes here. The aristocratic hierarchy, at least in its origin, was Catholic and rooted in a fairly mythical history that had endured for centuries. The bourgeois order, for its part, was more or less secular or Protestant. Calvinism gave it a religious backing by seeing social success as a sign of divine election and grace. Meritocracy could thus be viewed as resting not only on biological superiority, but also on a divine election of this kind (Galton's theories, indeed, have been described as 'a biological Calvinism in which nature and inherited potentials assign each person their due place in society'[31]). Divine election is manifest in and through biological selection, as much as in social success. The natural law is thus a divine law, but from a quite

31 G. Lemaine and B. Matalon, *Hommes supérieurs, hommes inférieurs? La controverse sur l'hérédité de l'intelligence* (Paris: Armand Colin, 1985), p. 25.

different perspective from that of the aristocracy. (See, on p. 37 above, Novicow's quotation from von Moltke on this subject: there is a conformity between the order of things established by God and that resulting from Darwinian natural laws.)

Just as social Darwinism replaced aristocratic descent by biological heredity, the aristocratic order, in essence Catholic, was replaced by a kind of Calvinistic biological order.[32] Here again, analysts have often created a historical short circuit by confusing the two things.

These confusions were all the more convenient for cleansing social Darwinism (and the scientists who supported or even invented it) of its consequences (those of the Nazis in particular) by attributing these to the old aristocratic model that had allegedly tried to re-establish itself in opposition to the bourgeois meritocracy. The way in which Gobineau was represented as the source of all racism (if not of eugenics and Nazism as well), whereas Haeckel became a hero of modern biology, is a good example of this. Historical analysis, by a curious short circuit, proceeded directly from Gobineau to Nazism, skipping over the social Darwinism of Haeckel and other eminent biologists or presenting it as an unimportant diversion.[33] We shall return to this in Part Two.

For the time being, we can conclude by discussing the criticisms that Novicow and other writers made of the biologization of society, and more specifically its Darwinization. With these authors, as also with those who championed this biologization, there is a great heterogeneity in the arguments presented; but on both sides, these are rarely very interesting and often display an intellectual mediocrity. Novicow was right to insist on the simplistic character of the social Darwinist claims, but the arguments he opposes to them are themselves very far from high-flying – whether by their own nature or as a reflection of the mediocrity of the arguments he criticizes.

32 From the 1890s onwards, this process could appeal to genetic theories such as Weismann's theory of the germ-plasm, which clearly recapitulated the seventeenth-century theories of a preformation of the living being in the germ (ovum or spermatozoon). At that time, these theories had been the biological counterpart of the Protestant and Jansenist doctrines of predestination. With Weismann's conception of an omnipotent inheritance materialized in the germ-plasm, the living being was 'biologically predestined' (today it is 'biologically programmed'), in perfect agreement with the 'biological Calvinism' of Galton and his followers.

33 I have no more sympathy for Gobineau's ideas than for those of Carrel, but both Gobineau and Carrel have manifestly been used as Aunt Sallys.

In certain cases, these arguments, for or against, refer to stages in the history of evolutionary biology and genetics that have today been superseded; only a good understanding of this history makes it possible to grasp their exact meaning. For example, Novicow was correct in counterposing De Vries's mutationism to Darwinism, as in his day there was indeed such an opposition in biology. Today, however, this opposition no longer makes any sense (it is even incomprehensible to those unaware of the history of genetics), because Darwinism and mutationism have both been modified to bring them into harmony. In other cases, there is clearly incomprehension as to what biologists were trying to say, both pro and con.

Finally, the majority of these arguments, on both sides, strike us as rather unoriginal, for with few exceptions they are the same ones that sociobiology and its adversaries use today (no doubt unaware of their origins). This is a further characteristic of this current of thought: the same ideas periodically return in a kind of circular rambling, from which their protagonists find it impossible to escape.

Here I shall just note two points that seem fundamental in the critique of the biologization of society, and specially its Darwinization. They have both been noted for a very long time, but perhaps have not always been given their due.

The first of these – the false naturalization of society – has already been mentioned. Darwinism imported sociological ideas into biology, and sociology took them back again once they had been 'naturalized'. The second, whose exact origin I do not know, is what Novicow (who perhaps coined the phrase) called the 'anthropological fable'. The anthropological fable is the foundation of Darwinian sociology, and the best way to demonstrate it is by a further extract from Novicow's text. He quotes and criticizes here a thesis bearing on competition for food that had been proposed by the Darwinian sociologist G. Ratzenhofer (another quotation from whom can be found in the box on p. 23). Ratzenhofer was, together with Gumplowicz, one of the first and principal theorists of the war of races:

> Let us then speak about food ... Here I would like to explain the point of view of the Darwinian sociologists. 'As foodstuffs become increasingly rare with the increase in population,' says G. Ratzenhofer (Sociologische Erkenntnis, p. 245), 'individuals are forced to struggle for existence. Two paths are then open to men: either they work to provide themselves with

> the means of subsistence, and, by an improved organization, make more of their original habitat despite the increase in population – in other words, the beginnings of civilization – or on the other hand, they attack their own kind and force them into servitude, in order to gain a greater share of the means of subsistence, i.e., violent struggle and policies of compulsion. It is environmental conditions that propel societies into one or the other of these directions. Initially, only men whose surroundings offered great advantages could opt for civilization; those who found themselves in unfavourable conditions were obliged to choose war and violence.
>
> Before all else, a first remark, which will also apply to all the examples I introduce below. Ratzenhofer was not present at the time when men took sides: some for work, others for violence. No individual has left us the least evidence for this of any kind. Ratzenhofer concludes without the slightest positive proof, solely by his own mental reasoning, that things took place in the way that he describes. He thus offers a work of pure imagination. This is why I give all these retrospective descriptions of the primordial condition of the human species the name of anthropological fables. They are pure fables in the full sense of the word, being based on no fact whatsoever ... Ratzenhofer's arguments, like those of all Darwinians, are based simply and solely on the deductions of their minds. These deductions can therefore be examined only from the standpoint of logic, and can only be rejected in the same manner.[34]

Novicow's text is perfectly clear: the anthropological fable is a work of imagination, a historical scenario, generally relating to the earliest age of humanity, yet offered as explanation of one or another social phenomenon of either that time or our own. It is a kind of reverse science fiction, situated in the past rather than in the future. Rosny's novel *Quest for Fire*, published a year after Novicow's book in 1911, was in fact rather more talented.

What claim can this kind of historical fiction make to be scientific? It simply cannot, even in the loosest sense of science. It is just that the anthropological fable appeals to ideas of competition, struggle, selection, etc., ideas of Darwinian biology – or rather, socio-economic ideas that Darwinism borrowed and naturalized, thus giving them a scientific backing. Returned to the sociology whence they came, they are endowed with a kind of scientific aura, and their use in anthropological fables confers on the latter a dignity to which they have no right.

---

34 J. Novicow, *La Critique du darwinisme social*, pp. 208–9.

The problem is that Darwinism, properly speaking, resorts to just this kind of historical scenario in its explanation of the origin of species. The simplest of these scenarios, in its modern form, sees a certain characteristic as appearing by chance mutation and, once shown to be favourable to its individual bearer, being preserved by natural selection. This basic model can be given added sophistication, mathematical for example, but the fact remains that the Darwinian explanation still consists in imagining a historical scenario constructed on the basis of a number of notions (variation, competition, selection, to which are added geographical isolation, fertility rate, or even natural catastrophe such as the meteorite supposedly responsible for the disappearance of the dinosaurs, etc.). To criticize the explanatory principle that the anthropological model provides in social Darwinism is equally to criticize the Darwinian principle that explains the evolution of species by reconstructing historical scenarios. It thus amounts to an attack on science (since Darwinism is deemed scientific, at least among biologists), whereupon one joins the category of 'fixists', religious creationists, Lysenko-ists, etc. – the 'time-servers and retrogrades', as Novicow had already ironically called them in 1910 (see quotation on p. 39).

Moreover, contrary to what one may initially believe, it is very hard to counter an anthropological fable with rational arguments. Novicow, for example, offers the following reason against the idea of a competition for food such as Ratzenhofer presented: 'When we consider that the population density of the globe was for a long time only one individual per hundred square kilometres, we can imagine how a struggle between tribes to monopolize fruit trees was hardly likely.'[35] The argument may seem perfectly reasonable, but it is more or less ineffective against Ratzenhofer's anthropological model, as it can always be neutralized by imagining a counter-argument (famine, local overpopulation, etc.). And so on ad infinitum. Such is its vagueness that it cannot even be called an irrefutable theory in the Popperian sense, and Rovicow's term 'anthropological fable' fits it perfectly.

Despite everything, it was in the guise of science that this kind of doctrine was able to impose itself:

> Social Darwinism has a great prestige today, precisely because it is dressed up in scientific guise. This theory existed in an unconscious state

35 Ibid., p. 261.

> for centuries. But only in the second half of the nineteenth century was it formulated in a precise fashion by scholars, biologists and sociologists, professional men, individuals with a high position in science. Coming from above, it enjoyed the very merited respect that science inspires; it showed itself, so to speak, in the aureole of its majesty, and spread among people as if surrounded by a halo of light.[36]

The 'scientific' forms of this doctrine are variable enough, the principal form being that which is today called sociobiology. We shall now present some elements of its history and epistemology.

36 Ibid., pp. 379–80.

3

Evolutionary Altruism

It is uncertain whether there can be a history and epistemology of sociobiology in the strict sense of these terms, given how shapeless the discipline and how bizarre its foundations. The preceding pages have given some elements of its prehistory and its ideological presuppositions. Nor are its history and philosophy any more clear. In any case, a detailed study of these is not our object here. As an expedient, therefore, I shall take a recent example and analyse its genealogy, methodology and philosophy. At first sight, this may seem no more than anecdotal, but we shall quickly perceive that it occupies a central position, and that the other aspects of sociobiology come a clear second in relation to it – which is why I have chosen it.

Let me also make clear from the start that I understand the word 'sociobiology' here in a broad sense. There exist in fact a whole range of disciplines that prefer to call themselves 'behavioural genetics', 'evolutionary ethology', 'behavioural ecology', 'evolutionary ecology', etc. These are all disciplines uncertain of themselves, which seek to assert themselves by assuming scholarly names, pompous and vaguely Greek, but are actually no more than variants of sociobiology. We can understand how, given sociobiology's past and reputation, they baulk at acknowledging themselves as such, but they nonetheless share the essentials of its ideas and methods.

The example I am going to start with comes from a fairly recent article, 'Evolution of indirect reciprocity by image scoring', by Martin A. Nowak and Karl Sigmund, in which the authors propose a mathematical model that accounts for altruism in a Darwinian perspective.[1] In the issue of *Nature* in which it was published, this

---

1 M. A. Nowak and K. Sigmund, 'Evolution of indirect reciprocity by image scoring', *Nature*, 1998, 393, pp. 573–7.

piece was presented and commented on in the following terms by Régis Ferrière:

> How do moral systems evolve? The common view, rooted in game theory, is that cooperation and mutual aid require tight partnerships among individuals, or close kinship. But this dogma is now being shaken ... Nowak and Sigmund report a new mathematical model to show that cooperation can become established even if recipients have no chance to return the help to their helped. This is because helping improves reputation, which in turn makes one more likely to be helped.[2]

In other words, the point is to show that altruistic behaviour may have a higher selective value than egoistical behaviour, and therefore a hereditary advantage preserved by evolution, even if the individuals affected are neither closely related nor even associated in such a way as to lead to an obligatory reciprocity. The solution offered is that such behaviour confers on the individual a positive 'brand image', which makes him or her susceptible to being helped in turn. The authors then present a mathematical model of this conception, using game theory and a computer simulation.

At first sight, faced with a publication such as this (no worse indeed than hundreds of others of the same kind), a normally trained epistemologist would just laugh, and say that if these authors focus on altruistic behaviour, it is because this is far more 'politically correct' than the commonly invoked genes for crime, alcoholism, homosexuality, etc., which have long formed the essence of this kind of study and are today in a threadbare state. In a similar way, the resort to rather sophisticated mathematical models is explained by the fact that the usual statistical studies, particularly those based on twins, have been too widely used to be credible (even leaving out frauds and falsifications such as those of Cyril Burt, which completely fabricated their 'experimental evidence' of heredity and intelligence).

It is easy to make this kind of remark, and scarcely epistemological, but to my mind it is nonetheless fully justified. Yet it is still somewhat insufficient, so I shall now try to flesh it out.

This question of 'altruism' may seem simply anecdotal, but in order to understand the stakes involved, we need to

---

2 R. Ferrière, 'Help and you shall be helped', *Nature*, 1998, 393, pp. 517–19.

know that so-called altruistic behaviour poses a problem for Darwinism, since this postulates that the motor of evolution is a struggle between individuals, in which only the victors survive (or leave a sufficiently great number of descendants). How then can altruistic behaviour, where cooperation takes the place of competition, have been selected for in evolution? Various solutions have been imagined, but all a bit shaky. In general they appealed more or less directly to 'group selection', that is, a selection applied to competing groups of individuals rather than to individuals themselves. The difficulty comes from the fact that a selection of this kind is incompatible with classical Darwinism, because for a group the notion of survival is simply a metaphor, and the perpetuation of the group in question rests in reality on the survival of the individuals who compose it, these being the 'objects' to take into consideration in the context of natural selection. Hence the resort to various subterfuges (kinship, reciprocity, and so on), charged with extending to the group a property that is valid only for its component elements.

The importance of the question raised by altruism is thus that its solution determines the compatibility of social behaviours (altruistic by definition, since they associate individuals) with Darwinian evolutionism. And it therefore determines the possibility of bringing the explanation of societies, animal or human, into the framework of this Darwinian biology. In other words, on this solution depends the possibility or impossibility of a biological sociology of Darwinian inspiration. Far from being as anecdotal as it seems at first sight, evolutionary altruism is in a very real sense fundamental to this discipline (and others of its kind), and was indeed already imagined very soon after the publication of *The Origin of Species* in 1859.

The Origin of Species does not discuss the human species, and it was only in 1871 that Darwin tackled this question, in *The Descent of Man, and Selection in Relation to Sex*. A decade earlier, in 1862 (though not published until 1864), Thomas Henry Huxley (1825–95) had proposed extending Darwinian principles to moral values.[3] According to Huxley, the moral faculties, like all others, were susceptible to an evolution subject to natural selection, in

3 T. H. Huxley, 'Six lectures to working men "On our knowledge of the causes of the phenomena of organic nature"' [1862–4], *Collected Essays*, II (Darwiniana) (London: Macmillan, 1893), pp. 303–475 (quote from pp. 471–5).

such a way that, thanks to such selection, man improved in his morals as well as in his physical qualities. The principle of this was very simple: 'There is not a single faculty – functional or structural, moral, intellectual, or instinctive – there is no faculty whatsoever which does not depend upon structure, and as structure tends to vary, it is capable of being improved.'[4]

Later Huxley changed his opinion and became one of the few major Darwinian biologists to reject this conception of morality, and even combat it, as we shall see later (quotation on pp. 95–96). At all events, however, this was the first allusion to an 'evolutionary morality', and a very general one.

A more precise idea of evolutionary altruism was developed by Alfred R. Wallace, the co-inventor of Darwinism, in a famous paper published in 1864.[5] After presenting competition and pitiless natural selection among the animals, Wallace acknowledges a certain virtue in co-operation among men. This cooperation modified the principles of selection, which thereby typically became a 'group selection': struggle between tribes replaced struggle between individuals, and the survival of the tribe rested on the cooperation of its members:

> In the rudest tribes the sick are assisted, at least with food; less robust health and vigour than the average does not entail death. Neither does the want of perfect limbs, or other organs, produce the same effects as among animals. Some division of labour takes place; the swiftest hunt, the less active fish, or gather fruits; food is, to some extent, exchanged or divided. The action of natural selection is therefore checked; the weaker, the dwarfish, those of less active limbs, or less piercing eyesight, do not suffer the extreme penalty which falls upon animals so defective.
>
> In proportion as these physical characteristics become of less importance, mental and moral qualities will have increasing influence on the well-being of the race. Capacity for acting in concert for protection, and for the acquisition of food and shelter; sympathy, which leads all in turn to assist each other; the sense of right, which checks depredations upon our fellows; the smaller development of the combative and destructive propensities; self-restraint in present appetites; and that intelligent foresight which prepares for the future, are all qualities, that from their earliest appearance must have been for the benefit of each community, and would, therefore,

---

4 Ibid., pp. 471.

5 A. R. Wallace, *Contributions to the Theory of Natural Selection* (London: Macmillan, 1871).

> have become the subjects of 'natural selection'. For it is evident that such qualities would be for the well-being of man; would guard him against external enemies, against internal dissensions, and against the effects of inclement seasons and impending famine, more surely than could any merely physical modification. Tribes in which such mental and moral qualities were predominant, would therefore have an advantage in the struggle for existence over other tribes in which they were less developed, would live and maintain their numbers, while the others would decrease and finally succumb.[6]

Wallace goes on to explain how, among men, intelligence and morality were able to substitute for the possession of specialized organs and brute force, and how intellectual and moral values thus became factors of adaptation:

> From the time, therefore, when the social and sympathetic feelings came into active operation, and the intellectual and moral faculties became fairly developed, man would cease to be influenced by 'natural selection' in his physical form and structure. As an animal he would remain almost stationary, the changes of the surrounding universe ceasing to produce in him that powerful modifying effect which they exercise over other parts of the organic world. But from the moment that the form of his body became stationary, his mind would become subject to those very influences from which his body had escaped; every slight variation in his mental and moral nature which should enable him better to guard against adverse circumstances, and combine for mutual comfort and protection, would be preserved and accumulated; the better and higher specimens of our race would therefore increase and spread, the lower and more brutal would give way and successively die out, and that rapid advancement of mental organization would occur, which has raised the very lowest races of man so far above the brutes (although differing so little from some of them in physical structure), and, in conjunction with scarcely perceptible modifications of form, has developed the wonderful intellect of the European races.[7]

The reader will excuse me for imposing these texts of fathomless stupidity, but this is what the historian of science is regularly faced with in this field, and the simplest and quickest way to demonstrate this is to cite its most explicit passages. We are in

6 Ibid., pp. 312–13.
7 Ibid., pp. 316–17.

the full flood of what Novicow called 'anthropological fables', a variant of Rosny's *Quest for Fire*.

We can see in any case that the object is to explain how altruism, along with intelligence, can be brought into the framework of natural selection. In some sense the point is to civilize and moralize such selection. Yet this is a very relative moralization, as Wallace immediately goes on to make clear that its direct consequence is the extinction of 'inferior' races, which are not competitive in the group selection that has replaced individual selection:

> It is the same great law of 'the preservation of favoured races in the struggle for life', which leads to the inevitable extinction of all those low and mentally undeveloped populations with which Europeans come into contact. The red Indian in North America, and in Brazil; the Tasmanian, Australian, and New Zealander in the southern hemisphere, die out, not from any one special cause, but from the inevitable effects of an unequal mental and physical struggle. The intellectual and moral, as well as the physical, qualities of the European are superior; the same powers and capacities which have made him rise in a few centuries from the condition of the wandering savage with a scanty and stationary population, to his present state of culture and advancement, with a greater average longevity, a greater average strength, and a capacity of more rapid increase, – enable him when in contact with the savage man, to conquer him in the struggle for existence, and to increase at his expense, just as the better adapted, increase at the expense of the less adapted varieties in the animal and vegetable kingdoms, – just as the weeds of Europe overrun North America and Australia, extinguishing native productions by the inherent vigour of their organization, and by their greater capacity for existence and multiplication.[8]

This is no longer a 'fire war' – *La Guerre du feu* was Rosny's more militaristic original title – but rather a race war. We might of course remind Wallace that many American and African plants were introduced to Europe and prospered (which is more than people from these lands ever did), but the suspicion is that the purpose of the botanical comparison is simply to naturalize the extermination of indigenous Americans and Australians, reducing this to as natural a phenomenon as the disappearance of a plant in a habitat colonized by another. (Seventy years later, Julian Huxley and A. M. Carr-Saunders still spoke of 'sweeping aside' indigenous populations; see p. 14 above.)

8 Ibid., pp. 318–19.

At all events, it should be stressed that the altruism Wallace discusses here is practised only within a particular human group (one for which it constitutes a selective advantage), and that the struggle for existence between individuals is now transformed into a struggle between groups (and even between races). The moralization of natural selection by altruism is thus highly relative, and might just as well pass for a 'scientific' justification of racism, or even for the extermination of the colonized 'inferior races', an extermination reduced to a natural process integrated into the evolution of species.

We often encounter this idea of a struggle between races in the writings of Darwinian sociologists. It is clear here that this was not their invention, but was originally imagined by the biologists themselves in their application of Darwinian theses to man. And it was precisely this application that was used as evidence for biological Darwinism, given that no comparable examples could be adduced in the animal kingdom – there is not a single description of animal evolution by natural selection in Darwin's *The Origin of Species*.

This way of bringing intelligence and altruism into natural selection was taken up by Darwin in 1871, in almost identical terms, in *The Descent of Man, and Selection in Relation to Sex* (Chapter 5: 'On the development of the intellectual and moral faculties during primeval and civilized times'). Darwin, moreover, cites Wallace's article in glowing terms, and accepts its conclusions – with quite comparable remarks on the colonizing virtues of the English. All the same, he seems confusedly to perceive the difficulty posed by group selection, since a characteristic that is advantageous for a group is not necessarily so for the individual that carries it. How, then, can natural selection, exercised on individuals whose survival and descendants it regulates, ensure the perpetuation of this characteristic? Darwin resolves the problem by proposing that factors such as reasoning, foresight, help being repaid in kind, and the taste for praise ended up by making cooperation a kind of selective advantage:

> But, it will best asked, how within the limits of the same tribe did a large number of members first become endowed with these social and moral qualities, and how was the standard of excellence raised? It is extremely doubtful whether the offspring of the more sympathetic and benevolent parents, or of those who were the most faithful to their comrades, would be reared in greater numbers than the children of selfish and treacherous

> parents belonging to the same tribe. He who was ready to sacrifice his life, as many a savage has been, rather than betray his comrades, would often leave no offspring to inherit his noble nature. The bravest men, who were always willing to come to the front in war, and who freely risked their lives for others, would on an average perish in larger numbers than other men. Therefore it hardly seems probable, that the number of men gifted with such virtues, or that the standard of their excellence, could be increased through natural selection, that is, by the survival of the fittest; for we are not here speaking of one tribe being victorious over another.
>
> Although the circumstances, leading to an increase in the number of those thus endowed within the same tribe, are too complex to be clearly followed out, we can trace some of the probable steps. In the first place, as the reasoning powers and foresight of the members became improved, each man would soon learn that if he aided his fellow-men, he would commonly receive aid in return. From this low motive he might acquire the habit of aiding his fellows; and the habit of performing benevolent actions certainly strengthens the feeling of sympathy which gives the first impulse to benevolent actions. Habits, moreover, followed during many generations probably tend to be inherited.
>
> But another and much more powerful stimulus to the development of the social virtues, is afforded by the praise and the blame of our fellow-men. To the instinct of sympathy, as we have already seen, it is primarily due, that we habitually bestow both praise and blame on others, whilst we love the former and dread the latter when applied to ourselves; and this instinct no doubt was originally acquired, like all the other social instincts, through natural selection. At how early a period the progenitors of man in the course of their development, became capable of feeling and being impelled by, the praise of blame of their fellow-creatures, we cannot of course say. But it appears that even dogs appreciate encouragement, praise, and blame. The rudest savages feel the sentiment of glory, as they clearly show by preserving the trophies of their prowess, by their habit of excessive boasting, and even by the extreme care which they take of their personal appearance and decorations; for unless they regarded the opinion of their comrades, such habits would be senseless.[9]

Just like Wallace, Darwin himself shows an astonishing (and very Victorian) mixture of religious moralism and intellectual poverty, along with a colonialist racism quite lacking in soul. The 'altruistic' thesis that Darwin proposes, in fact, no longer requires

9 C. Darwin, *The Descent of Man, and Selection in Relation to Sex* (Harmondsworth: Penguin, 2004), p. 156.

a transposition of the struggle between individuals into a struggle between groups or races. But Darwin does not notice this, and reasons more or less as Wallace did on this subject. There is in fact no coherence in his argument. And no more would there be with those authors who took up the problem of altruism later on, in different ways and with different intentions.

What is of particular interest for us here is that Darwin had already more or less resolved the problem of group selection, and his 'solution' was very close to that imagined in 1998 by Nowak and Sigmund. Where the latter speak of 'brand image', Darwin speaks of experience teaching the individual that 'if he aids his fellows, they will aid him in turn', or again of 'the approval or blame of our fellows'. If this is not quite identical, it is very close to a brand image, and the slippage is easily effected from one to the other. Contrary to what Ferrière claims, therefore, there is no great novelty in the approach of Nowak and Sigmund. They did little more than simply graft a mathematical treatment onto a Darwinian conception already 130 years old. It is possible, even probable, that Nowak and Sigmund were unaware of this text of Darwin's; at least their bibliography makes no reference to it. In this case, they may have reinvented the idea for themselves, which would not be surprising given how widespread this kind of simplistic reasoning is.

Yet the 130 years between Darwin and Nowak-Sigmund was not just a great void. Throughout the period, a large number of similar reflections can be found. The compatibility of altruism with Darwinism has always been one of the most debated questions. It has been used in all kinds of ways, and with all kinds of embellishment. Shortly after Darwin's own day, for example, it served as the foundation for the 'monist' sociology and morality of Haeckel, in which altruism, the Kantian imperative and love of one's neighbour were explained biologically in terms of social instincts that included the necessity of group survival, if need be at the expense of the egoistic interest of the individual. More precisely, Haeckel imagined a dialectic between egoism and altruism, both of these being equally indispensable. This theory was presented in very clear fashion in 1899, in Chapter 19 of *The Riddle of the Universe* (compare this with the quotation from Ammon on p. 15 above, a text dating from the same era):

> Man belongs to the social vertebrates, and has, therefore, like all social animals, two sets of duties – first to himself, and secondly to the society

> to which he belongs. The former are the behests of self-love or egoism, the latter of love for one's fellows or altruism. The two sets of precepts are equally just, equally natural, and equally indispensable. If a man desire to have the advantage of living in an organized community, he has to consult not only his own fortune, but also that of the society, and of the 'neighbours' who form the society ...
>
> (1) Both these concurrent impulses are natural laws, of equal importance and necessity for the preservation of the family and the society; egoism secures the self-preservation of the individual, altruism that of the species which is made up of the chain of perishable individuals. (2) The social duties which are imposed by the social structure of the associated individuals, and by means of which it secures its preservation, are merely higher evolutionary stages of the social instincts, which we find in all higher social animals (as 'habits which have become hereditary').[10]

This seems to be the first use of the terms 'altruism' and 'egoism' in biological sociology. They do not figure explicitly in Darwin or Wallace, and I do not know of any other writer who used them before Haeckel (apart from Ammon, whose text is exactly contemporary). An individual is altruistic when the egoism that characterizes him in the theory of natural selection is transferred to a higher level, that of the social group whose survival takes precedence over his own.[11]

The altruism in question thus refers to behaviours ensuring the existence and perpetuation of the social group, sometimes at the expense of that of the individual. It is a generic designation of behaviours that may be very different in kind (social, parental, filial, alimentary, etc.) but have in common the fact that they require individuals to co-operate. In other words, they have the common property of posing a problem in terms of explanation by competition and natural selection, since these tend to favour

10 E. Haeckel, *The Riddle of the Universe* (Prometheus Books: New York, 1992), pp. 350–1.

11 The same 'egoism' is used today in Richard Dawkins's theory of the 'selfish gene'. In this theory, instead of explaining individual altruism by group egoism (the altruism of the individual being useful to the group), it is explained by the egoism of the gene (the altruism of the individual is useful to the gene, by assuring its transmission and multiplication within the population). The two theories are symmetrical, and each is as absurd and ill-founded as the other. They differ only in that, in the one, individual egoism (as postulated by the theory of natural selection) is transferred to a higher level (the group), and in the other, to a lower level (the gene). Richard Dawkins, *The Selfish Gene* (Oxford: OUP, 1976).

behaviour that is 'egoistic' in that it involves the survival of the individual and not that of the group.

This way of grouping different kinds of behaviour under the same generic label of altruism clearly reveals that the aim of this kind of theory is not to explain the behaviours in question in their diversity and particularity, but rather to make them compatible with Darwinism, bringing them into the framework of competition and natural selection. Once this is done, these behaviours, no matter how varied, are considered explained, as they now fall under the common Darwinian law. We may note, moreover, that no such explanation is sought for 'egoistic' behaviours, since these automatically enter into the Darwinian framework.

The supposedly hereditary character of altruistic behaviours, moreover, is conceived in a very curious fashion. By virtue of their different physical forms and diverse natures (social, parental, alimentary), these behaviours cannot derive from the same genetic substratum. It is also hard to imagine a gene for altruism that would be involved in each of these and be common to them. Altruism is not a behaviour in itself, rather a quality that the observer attributes to a behaviour, just as there is no gene for egoism that is involved in the substratum of all egoistic behaviours. What is considered in these behaviours is not really the genetic substratum, this being as varied as the behaviours themselves – assuming that these really are hereditary, and no knowledge of this substratum is invoked to declare them so. Neither is there any empirical or experimental data. They are only declared hereditary so that the social Darwinian fable can function. The only thing that is taken into account is their 'altruistic' aspect. This alone is what Darwinism seeks to explain, since once this is done, competition and selection take care of the rest.

On top of all this, the terminology itself gives the behaviours in question a moral connotation: they are altruistic even if it is not humans that are involved, but animals or even plants. The morality in question is then judged by the criterion of group interest; it is a social morality and not a private morality, and besides, a utilitarian morality rather than one based on a notion of good having an intrinsic value.

This moral aspect is clear in the above quotation from Haeckel. He not only proposes to explain altruistic behaviours but evidently seeks to moralize the rather brutal aspect of the Darwinist struggle for existence. He thus develops an entire, supposedly naturalistic ethics that he opposes to Christian morality (sometimes in a curious manner), relating this to Oriental traditions, and Buddhist ones in particular. On the love of animals, for example:

> Christianity has no place for that well-known love of animals, that sympathy with the nearly related and friendly mammals (dogs, horses, cattle, etc.) which is urged in the ethical teaching of many of the older religions, especially Buddhism. Whoever has spent much time in the south of Europe must have often witnessed those frightful sufferings of animals which fill us friends of animals with the deepest sympathy and indignation ...
>
> Darwinism teaches us that we have descended immediately from the primates, and, in a secondary degree, from a long series of earlier mammals, and that, therefore, they are 'our brothers' ... No sympathetic monistic scientist would ever be guilty of that brutal treatment of animals which comes so lightly to the Christian in his anthropistic illusion – to the 'child of the God of love'.[12]

It is amusing to see Haeckel disguise himself as St Francis of Assisi, especially after he has classified human races in a pitiless evolutionary hierarchy. This very quotation bears the trace of this, in its reference to the Catholic peoples of southern Europe. In actual fact, Haeckel's hatred of Christianity is above all anti-papism, bound up in his case with a pan-Germanism that is both political and racial, in particular asserting the superiority of the Indo-Germanics (the term used at that time, rather more restrictive than the more classical 'Indo-European').[13] The Monist League that Haeckel founded to propagate his doctrine is today viewed as one of the nurseries in which the ideas that

---

12 E. Haeckel, *The Riddle of the Universe* (London: Watts, 1904), pp. 355–6.

13 Haeckel was a supporter of Bismarck, who viewed the Catholic Church as a foreign power meddling in German affairs, and his pan Germanism included a pronounced anti-Catholic component. On the specifically political level, this Bismarckian anti-Catholicism was translated into a secularization of society, for example by withdrawing the church's budget. On the ideological level, a Kulturkampf combined support for Protestant culture (understood as genuinely German) with a scientistic culture (opposing scientific, technical and industrial progress to a Catholicism seen as rather retrograde in these matters). Haeckel was the chief representative of this current, influencing both left-wing ideologists seduced by his materialism, and those of the right, who particularly took up his racial theories.

would develop into Nazi biological-political doctrine were elaborated.[14]

On the question of love of animals, Haeckel had a celebrated predecessor. Darwin himself expressed the same kind of reflections – grotesque enough – when in 1871 he resigned himself to admitting the consequences of his theory that all the world except he himself had already drawn. He then wrote:

> The main conclusion arrived at in this work, namely that man is descended from some lowly organized form, will, I regret to think, be highly distasteful to many ... For my own part I would as soon be descended from that heroic little monkey, who braved his dreaded enemy in order to save the life of his keeper, or from that old baboon, who descending from the mountains, carried away in triumph his young comrade from a crowd of astonished dogs – as from a savage who delights to torture his enemies, offers up bloody sacrifices, practises infanticide without remorse, treats his wives like slaves, knows no decency, and is haunted by the grossest superstitions.[15]

I do not cite these texts simply for comic effect, nor even to emphasize the scientistic stupidity of their authors. They are here above all to attest to the extreme ambivalence of these naturalistic and evolutionary sociologies and moralities, and their capacity simultaneously to justify love of animals and the extermination of the 'inferior human races' (see on p. 37 Novicow's comment on the double aspect of Darwinism, able to please both enlightened and brutish minds). Rather than 'naturalistic' we should say 'pseudo-naturalistic', as we have seen how in this kind of theory nature is understood after the social model of the nineteenth-century Industrial Revolution.

Haeckel's *Riddle of the Universe*, from which these quotes are taken, was translated into many languages and sold four hundred thousand copies in Germany alone: a large enough figure today, and far more so in the late nineteenth century.[16] We are thus not talking here about marginal theorists like Vacher de Lapouge, who lacked any academic recognition, but doctrines that if not

14 J. Roger, 'Darwin, Haeckel et les français', in *Pour une histoire des sciences à part entière* (Paris: Albin Michel, 1995), p. 387. More generally on the role of Haeckel's theories in the birth of fascism, see D. Gasman, Haeckel's *Monism and the Birth of Fascist Ideology* (New York: Lang, 1998).

15 C. Darwin, *The Descent of Man, and Selection in Relation to Sex*, p. 689.

16 J. Roger, 'Darwin, Haeckel et les français', p. 373.

scientific were at least completely accepted, and even propagated, by established science, and popularized on a large scale.

At around the same time, Kropotkin adopted this conception of altruism (which he ascribed to Darwin) to praise solidarity and mutual aid, criticizing Spencer for having failed to understand that the good of the individual and the good of the species coincide:

> With all due respect to the popularizers of Darwin, who ignore everything in his work that he did not take over from Malthus, the sentiment of solidarity is the predominant feature in the life of all animals that live in societies. The eagle devours the sparrow, the wolf devours the marmot, but both eagles and wolves aid their kind in hunting, and sparrows and marmots support their fellows so well against their predators that only the clumsy individuals are caught. In any animal society, solidarity is a law of nature, i.e., a general fact, infinitely more important than this struggle for existence whose virtue the bourgeois sing in every key, the better to stupefy us ... We shall one day return to this subject, in order to demonstrate with abundant proof how, in the animal world as in the human, the law of mutual support is the law of progress, and how mutual aid, just as the individual courage and initiative that flow from it, assure victory for the species that is best able to practise these ...
>
> And when Spencer foresees a time when the good of the individual is identified with that of the species, he forgets one thing: that if the two had not always been identical, the very evolution of the animal kingdom could not have been accomplished.[17]

Where Darwinism projected on the animal world the notions of Malthusian sociology, Kropotkin projected the values of solidarity and mutual aid. He offers this as an explanation of the evolution of species, and wrongly attributes it all to Darwin, whose work was supposedly betrayed by his followers. (This betrayal, completely imaginary, became a leitmotif for those left-wing Darwinians always rather embarrassed by the master's Malthusian references, and by social theories that appealed to Darwin.)

Kropotkin's procedure is clearly more sympathetic than those previously discussed, but it is not innocuous for all that. In fact, it falls back on the transposition already noted of the struggle between individuals onto a struggle between groups (between races, with Wallace and Darwin). As far as the political consequences of this

17 P. Kropotkin, *La Morale anarchiste* (Paris: Groupe libertaire Kropotkin, 1969), pp. 15–16, 29.

'evolutionary altruism' are concerned, Kropotkin was no more lucid than those he criticized.

Jean-Louis de Lanessan had already proposed a similar conception in 1881, highlighting co-operation and, more specifically, co-operation in a struggle against the environment or other groups.[18] In 1910 Novicow also appealed to mutual aid and co-operation in his critique of social Darwinism.[19] Bouglé did the same in the 1920s.[20] Novicow and Bouglé both cite Kropotkin, but without continuing his attempt to reconcile co-operation with Darwinian struggle.

The political and philosophic commitments of the Russian anarchist were hardly compatible with those of the Bismarckian Haeckel. Nor with the very Victorian Darwin. All these authors, however, concurred in discovering in science – or, more precisely, in what they believed to be scientific – a justification for their social and moral ideas. As well as this scientism, we may note the astonishing adaptability of Darwinian doctrine, to which a range of completely different ideologies could appeal. With these biological social theories, in fact, we enter a domain of no matter what justifies no matter how – but always by reference to Darwinism, which can be made to say whatever one wants. It should be recognized, indeed, that Darwin, by constantly contradicting himself in his writings, opened wide the gate to these abuses.

Still in the early years of the twentieth century, the American psychologist James Mark Baldwin made the same use of altruism in criticizing Thomas Henry Huxley for having claimed that evolution was incompatible with morality. As discussed above (p. 54), Huxley was one of the first, in 1862, to conceive an evolution of morality by natural selection, but he subsequently changed his views, which Baldwin saw as a kind of betrayal of Darwinism:

> [Huxley] did not take account of the fact that there are transition stages between biological struggle and social rivalry, between the physical aptitude required by the one and the social aptitude required by the other. As soon as we grasp that the aptitude of a group demands an internal organization,

18 J.-L. de Lanessan, *La Lutte pour l'existence et l'association pour la lutte* (Paris: Doin, 1881).

19 For example, J. Novicow, *La Critique du darwinisme social*, pp. 205–6.

20 C. Bouglé, *La Démocratie devant la science*, pp. 221ff.

> which in its turn demands socialization rather than individual egoism, then the difficulty disappears. Collective utility presupposes self-control and altruism on the part of the individual. By extending the application of natural selection to groups, instead of restricting it to individuals, the origins of morality are explained. The Darwinian principle is thus preserved. The theatre of competition is the social order, not the physical environment; in this order, it is social interests that are primordial and essentially useful. Morality arises because it is socially useful; this is the Darwinian explanation. An intelligent altruism expresses a higher type of social life, rather than a powerful egoism: the best type prevailed and must prevail.[21]

We find yet again here the method of transposing the struggle between individuals into a struggle between social groups (or even races), with the object of rescuing morality and altruism within these groups. It is impossible to determine whether Baldwin did not understand the difficulties that group selection poses for Darwinism, or whether he simply adopted the 'solution' that Darwin himself had conceived. To my mind, this is not a very important matter, given that we are led here into ever vaguer theories. It should simply be noted that here again, and in this case formulated very explicitly, morality is born and develops by Darwinian evolution because it is socially useful.

In the 1930s, it was still the same idea that persisted or returned. There now began to appear 'non-formalized models' that sought to resolve the problem of group selection. J. B. S. Haldane, for example, in 1932, conceived a kind of parental selection of small groups of individuals.[22] These ideas continued in the 1940s. An American biologist even saw in them a biological foundation for ethics, by assimilating society to a living being whose component parts were differentiated and specialized. That was hardly anything original. Here is an extract from his text – the last that we shall cite on this subject, as it synthesizes the characteristic ideas of the first half of the century:

---

21 J. M. Baldwin, *Darwin and the Humanities*, retranslated from the French edition, *Le Darwinisme dans les sciences morales* (Paris: Alcan, 1911), pp. 83–4. The book that Baldwin criticizes here is T. H. Huxley, *Evolution and Ethics, and Other Essays* (London: Macmillan, 1893). See pp. 95–6 below for a quotation from this work.

22 J. B. S. Haldane, *The Causes of Evolution* [1932] (Ithaca: Cornell University Press, 1966).

> A rational, or scientific, ethics can only be on the public level of intercommunication and agreement. It cannot be circumscribed by or built upon any fiats of innate a priori knowledge of right and wrong or of anything else. It cannot accept any intuitive certainty, any categorical imperative of duty, any revealed will of God, or love for Him, as its base or end. It cannot rely on a reward in the hereafter for virtue in the here-now. It must be empirical. It must seek ever wider and deeper roots in biology as the store of physiological and psychological knowledge grows ...
>
> Any man in a communal setting, then, is in part a complete individual, in part a unit in an individual of next higher order. The many antinomies of selfish and altruistic, private and public, individual and group, must be reconsidered from this viewpoint. Huxley's dilemma [the opposition between evolution and morality], for example, is easily resolved. The conflict between his 'ethical process', governing society by the rule of the common good, and his 'cosmic process', of competition for survival and evolution, is a typical instance of the relation between any organized system and its units. Man as an organism competes with every other man, but men as units in an epiorganism cooperate for the welfare of their society in its struggle with its environment – biotic or physical ... The direction of evolutionary change is consistently from the more homogeneous, with different structural regions performing like functions in 'competition' with one another, to the more differentiated and reintegrated, with different regions specialized for separate functions, division of labour, and cooperation of all parts in terms of the whole. This is as true at the levels of protoplasm, cell, tissue, organ, or multicellular individual as it is at the levels of family, hive, herd, clan, community, state, or whole species ...
>
> Selfish and altruistic impulsions are thus ancient and both have evolved progressively. The latter, however, have gained relatively to the former and in more recent animal history have increased notably. They have flowered with the actual growth of the cerebrum, so dominant in man, and they have been fostered by all social milieus for they are necessary to group living.[23]

It is needless to give further details on this question, as it is always the same notions that recur. All that varies is the sauce with which they are seasoned (more or less philosophical, more or less mathematicized, more or less democratic, more or less racist, spiced or bland, etc., according to time, place and fashion).

I have not carried out in-depth research on the subject, but simply cited the principal texts I am familiar with. However, given

23 R. W. Gerard, 'A biological basis for ethics', *Philosophy of Science* 1942, 9, pp. 92–120 (extracts from p. 103, 107–8, 114–15).

the frequency of these kinds of ideas, whether explicitly developed or simply mentioned, it is highly likely that a far greater number of variations on the theme could be found, among a whole range of writers of this period.[24] This magical Darwinian altruism is in fact a cliché of biological sociology, and has been periodically revisited (or reinvented) in the context of supposed evolutionary ethics.

24 One of these variants worth mentioning here is the work of F. Le Dantec, *L'Égoisme, base de toute société* (Paris: Flammarion, 1911), in which society is seen as a 'collection of egoists', but where morality is finally saved and men end up finding cooperation advantageous as 'egoistic work can be useful for all'.

4

Sociobiology Today

After the Second World War, ideas of this kind became more rare, or, more precisely, lost their theoretical importance. They did not disappear, but less commonly appeared in the foreground; they also seemed to be concentrated among Anglo-Saxon authors, who now maintained the tradition. At the same time, informal theories of group selection made way for more mathematicized approaches, even mathematical modelling – especially, in the 1960s, with the work of W. D. Hamilton.[1]

These changes can be understood in the following way. First of all, there was the rise of molecular genetics. This occupied the high ground, and made redundant the approach to heredity used in the altruistic theories, which now looked old-fashioned. It became clear that these lacked any serious foundation, and they were terribly marked by pre-war ideology. They lost credibility and disappeared from the stage. But they profited from this effacement, which afforded theorists the opportunity to try and give these approaches a rather more scientific appearance and thereby regain ground. To reduce the aura of the anthropological fable, the theories became formalized by way of mathematical modelling – based on a combination of population genetics and economic models. This mathematical veneer masked the poverty of their biological reasoning and ensured their survival, if not much more.

1 W. D. Hamilton, 'The genetical theory of social behaviour, I, II', *Journal of Theoretical Biology*, 1964, 7(1), pp. 1–52; 'Selfish and spiteful behaviour in an evolutionary model', *Nature*, 1970, 228 (5277), pp. 1218–20; 'Geometry for the selfish herd', *Journal of Theoretical Biology*, 1971, 31 (2), pp. 295–311; 'Selection of selfish and altruistic behaviour in some extreme models', in J. F. Eisenberg and W. S. Dillon (eds), *Man and Beast: Comparative Social Behavior* (Washington: Smithsonian Institution Press, 1971), pp. 57–91; 'Altruism and related phenomena, mainly in social insects', *Annual Review of Ecology and Systematics*, 1972, 3, pp. 193–232.

In the mid 1970s, this passage through the desert came to an end, and the idea of Darwinian altruism returned to the front of the stage, with E. O. Wilson's sociobiology and Richard Dawkins's book *The Selfish Gene*. These doctrines had no better claim to scientific status than those of the pre-war era or the 1960s, but the ideology underlying them had regained lost ground (see below). The article quoted above is an example of the new current initiated at this time.

We shall focus first of all on E. O. Wilson, as it was his doctrine that gave its name to the new discipline.[2] I shall use Dawkins's *Selfish Gene* only by way of comparison[3] – first of all, because it was published after Wilson's magnum opus; then because it is closely equivalent in its principles (all that interests us here); and finally, because it was written in the style of an idiotic hawker ('genetics and society explained to clever children and rather simple adults'), a style no doubt needed to hide the incoherence of the theory, but which rapidly makes it insufferable reading.

The general principle is simple. With both authors, this consists in transposing to the level of the gene the egoism that characterizes the individual in the theory of Darwinian selection. (The theories of altruism discussed above transposed this egoism to the level of the group: altruistic behaviour on the part of the individual was justified by serving the group interest.) In sociobiology, it is no longer the survival of the individual that forms the axis on which the theory turns, nor the survival of the group, but rather the 'survival' of the gene and its replication.

Altruistic behaviour is then very easy to explain. It suffices to show that this behaviour, even if not favourable to the individual itself, is favourable to its genes. The first thing necessary here is that this behaviour should improve the survival and multiplication of individuals who have genes in common with the individual in question; the second, that the survival of these individuals should be more favourable to the perpetuation and multiplication of these genes than the survival of the individual itself. The persons concerned are evidently the close kin of the altruistic individual: brothers or sisters, cousins, etc., who share certain genes in varying proportions and will be likely to benefit from this altruism to a corresponding degree. In his or her

2 E. O. Wilson, *Sociobiology, the New Synthesis* (Cambridge, MA: Belknap/Harvard University Press, 1975).

3 R. Dawkins, *The Selfish Gene*.

altruistic behaviour, the individual will therefore prefer a sibling to a cousin, as they have more genes in common; but they will prefer a cousin to a foreigner, and a foreigner of the same race to one of a different race.

All behaviour, and altruistic social behaviour in particular, is thus a means by which genes perpetuate themselves and multiply by way of a network of related individuals. The perpetuation and multiplication of his genes are the individual's essential motivations, rather than his own egoistic survival:

> In a Darwinian sense the organism does not live for itself. Its primary function is not even to reproduce other organisms; it reproduces genes, and it serves as their temporary carrier. Each organism generated by sexual reproduction is a unique, accidental subset of all the genes constituting the species. Natural selection is the process whereby certain genes gain representation in the following generations superior to that of other genes located at the same chromosome position ...
>
> In the process of natural selection, then, any device that can insert a higher proportion of certain genes into subsequent generations will come to characterize the species ...
>
> This brings us to the central theoretical problem of sociobiology: how can altruism, which by definition reduces personal fitness, possibly evolve by natural selection? The answer is kinship: if the genes causing the altruism are shared by two organisms because of common descent, and if the altruistic act by one organism increases the joint contribution of these genes to the next generation, the propensity to altruism will spread through the gene pool. This occurs even though the altruist makes less of a solitary contribution to the gene pool as the price of its altruistic act.[4]

Here, for comparison, is the formulation adopted by Dawkins, which is perhaps a bit clearer:

> What is the selfish gene? It is not just one single physical bit of DNA ... it is all replicas of a particular bit of DNA, distributed throughout the world. If we allow ourselves the licence of talking about genes as if they had conscious aims ... we can ask the question, what is a selfish gene trying to do? It is trying to get more numerous in the gene pool. Basically it does this by helping to program the bodies in which it finds itself to survive and to reproduce. But now we are emphasizing that 'it' is a distributed agency, existing in many different individuals at once. The key point of this chapter

4 E. O. Wilson, *Sociobiology, the New Synthesis*, pp. 3–4.

> is that a gene might be able to assist replicas of itself which are sitting in other bodies. If so, this would appear as individual altruism but it would be brought about by gene selfishness.[5]

In both Wilson and Dawkins, altruistic behaviour is a ruse employed by the genes to perpetuate and multiply themselves, if need be by sacrificing the individual who bears them, to the benefit of the survival of other individuals who are equally their bearers but are in a better position to ensure this perpetuation and multiplication.

The importance of the degree of kinship of the individuals towards whom the altruist displays this characteristic behaviour (going as far as self-sacrifice) is underlined by Dawkins in a more direct manner than by Wilson: from the point of view of his genes, the life of the altruistic individual is worth that of two of his or her children, four nephews, eight cousins, etc. From this point of view, he or she should therefore sacrifice him- or herself if this will ensure the survival of more than two children, more than four nephews, more than eight cousins, etc:

> Now we are in a position to talk about genes for kin-altruism much more precisely. A gene for suicidally saving five cousins would not become more numerous in the population, but a gene for saving five brothers or ten first cousins would. The minimum requirement for a suicidal altruistic gene to be successful is that it should save more than two siblings (or children or parents), or more than four half-siblings (or uncles, aunts, nephews, nieces, grandparents, grandchildren), or more than eight first cousins, etc. Such a gene, on average, tends to live on in the bodies of enough individuals saved by the altruist to compensate for the death of the altruist itself.[6]

It goes without saying that, contrary to what Wilson and Dawkins make out, all this has only a distant relationship with Darwin's original proposition, and even with Darwinism as this subsequently developed. We can distinguish the influence of Weismann's genetics, but Weismann read in a very particular fashion – a point we shall return to later on.

There is a manifest anthropomorphism here, a kind of personification of genes whose 'interest' replaces that of the individual in natural selection. Dawkins, in particular, constantly

5 R. Dawkins, *The Selfish Gene*, p. 95.
6 Ibid., p. 100.

speaks of genes as if they were living beings, and as if the fundamental question were their 'survival' in the face of natural selection:

> It will be remembered that the 'central theorem' of the selfish organism claims that an animal's behaviour tends to maximize its own (inclusive) fitness ... An animal's behaviour tends to maximize the survival of the genes 'for' that behaviour, whether or not those genes happen to be in the body of the particular animal performing it.[7]

This personification of genes even borders on infantile farce in some of his works of popularization, where genes become happy fellow-travellers passing down through the generations – though in this field, it is hard to draw any firm line between 'science' and vulgarization:

> I have spoken of a river of genes, but we could equally well speak of a band of good companions marching through geological time. All the genes of one breeding population are, in the long run, companions of each other. In the short run, they sit in individual bodies and are temporarily more intimate companions of the other genes sharing that body. Genes survive down the ages only if they are good at building bodies that are good at living and reproducing in the particular way of life chosen by the species.[8]

Apart from this kind of personification of genes, there are two problems that must be settled. The first is to explain how the individual can perform behaviours that are not justified by his or her own survival but by that of individuals who carry genes identical to their own. The response Wilson gives is that behaviour is controlled by a part of the nervous system (the limbic system) that is programmed for this purpose. Nothing precise is evidently said about the manner in which this programming takes place; it is simply ascribed to Darwinian evolution. At all events, the limbic system programmed in this way is supposed to calculate and command the behaviour adequate to the best preservation and multiplication of the genes, even if this involves the death of the individual:

7 R. Dawkins, *The Extended Phenotype* (Oxford: OUP, 1989), p. 233: his emphasis.
8 R. Dawkins, *River Out of Eden* (London: Weidenfeld & Nicolson, 1995), p. 5.

> The hypothalamic-limbic complex of a highly social species, such as man, 'knows', or more precisely it has been programmed to perform as if it knows, that its underlying genes will be proliferated maximally only if it orchestrates behavioural responses that bring into play an efficient mixture of personal survival, reproduction, and altruism. Consequently the centres of this complex tax the conscious mind with ambivalences whenever the organisms encounter stressful situations. Love joins hate, aggression, fear, expansiveness, withdrawal, and so on, in blends designed not to promote the happiness and survival of the individual, but to favour the maximum transmission of the controlling genes.
>
> The ambivalences stem from counteracting pressures on the units of natural selection.[9]

Why the limbic system? Because this is the part of the nervous system which, it is believed, controls emotions and feelings. It is these emotions and feelings that are seen as underlying altruistic behaviour, which may possibly put the life of the individual concerned at risk. This presupposes that egoistic behaviours, which aim at the survival of the individual, are for their part under the control of rationality, and that this rationality has to be submerged by emotion and feeling in order for the altruistic component to triumph over egoism. This is perfectly in keeping with the most immediate common sense – egoism is always calculating (hence rational), whereas altruism is emotional and sentimental:

> The biologist, who is concerned with questions of physiology and evolutionary history, realizes that self-knowledge is constrained and shaped by the emotional control centres in the hypothalamus and limbic system of the brain. These centres flood our consciousness with all the emotions – hate, love, guilt, fear, and others – that are consulted by ethical philosophers who wish to intuit the standards of good and evil. What, we are then compelled to ask, made the hypothalamus and limbic system? They evolved by natural selection.[10]

But it still has to be made clear that this emotional and sentimental aspect operates only at the level of the individual, not at that of the genes, where calculating egoism takes the upper hand. Individual feeling and emotion are simply a Darwinian ruse of the

9 E. O. Wilson, *Sociobiology, the New Synthesis*, p. 4.
10 Ibid., p. 3.

genes: 'When altruism is conceived as the mechanism by which DNA multiplies itself through a network of relatives, spirituality becomes just one more Darwinian enabling device.'[11]

Once it is accepted that the limbic system is programmed to calculate the behaviour that will ensure the best preservation and multiplication of the genes, it still remains to be seen how the individual knows the genetic composition of the partners to whom she devotes herself (which is evidently necessary to calculate behaviour, if the aim of this is the best survival of the genes). Wilson is clearly unable to resolve this difficulty, and for good reason. Marshall Sahlins made fun of this with the example of a Hawaiian woman who explained her relationship to the daughter of her adoptive brother as follows:

> Kealoha is my brother's child – of course my brother isn't really my brother as both he and I are hanai [adoptive] children of my father. I guess my father isn't really my father, is he? I know who my real mother is, but I didn't like her and I never see her. My hanai brother is half-Hawaiian and I am pure Hawaiian. We aren't really any blood relations I guess, but I always think of him as my brother and I always think of my [adoptive] father as my father. I think maybe Papa [her adoptive father] is my grandfather's brother. So I don't know what relation Kealoha really is, though I call her my child.[12]

Let us hope that this woman's limbic system has a clearer idea of her relations of genetic kinship, if it has to calculate her behaviour in order to maximize the propagation of her genome.

It goes without saying that sociobiological theory is no more than a cock-and-bull story. In fact, it is not even developed in Wilson's work, but simply occupies a few pages in the first chapter. The rest of the book's several hundred pages are devoted to explaining all kinds of questions regarding social behaviour, both in a whole range of animals and in humans. This is all more or less independent of the theory presented at the start of the book, which is only referred to occasionally, and even then, simply for illustrative purposes.

The reason for this is easy to understand. On the one hand, the thesis is so shaky that it is impossible to give it any kind of

11 Ibid, p. 120.

12 M. Sahlins, *The Use and Abuse of Biology* (Ann Arbor, MI: University of Michigan Press, 1976), p. 52.

coherent theoretical development. On the other hand, contrary to its pretensions, sociobiology (like the theory of the selfish gene) is not designed to provide an explanation of (altruistic) social behaviour but simply to make it compatible with Darwinism. Once this compatibility is 'established', the theory can be forgotten – it is idiotic – and the various behaviours can be explained as if their compatibility with the Darwinian framework were already resolved. Which is why Wilson, apart from the first twenty pages presenting his theory of sociobiology, devotes the rest of his book essentially to descriptions of animal and human behaviour.

Wilson's book was strongly criticized on its first publication in 1975. The French translation did not appear until 1987, and this was in a collection devoted to esoterica (among titles such as *L'Alchimie de la vie*, *La Science et les pouvoirs psychiques de l'homme*, and even *L'Énergie micro-vibratoire et la vie)*. This indicates the scientific value it was accorded; Wilson is certainly a genuine scientist, but should perhaps not have abandoned himself to the perverse pleasure of general theories. As for its political connotation, it is sufficient to note that this French translation was edited by a journalist generally associated with the far right. In itself, however, Wilson's book has a moderate tone, including the chapter devoted to human behaviour, which corresponds completely to what might be expected of an Anglo-Saxon Darwinian – no more (or hardly), no less. The far-right connotation, however, is not completely unjustified, as we shall soon see.

The Selfish Gene perhaps includes more theoretical developments than does Wilson's book. But, as we have said, it is written in an infantile style in which genes are almost equated with living beings – so much so that the theoretical developments presented can scarcely be taken seriously. You can demonstrate anything you like in this kind of style. It is undoubtedly because of its rather stupefying popularization that Dawkins's book was translated into French earlier than Wilson's, as early as 1976.

The theories of Wilson and Dawkins, however poorly received at first, nevertheless steadily managed to win recognition and establish themselves in the scientific landscape – very largely because of the indifference that surrounded them once the initial scandal had passed. This enabled a certain number of third-rate biologists to specialize in this field, buoyed up by mathematicians whose main talent lay in applying theories invented by others. This was no doubt because, as we shall see, the theoretical difficulties of molecular genetics were to revive ways of thinking that dated from before the Second World War; as well as because of the very particular nature of this theory.

We should not in fact see sociobiology as simply an extension of the biological sociologies of the first half of the twentieth century, social Darwinism and others. There are obviously certain common elements, as the definition of sociobiology given by Wilson clearly indicates: in his words, the object is to explain society in terms of its biological base, and thus articulate it to the 'modern synthesis' – i.e., the 'synthetic theory' put forward in the late 1930s to bring together genetics and evolutionism.[13] What Wilson sought to achieve was a new synthesis, combining this 'synthetic theory' (genetics and evolutionism) with sociology, the whole under the sign of a Darwinism that was both biological and social:

> Sociobiology is defined as the systematic study of the biological basis of all social behaviour. For the present it focuses on animal societies, their population structure, castes, and communication, together with all of the physiology underlying the social adaptations. But the discipline is also concerned with the social behaviour of early man and the adaptive features of organization in the more primitive contemporary human societies ... One of the functions of sociobiology, then, is to reformulate the foundations of the social sciences in a way that draws these subjects into the Modern Synthesis.[14]

This scientism, the desire to naturalize sociology, has its counterpart in the field of ethics, which should now pass to the hands of biologists: 'Scientists and humanists should consider together the possibility that the time has come for ethics to be removed temporarily from the hands of the philosophers and biologized.'[15]

That is perfectly clear, and perfectly in the lineage of the pre-war Darwinian sociologies. But we should not trust Wilson's statements too much, as his project, like that of Dawkins, goes well beyond the simple framework of a naturalization of sociology (or ethics). As Sahlins stressed as far back as 1976, with sociobiology we enter the domain of a new theory that Sahlins

13 The title of Wilson's book, *Sociobiology, the New Synthesis*, is a reference to Julian Huxley's book *Evolution, the Modern Synthesis* (London: Allen and Unwin, 1942), which coined the description of a 'synthetic theory'.

14 E. O. Wilson, *Sociobiology, the New Synthesis*, p. 4.

15 Ibid., p. 562.

called 'genetic capitalism'.[16] I shall keep this description, even though I understand the phenomenon slightly differently.

As against the biological sociology of the pre-war era, sociobiology is biology before being sociology; but it is a biology injected with the principles of Darwinian sociology. The difference lies in the change of label: in the early twentieth century the term was 'bio-sociology' (sociology with a biological dimension); today it is sociobiology (a biology that has integrated the social dimension). The origin of this sociobiology thus presents the following stages: 1) Darwinism imported sociological principles into biology; 2) Darwinian sociology recuperated these sociological notions, naturalized by their transit through biology; 3) sociobiology took up these notions from Darwinian sociology in order to reintroduce them into biology.

Conceived in this way, sociobiology is not really a biological explanation of society, nor even of social behaviour among animals. This is why it matters little that the explanation it gives is totally ridiculous. All this is just a façade, and the real stake is something quite different: sociobiology is first and foremost a new definition of life, a definition that integrates the notions of Darwinian sociology.

In point of fact, sociobiology does not continue the old tradition that sees life as the egoistic preservation of the individual being – a definition still operative in the Darwinian theory of the struggle for existence. It sees life rather as the maximization of genetic inheritance, on the economic model of the maximization of capital; the modelling used by sociobiology is in fact copied from that of economics. Far indeed from the individualistic egoism of the traditional definition of life, it considers individuals as negligible quantities, simple supports for a genetic inheritance whose maximization they must ensure. (Just as economic individuals are negligible quantities whose value lies only in their role in the maximization of capital – a parallel that is enlightening as to the origins of the theory.) This attitude towards the individual goes back to that of Darwinian sociologists such as Gumplowicz, who likewise deemed individual life cheap (see the quotation on p. 20).

The result is that sociobiology does not so much biologize society, but rather 'socializes' life by giving it an end that goes beyond individual destiny and is attached to the destiny of the

16 M. Sahlins, *The Use and Abuse of Biology*, p. 72.

genome (that is, the destiny of the population sharing this genome, by and large the destiny of the race).

It is undoubtedly this conception of life, far more than the ridiculous biological explanation it gives of society, that has given sociobiology a certain success. And it is also this that is responsible for its far-right political reputation. In fact, if this idea has a vague connection with Darwin (orthodox Darwinism sees success in competition falling to whoever leaves the most descendants, which can already be understood as a maximization of genetic inheritance), and likewise with Weismann (for whom the germ-plasm was a kind of immortal entity from which mortal individuals temporarily developed), this origin has now become very remote. Neither Darwin, nor Weismann, nor even Darwinism and Weismannian genetics present the individual and its survival as overshadowed in this way. Where this overshadowing can be found is in Nazi biology, where the mysticism of the germ-plasm hinted at by Weismann finds its full development, either in this form of genetics or in the glorification of the race that has this plasm (or plasma) as its physical substratum. Whether they are aware of it or not, we can find formulations in both Wilson and Dawkins that closely echo Nazi biologists such as Otmar von Verschuer. Here is an extract from the latter's *Manuel d'eugénique et d'hérédité humaine*, followed by a fragment from a speech by Hitler:

> In the ethnic, National Socialist state, we understand by 'people' and 'ethnic group' a spiritual and biological unity. The spiritual unity of a people becomes an experience of each member of the ethnic group at the great historical moments in the life of the nation. But the 'people' is also a biological unity. Each person is linked by ties of blood to his parents, grandparents, children, brothers and sisters, cousins, etc., and one family is linked to another ... This biological unity of the people is the foundation of the ethnic body, an organic structure with a totalitarian character, whose various fractions are nonetheless components of the same whole. The essential, what is constant and durable in the ethnic body, is not for us a sum of individuals, but rather the hereditary patrimony, which, like a river, flows from one generation to another and represents, in each of these, a particular entity linked to the whole by its parental abilities. If we start from this notion of the 'people', demographic policy is a policy of protection of the ethnic body, by the maintenance and improvement of the healthy heritage, the elimination of its diseased elements, and the preservation of the specific racial

character of the people – in other words by eugenics, cultivation of the hereditary patrimony and racial hygiene.[17]

The fundamental pillar of National Socialism is the abolition of the liberal concept of the individual along with the Marxist concept of humanity, substituting for these the community of the Volk, rooted in its soil and united by the links of a common blood.[18]

Verschuer makes very clear the pre-eminence of genetic inheritance, comparing this with a river traversing individuals and generations, and the importance of ties of biological kinship in the preservation of this genetic patrimony – in exactly the same fashion as Wilson's sociobiology and Dawkins's 'selfish gene'. The only difference is that Verschuer draws political conclusions, whereas Wilson and Dawkins carefully abstain from doing so, leaving this concern to the ideologists of the far right that gravitate around their theses.

Here is a further example, this time concerning the manner in which this politics of 'blood' was put into practice. This is an extract from the notorious speech Himmler gave on 4 October 1943 to SS forces charged with operations on the Russian front:

For the SS man, one principle must apply absolutely: we must be honest, decent, loyal, and comradely to members of our own blood, and to no one else. What happens to the Russians, the Czechs, is totally indifferent to me. Whatever is available to us in good blood of our type, we will take for ourselves, that is, we will steal their children and bring them up with us, if necessary. Whether other races live well or die of hunger is only of interest to me insofar as we need them as slaves for our culture ... That is how I would like to indoctrinate this SS; and, I believe, have indoctrinated, as one of the holiest laws of the future: our concern, our duty, is to our people, and to our blood. That is what we must care for and think about, work for and fight for, and nothing else ... We have arisen through the law of selection. We have selected from the average of our people. Our people arose through the dice game of Fate and history in long primaeval times, over generations and centuries ... The moment we forget the law of the racial foundation of

17 O. von Verschuer, *Manuel d'eugénique et d'hérédité humaine* (Paris: Masson, 1943), p. 114.

18 Hitler speech of 30 January 1937, cited by Y. Ternon and S. Helman, *Les Médecins allemands et le National-socialisme, les métaphorphoses du darwinisme* (Tournai: Casterman, 1973), p. 22.

> our people, the law of selection and severity with regards to ourselves, then the germ of death will lie within us ... For that reason, it is our duty ... to remember our principle: blood, selection, severity.[19]

Here again we have the same theme, which is not taken from either Darwin or Weismann: the pre-eminence of 'blood' (the genetic inheritance) over the individual, and the egoism of 'blood' that links related individuals and separates them from individuals of a different (or even opposing) 'blood' – and this across nationalities, as what matters is not nationality but 'blood', race.

A second factor in the success of sociobiology, along with its political connotation, comes from its irrational – or better, irrationalist – aspect, which has no difficulty in combining with a very pronounced reductionist scientism.[20] This irrationalist aspect opened the door wide to a kind of mysticism, and seduced a large following, including a number of scientists, attracted more by its easy analogies than by the difficulty of its arguments. It is thus no accident that the French translation of Wilson's *Sociobiology* was published in a collection of esoterica.

The importance ascribed to the genome has made this a kind of mystical entity traversing the generations, the 'holy grail' of life or even race, whose transmission must be preserved at all costs. With Dawkins, this importance is explicitly compared with the worship of ancestors, and its passage through successive generations becomes a kind of epic:

> All peoples have epic legends about their tribal ancestors, and these legends often formalize themselves into religious cults. People revere and even worship their ancestors – as well they might, for it is real ancestors, not supernatural gods, that hold the key to understanding life.[21]

---

19 Speech of Reichsführer-SS Heinrich Himmler at Posen, 4 October 1943. Document no. 1919-PS, Nuremberg Trial (www.codoh.com/incon/inconhh.html).

20 Here, too, sociobiology continues an old tradition, as Haeckel's monism was not a simple materialism, but derived from a kind of pantheistic religion shared by a very varied range of esoteric currents (including Éliphas Lévi, Annie Besant, Mme H. P. Blavatsky, Rudolf Steiner, etc. – É. Schuré saw Haeckel himself as a great prophet), as well as by Nazism in its pseudo-mystical aspects (see D. Gasman, Haeckel's *Monism and the Birth of Fascist Ideology*, pp. 69ff.). We can add to this list the so-called New Age, as at the end of his life Haeckel wrote a book on the subject of the energy at work in crystals: Krystalseelen. *Studien über das anorganische Leben* (Leipzig: Kröner, 1917).

21 R. Dawkins, *River Out of Eden*, p. 1.

This sacralization of the genome is nothing new. We can already find an explanation of it in Vacher de Lapouge, along with the religious aspect that often characterizes doctrines bent on the preservation or improvement of the genetic inheritance, racism and eugenics (Galton termed eugenics 'the religion of the future', an expression that Julian Huxley adopted). According to Vacher de Lapouge, as we see below, the germ-plasm had taken the place of the soul in terms of immortality. It was no longer the soul that was immortal (no biologist still believed in the soul), but the support for heredity that traversed the generations certainly was so. This supposed immortality made the physiological support for heredity the essence of life. Hence a kind of mysticism of life – not individual life, but life as represented by the germ-plasm, along with a mysticism of fertility (Lapouge also saw life as the means by which the universe becomes conscious of itself):[22]

> Scarcely anyone still believes in the immortality of the soul. The most idealistic and religious minds are seized by doubt. Science no longer needs the soul in order to explain life. In former times, it was said that the soul was immortal, because the idea of the soul and that of immortality were inseparable ... But immortality is not a mere deception ... It is almost the opposite of what people were searching for. What is immortal is not the soul, a doubtful and probably imaginary entity; it is the body, or rather the germ-plasm. The being that does not die is the material of the reproductive cells ... The germ-plasm is like the underground stem, the invisible rhizome of which individuals are simply the visible shoots ... This relative immortality, which lasts as long as the rituals of fertilization are repeated, is the only immortality: any other kind is illusory. It is through this immortality that the being returns to life, with its form, its behaviour, its instincts, its thoughts. A poet would call this the rebirth of the phoenix. A metaphysician would say that physical and mental heredity are the effort of the universe to preserve self-consciousness. Living beings really are the consciousness

22 An implication of this is that the scientist, aware of this fact, is the representative of God: 'God has consciousness through the hierarchy of beings that feel and think, from the moneron in which the soul has just awakened through to the scientist who knows the infinitely great and the infinitely small ... For this reason the scientist is the partial avatar of God, and human moral purpose is the expansion of consciousness.' G. Vacher de Lapouge, preface to his translation of E. Haeckel, *Le Monism, lien entre la religion et la science* (Paris: Schleicher, 1897), p. 8.

> of the world. This consciousness rises from the lower animals to man, and from man to the scientist ... It is in this way that science reawakens an entire profound metaphysics of the call of life, conceived many times by the sages of Asia, which predated Christianity and will survive it, if metaphysics is still needed. Immortality, cosmic awakening, all this has fertilization as its base. This is why the sexual act is not creative simply in its evocation of a new being, the accomplishment of the absolute precondition for immortality: it radiates a divine character because it is the transmission of the consciousness of the world, and thus becomes a theogony. This is why the absolute sin is infertility.[23]

Vacher de Lapouge goes on to contrast this mysticism to Christianity, which sees chastity as a virtue.

All this is obscurely present in the kind of genetic mysticism reintroduced by Wilson and Darwin. We might add that the role Wilson ascribes to the limbic system – as a means by which genes find expression in social behaviour, beyond the rationality of egoistic behaviour – confers on emotion and irrational intuition a superiority over rationality. This emotion and irrational intuition express the higher interest of the genes (and of Life, which they represent), whereas rationality simply pursues the egoistic interest of the individual. The emotional irrationality of behaviour commanded by the limbic system as a function of the interest of the genes thus becomes a kind of 'voice of blood'. Genetic self-interest and the voice of blood easily become the interest and voice of the race.

Here, too, we can trace, if not an origin, then at least a readily identifiable predecessor. This was the Anglo-German racist theorist Houston Stewart Chamberlain (1855–1927), forerunner of Nazism. In order to paper over the difficulties of an objective definition of race (a perennial problem, as we shall see), Chamberlain conceived a subjective definition: people supposedly had an immediate and intuitive knowledge, rather than a rational one, of their own race and that of others, in the same way that sociobiology supposes that individuals, in order to calculate their altruistic behaviour, grasp in an intuitive and immediate fashion their links of genetic kinship with this or that other individual (hence the difficulties faced by Sahlins's Hawaiian). Chamberlain opposes life ('whose roots plunge to a depth that knowledge will never attain') to the rational approach, just as feeling and emotion

23 G. Vacher de Lapouge, *Les Sélections sociales*, pp. 306–7.

governed by the limbic system are opposed to the rationalism of egoistic behaviour:

> Pure science (in contrast to industrial science) is a noble plaything; its great intellectual and moral worth rests in no small degree upon the fact that it is not 'useful' ... Life, on the other hand, purely as such, is something different from systematic knowledge, something much more stable, more firmly founded, more comprehensive; it is in fact the essence of all reality, whereas even the most precise science represents the thinned, generalized, no longer direct reality. Here I understand by 'life' what is otherwise also called 'nature' ... Nature is in fact what we call 'automatic', its roots go very much deeper than knowledge will even be able to follow ... Though science leaves us in the lurch at many points, though she, fickle as a modern parliamentarian, laughs today at what she yesterday taught as everlasting truth, let this not lead us astray; what we require for life, we shall certainly learn ... The very fact that we are living beings gives us an infinitely rich and unfailing capacity of hitting upon the right thing, even without learning, wherever it is necessary ... So it is, too, in regard to the question of the significance of race: one of the most vital, perhaps the most vital, questions that can confront man ...
>
> Nothing is so convincing as the consciousness of the possession of Race. The man who belongs to a distinct, pure race, never loses the sense of it.[24]

Much later in the same work, Chamberlain writes:

> Darwin himself, who worked all his life with compass, ruler and weighing machine, is always in his studies on artificial breeding calling attention to the fact that the eye of the born and experienced breeder discovers things of which figures give not the slightest confirmation, and which the breeder himself can hardly ever express in words; he notices that this and that distinguishes the one organism from the other, and makes his selection for breeding accordingly; this is an intuition born of ceaseless observation ... The hieroglyphs of nature's language are in fact not so logically mathematical, so mechanically explicable as many an investigator likes to fancy. Life is needed to understand life. And here a fact occurs to me which I have received from various sources, viz. that very small children, especially girls, frequently have a marked instinct for race. It frequently happens that children who have no conception of what 'Jew' means, or that there is any such thing in the world, begin to cry as soon as a genuine Jew or Jewess

24 H. S. Chamberlain, *The Foundations of the Nineteenth Century* [1899] (London: Lane, 1910), vol. 1, pp. 268–70.

> comes near them! The learned can frequently not tell a Jew from a non-Jew; the child that scarcely knows how to speak notices the difference.[25]

We see again here the extreme ambiguity of this kind of theory, and how easy it is to move from Darwinian altruism to theories that have nothing altruistic about them.

25 Ibid., pp. 536–7.

5

Darwinism, Society and Morality

Why was the idea of an evolutionary biological altruism so common in the first half of the twentieth century – a fact that attests to its importance, however trivial it seems at first sight? Why did it regress after the Second World War, to reappear later in sociobiology?

The primary reason for this evolutionary altruism, as we have seen, is the need to reconcile the existence of societies, both animal and human, with Darwinism. For a very long time, instincts were arbitrarily invoked to explain behaviour. Darwin himself extended his theory to such instincts: whatever they were, they had to be viewed from the standpoint of the advantage they conferred on the individuals possessing them. In theoretical terms, this was all well and good; it was sufficient to consider these instincts and behaviours as extensions of purely biological characteristics. The only problem was that presented by social instincts and behaviour, as these could not easily be fitted into the framework of competition and selection. This lack of fit between Darwinism and sociability was all the more marked, and all the more awkward, given how the competition for existence was deemed to exert itself, in the orthodox theory, between individuals of the same species occupying the same territory – in other words, precisely those who might be found grouped together in a society.

The most difficult case was that of the social insects – ants, bees, termites – traditionally depicted as models of order, discipline and hard work, which should be an inspiration to human societies that are disorganized by comparison. Among the insects, the social dimension includes a very well-defined division of tasks, even translated into different morphologies and physiologies of individuals with different 'specializations' (queen, workers, soldiers, etc.). These specialized morphologies and physiologies, being of use only for the group, fitted very poorly into the framework of natural selection. The sterility of the majority of the individuals concerned was still harder to explain. How could it be seen as advantageous for them, and how could that advantage be transmitted to future generations? Darwin already perceived this problem in *The Origin of Species*, and sought to resolve it by way of 'group selection': the characteristics in question, including sterility, were useful for the group rather than for the individual itself.[1] This explanation was revived on many occasions in different forms.

In a general sense, the difficulty arose from the fact that Darwinism had introduced into biology the war of all against all, making this not only the motor of an evolution understood broadly as progress, but also a universal mode of explanation (see above, pp. 21–34). We can recognize here certain particular philosophical, sociological and economic influences (Hobbes, Adam Smith, Malthus, and others), but where these theorists appealed to political power, a social contract, a 'hidden hand', a morality dependent on a transcendent being, or some other such mechanism, biology found itself quite unequipped to explain social phenomena, whether animal or human, whilst remaining within its naturalist field. To account for the existence of societies – and, more generally, of behaviour in which individuals co-operate rather than struggle – it was necessary to introduce a moderating factor, a counterweight to the war of all against all, and a biological counterweight at that: the evolutionary value of altruism.

Altruism is, then, the pendant of struggle in Darwinian sociology; the more struggle is invoked to explain social processes, the more frequent the resort to altruism to correct its negative effects and make society possible. Thanks to which, by a subtle combination of struggle and altruism, invoked simultaneously or successively, it is possible to explain anything. Altruism thus makes it possible to base animal or human society on a biological foundation compatible with competition and selection. It places society under the rule of Darwinian biology. In other words, it enables biological sociologies and sociobiology to exist; without altruism, they would not be possible (at least in the Darwinian framework that is their 'natural' habitat). This is the first reason. It remains to be explained why there is such a recurrence of theories along these lines.

---

1 C. Darwin, *The Origin of Species* (Harmondsworth: Penguin, 1985), pp. 234ff.

In itself, this recurrence is not surprising. We are dealing here, in fact, with a discipline that is intellectually very weak and working with a very limited number of ideas (pseudo-concepts such as randomness, competition, selection, survival, egoism, and altruism). It combines these ideas with one another, explores the possibilities opened up by a given combination, and after rapidly exhausting these, moves on to another. As there are so few of these ideas, their combinations are equally limited, and the same theories are therefore fated to return periodically, sometimes put forward by authors ignorant of their precursors or contemporaries. There are certainly variants of detail or ideological colour (as we have seen with Darwin, Haeckel and Kropotkin), but the principle remains the same.

These theories are all marked by a very high level of generality, imprecision, and artistic vagueness. They are anthropological fables – or zoological ones when they concern animals. None of them is very satisfactory, but the process goes on and on. Each theory fairly quickly loses its attraction; the attempt is made to improve it, but as the discipline is so limited intellectually, the search is soon begun for new theories that closely resemble the old ones, differing from them only in some unimportant details (or by a different ideological coloration). These small variations enable their authors to claim they have invented something new. Perhaps they believe this themselves, but in reality they take the same paths again and again without recognizing them. Over 140 years, more or less, variations on the same theory have been resurrected and 'improved'. There is no real development in these theories, but rather a perpetual return; hence it is almost impossible to trace their history.

The second reason for the omnipresence of evolutionary altruism is of a moral order. These scientists were in the main good Victorian or Bismarckian bourgeois.[2] They were scientists, even materialists, but they believed in morality. The application of Darwinian principles to human society made selection and the war of all against all a necessary part of the landscape. This raised certain problems of conscience, which the scientists in question had to resolve by civilizing and moralizing somewhat their original convictions. Hence altruism, which imparted a softening

2 France, which remained very Lamarckian, and the countries of 'Catholic Europe' in general, have been less affected, always being somewhat reticent about Darwinism.

Christian sentiment to the merciless sway of competition and selection.

It is traditionally accepted that Darwinism had a threefold effect on religion and morality. Two of these directly affected religion and thus, indirectly, morality; they resulted from the opposition to creationism, which meant rejecting the fixity of species and introducing chance into their formation, thus running against the idea of a predetermined plan of divine creation. The third was more a matter of morality than religion, bearing on the consequences of applying natural selection to humanity. Contrary to what is often maintained, it is the third problem that predominated, far more than those posed by the rejection of creationism.

In actual fact, the importance of the non-fixity of species and creationism in the nineteenth century has been much exaggerated: by no means were Darwin's opponents always fixists, any more than fixists were always creationists. The idea of a transformation of species was already long established, and had been formalized by Lamarck in 1809; it no longer shocked anybody. As for creationism, this had really never existed as a constituted biological doctrine. Contrary to what is often maintained, Cuvier was not a creationist but a fixist, and his arguments were not religious in nature but scientific, bound up with his principles of comparative anatomy and taxonomy. In his *Discourse on the Revolutions of the Surface of the Globe*, he did claim to explain the disappearance of fossil species as a result of geological catastrophes (including the biblical flood), but he never appealed to divine creation in order to explain the appearance of new species after such catastrophes. He simply did not offer any explanation.[3]

After the death of Cuvier in 1832, his theses lost influence with the spread of transformist ideas, but these were still unable to provide a satisfactory explanation for new species. Lamarck's own explanation, based on the typically mechanistic biology of the eighteenth century, was already outmoded. In the face of the rise of transformism, Cuvier's disciples attempted to perpetuate his ideas, and it was they who, in order to rescue these, resorted

3 G. Cuvier, A *Discourse on the Revolutions of the Surface of the Globe* (London: Whitaker, Treacher & Arnot, 1829). The first edition of 1812 has been republished under its original title: *Recherche sur les ossements fossiles de quadrupèdes, Discours préliminaire* (Paris: GF-Flammarion, 1992).

to one or more acts of creation – though even these acts were not necessarily divine, as the arguments raised against transformism were still based on principles of comparative taxonomy and anatomy rather than on the Bible.

At that time, moreover, the sciences in question (taxonomy and palaeontology) were being relegated to the background by the rise of physiology, cytology, biochemistry, and so on, and by the time Darwin's *The Origin of Species* was published, Cuvier's followers had been reduced to a rather weak current of thought. If there still were some creationists around (generally second-rate scientists, or already aged, encrusted withn the ideas of their Cuvierist youth), this does not mean that creationism existed as a scientific doctrine. Each creationist had his own theory, from divine intervention to the passage of comets that destroyed certain forms of life and replaced them by others. The names of Pierre Flourens (1794–1867) and Louis Agassiz (1807–73) are still well remembered, but who today, apart from a few specialists, knows anything of Henri Marie Ducrotay de Blainville (1777–1850), Charles Léopold Laurillard (1783–1853) and his comets, or the twenty-seven successive creations imagined by Alcide d'Orbigny (1802–57)?[4] It should also be noted, moreover, that of these five eminent creationists (the remainder being still more obscure), three were already dead by the year of Darwin's publication, and it was the ideas of Lamarck, Geoffroy Saint-Hilaire and the like that they opposed, rather than those of Darwin.

Based on Darwin's having written that he had admired William Paley's *Natural Theology* in his youth, biologists and historians of science often claim that pre-Darwinian biology was marked by that book.[5] This is totally wrong. Darwin's professed admiration for Paley only attests to the mediocrity of his intellectual references (Paley had already been treated as a 'weakness of the century' by Thomas de Quincey), but certainly not of the state of biology in the 1850s. On the one hand, this natural theology was characteristic of the eighteenth century rather than the nineteenth. On the other hand, the 'natural theologians' were authors who made use of scientific theories (the animal-machine served to demonstrate

4 For a presentation of their different theories, see G. Laurent, *Paléontologie et évolution en France, 1800–1860* (Paris: Éditions du Comité des Travaux Historiques et Scientifiques, 1987), pp. 249–319.

5 For example, F. Jacob, *The Logic of Life and the Possible and the Actual* (Harmondsworth: Penguin, 1989), p. 364.

the existence of the great watchmaker) rather than scientists concerned with theology. Finally, Paley, who was himself not a biologist but a theologian, had died in 1805; he had not read Lamarck's *Zoological Philosophy* of 1809, nor the major works of Cuvier, nor those of the Geoffroy Saint-Hilaires (father and son), not to speak of all the revolutions in biology between 1805 and 1859 (the cell theory, Magendie's experimental physiology, the beginnings of biochemistry and the physiology of Claude Bernard, the first works of Pasteur, and so forth). His ideas, close to those which the Abbé Pluche had propounded in France in the mid-eighteenth century in his *Spectacle de la nature*, could thus in no way be representative of the state of biology immediately before Darwin.

It is clear enough that nineteenth-century theology remained reticent about the reduction of man to an animal (a reticence that dates from the mid-eighteenth century, when Linnaeus had integrated man into his classification of animals), and preferred in general to stick to a creationist fixism, accepting perfectibility only for man, and even then only in the realm of the spirit, not that of the body. But this was a matter for theology, not for a biology that already in the nineteenth century was no longer dependent on religious principles.

This theology was far from homogeneous, as well as more flexible than is generally imagined. Contrary to the claim often made, it had begun to envisage, in the face of the rise of evolutionary ideas, a way of reconciling evolution with divine creation, even before the publication of Darwin's book in 1859. Here is, for example, what the *Dictionnaire des harmonies de la raison et de la foi* of Abbé Le Noir had to say in 1856:

> The first of these systems [transformism] is repugnant to human conscience; it encroaches on our dignity by reducing us to the level of the animal. Lamarck is one of those who championed this system most methodically; that is what he attempted in his Zoological Philosophy. Although this theory has hardly been supported by anyone except atheists, materialists and pantheists, it does not necessarily exclude genuine theism. It is very easy to conceive, a priori, a world created by God along these lines. One need only assume that He placed in the first organized being, or even in mineral nature, an immaterial principle that was the embryonic soul; that this principle had both the property of propagating itself and of perfecting itself by the gradual development of nature with the organism itself, in such a way as to generate by successive steps the various species through to

man. It would be wrong to maintain that God's creations do not include a creation after this model.[6]

Larmarck, who professed a rather vague deism, showed theology the way by writing in his *Zoological Philosophy*:

> [Linnaeus, sixty years ago,] assumed that each species was invariable and as old as nature itself, and that its particular creation was the work of the supreme Author of all that exists. It is indeed true that nothing exists but by the will of the sublime Author of all things. But can we assign him rules in the execution of his will, and fix the mode that he followed in this respect? Could His infinite power not create an order of things that successively gave existence to all that we see, as well as to all that exists and is unknown to us? Certainly, whatever His will might be, the immensity of His power is always the same; and in whatever manner this supreme will is executed, nothing can diminish its grandeur.[7]

As far as the American Protestant churches – that supposedly high ground of creationism – were concerned, Windsor Hall Roberts noted that, although there was still fairly strong opposition to evolutionism in the 1860s, it was steadily accepted in the 1870s, accompanied by the adaptation of theological doctrines (original sin, for instance, instead of being based on Adam and Eve's disobedience to divine will, became the mark of our animal nature – which virtue commanded us to resist).[8] There was certainly a creationist reaction in the United States, but this was little more than the work of fundamentalist sects and conservative politicians, and not representative of the general situation – neither very major nor very long lasting (even among religious people), and still less a moral problem.

The random nature of evolution, as opposed to the predetermined divine plan of creation, was not very awkward for

6 Abbé Le Noir, *Dictionnaire des harmonies de la raison et de la foi, ou Exposition des rapports de concorde et de mutuel secours entre le développement catholique, doctrinal et pratique, du christianisme et toutes les manifestations rationelles, philosophiques, scientifiques, littéraires, artistiques et industrielles, de la nature humaine individuelle et sociale* (Paris: Encyclopédie Migne, 1856), vol. 19, col. 1339.

7 J. B. P. A. Monet de Lamarck, *Philosophie zoologique* (Paris: GF-Flammarion, 1994), p. 102.

8 W. H. Roberts, *The Reaction of American Protestant Churches to the Darwinian Philosophy, 1860–1900*, University of Chicago Ph.D. dissertation, private edition, distributed by The University of Chicago Libraries, 1938.

original Darwinism either, being almost absent in Darwin's own writings. He originally ignored the idea of mutation. Instead, he assumed a large number of continuous small variations, of very varied and as yet unexplained origin, giving living beings a kind of infinite malleability; by the work of selection on this raw material, only the fittest forms were preserved.

It was only after the rediscovery of Mendel's laws and the development of the idea of mutation by Hugo De Vries, in the first decade of the twentieth century, that random variation was truly introduced into Darwinism. Galton's biometrics had opened the way to this in the 1880s but in a manner that was unusable, and had to be corrected in the Mendelian framework. After 1900, instead of an infinite malleability through many small continual variations, the material that selection had to work on was now sudden, rare and haphazard mutations. These thus became the limiting factor in relation to evolution. Selection intervened only subsequently, and only if necessary; it was no longer the determining factor that fashioned species as a function of the environment to which they had to be adapted. De Vries could therefore write that, in his theory, evolution depended on the rate of mutations, and thus their random character, whereas in Darwin's original conception it depended almost entirely on the pressure of selection:[9] selection, rather than random variation, was its essential parameter. Chance could clearly be reintroduced by way of the random character of the external conditions responsible for selection, but as far as I am aware, no one did so at this time. Besides, the random character of external conditions was already present in Lamarck's theory.

Before 1900, therefore, the random nature of evolution could not have posed any very great problems for morality and religion. There seem to have been no major authors who developed the idea of chance as a perturbation of a premeditated divine plan, most likely because a plan of this kind refers to divine creation whereas chance refers to evolution, the consequence of this being that the idea mixes two different conceptions and is only to be found in attempts to reconcile creation with evolution (as in Abbé Le Noir, cited above). This kind of question is visible only negatively, in attitudes such as that of the botanist Asa Gray, who was both a pioneer of evolutionism in the United States and a deeply religious

9 H. De Vries, *The Mutation Theory* [2 vols, 1901–3] (New York: Kraus Reprint Co., 1969), vol. 1, p. 69.

person, and who sought to reintroduce divine will by entrusting it with the task of directing evolution along certain beneficial lines by the play of variations, for which there was no explanation before the mutation theory of De Vries.[10]

What was embarrassing, in fact, in the (apocryphal) formula that 'man is descended from the ape' was less the incompatibility of Darwinism with an improbable biblical creationist taxonomy (Lamarck's theory scarcely caused a scandal in 1809), but rather the moral questions discussed above, and their resolution by a pseudo-naturalistic thesis that justified both the love of animals (our distant relatives) and the extermination of 'inferior races' (evidence of a state of humanity superseded by the European races and destined to disappear in the name of progress).

The moral – or rather, immoral – consequences of natural selection had been immediately understood and were formulated very explicitly. Clémence Royer, for example, could write in the preface to her French translation of *The Origin of Species*:

> But the law of natural selection as well, when applied to humanity, makes us see with surprise and pain how false our political and civil legislation has been up to now, as well as our religious morality. We need only adduce here one of the vices least often signalled, but not one of the least for all that. I mean that imprudent and blind charity towards badly constituted individuals, in which our Christian era has always sought its ideal of social virtue, and which democracy would like to transform into a source of compulsory solidarity, even though its most immediate consequence is to aggravate and multiply in the human race the very ills it claims to remedy. The result is to sacrifice the strong to the weak, the good to the bad, those well endowed in mind and body to the vicious and sickly. What is the end product of this unintelligent protection granted exclusively to the weak, the infirm, the incurables, even the wicked – indeed, all the defectives of nature? It is that the evils they are affected with tend to perpetuate themselves indefinitely; the evil increases rather than diminishing, and it increases ever more at the expense of the well. Whilst all the care and devotion of love and pity are considered the due of the declining or degenerate members of the species, nothing is done to aid newborn strength, to develop it, to multiply merit, talent or virtue.[11]

10 P. J. Bowler, *Darwin, The Man and His Influence* (Oxford: Blackwell, 1990), p. 215.

11 C. Royer, preface to her 1862 translation of Darwin, *L'Origine des espèces* (Paris: Flammarion, 1918), pp. xxxiv–v.

Royer's entire preface is a diatribe against Christianity and democracy, accused of running counter to the natural direction of evolution by protecting the weak. For the 'religion of the fall' (Christianity) she proposes to substitute the religion of progress under the auspices of Darwin – without, of course, any heed for evolutionary altruism, writing as she did before Wallace's pioneering article on this question.

If a figure like Clémence Royer was very happy with this situation, and scarcely bothered by the ethical problems that Darwinism raised (she was in fact half-crazy, her delirious ravings an embarrassment to Darwin, giving his doctrine unwelcome ethical and political connotations in France), this was not the case with more responsible biologists. They may well have shared the same political and social ideas, which, after all, Darwin had simply taken over and naturalized before biological sociologists later reimported them, but they did not like the idea of these ideas being expressed too crudely. At least in appearance, moral values had to be saved.

As I explained on pp. 53–4 above, T. H. Huxley was one of the first, as early as 1862, to apply Darwinism to morality, imagining a moral progress thanks to natural selection. This idea was taken up and developed, and very soon a whole variety of theses of this kind appeared.[12] Later, however, Huxley changed his mind and came to see evolution as incompatible with morality – in other words, presenting an opposition between nature and civilization. He was one of the few major Darwinians to have defended this idea, which Baldwin criticized him for in the quotation given on pp. 65–66 above:

> That which lies before the human race is a constant struggle to maintain and improve, in opposition to the State of Nature, the State of Art of an organized polity; in which, and by which, man may develop a worthy civilization, capable of maintaining and constantly improving itself, until the evolution of our globe shall have entered so far upon its downward course that the cosmic process resumes its sway; and, once more, the State of Nature prevails over the surface of our planet ...
>
> The science of ethics professes to furnish us with a reasoned rule of life; to tell us what is right action and why it is so. Whatever differences of

12 On Darwinian ethics following the publication of *The Origin of Species* and *The Descent of Man*, see K. F. Gantz, *The Beginnings of Darwinian Ethics: 1859–1871* (Ph.D. thesis, 1937), private edition, distributed by The University of Chicago Libraries (reprinted from The University of Texas Studies in English, 1939), pp. 180–209.

> opinion may exist among the experts, there is a general consensus that the ape and tiger methods of the struggle for existence are not reconcilable with sound ethical principles ...
>
> Thus, brought before the tribunal of ethics, the cosmos might well seem to stand condemned. The conscience of man revolted against the moral indifference of nature, and the microcosmic atom should have found the illimitable macrocosm guilty. But few, or none, ventured to record that verdict ...
>
> Let us understand, once for all, that the ethical progress of society depends, not on imitating the cosmic process, still less in running away from it, but in combating it. It may seem an audacious proposal thus to pit the microcosm against the macrocosm and to set man to subdue nature to his higher ends; but I venture to think that the great intellectual difference between the ancient times . . . and our day, lies in the solid foundation we have acquired for the hope that such an enterprise may meet with a certain measure of success.[13]

It would seem, therefore, that we have to admit that one of the reasons for the omnipresence of evolutionary altruism was the need to neutralize what in *The Origin of Species* was embarrassing to Christian morality. Equipped with a hereditary altruism (however tortuous this was, and no matter what problems were raised by group selection), Darwinism could then enter perfectly well into bourgeois society, sharing as it did all the values of that society, on the one hand those of competition and the war of all against all, and on the other hand those of Christian morality, or at least its appearance. The subversive character of Darwinism is a sweet illusion that certain biologists cultivate, peaceable characters who like to see themselves as bold revolutionaries of thought (and since this has been going on for 140 years, they are old revolutionaries indeed ...). Anyone who would doubt the nature of Darwinism can refer to the depiction that Dumont gave of the French conservatives in 1873 (see above, pp. 7–8).

This altruism was equally well suited to those who despised bourgeois society, like Kropotkin, even 'priest-eaters' such as Haeckel, enabling them as it did to anchor love of one's neighbour, generosity and mutual aid in nature and biology, rather than in a social order that they challenged, and from which nothing good

13 T. Huxley, *Prolegomena to Evolution and Ethics* [1894] in T. H. Huxley and J. Huxley, *Evolution and Ethics* [1947] (New York: Krauss Reprint Co., 1969), pp. 60, 64, 68, 82–3.

could come (Kropotkin's view), or else in some kind of divine commandment that affronted their materialism (Haeckel). In both cases, morality descended again to earth, came closer to men, and by a process of naturalization was able to become 'scientific' – an ideal widely shared at that time.

To sum up, evolutionary altruism was the source of a pseudo-naturalistic morality, which made it possible to reconcile the law of the jungle and the ideology of the noble savage. It animalized human society by biologizing it, and humanized animal society by anthropomorphically extending to it the psychological and moral dimension of human social behaviour. Hence the reversibility of its inherent conceptions, which could serve equally to justify the love of animals and the extermination of the 'inferior races'. It could lend itself to anything and everything, good business as well as good sentiments. That is why it is so frequently encountered in all kinds of writers of this time, in a whole range of contexts and to justify whatever they wanted, and it continues this brilliant career today, as evidenced by the article in *Nature* with which we began this discussion on pp. 51–2 above.

6

Methodological Issues

It is time now to turn to the methods used by biological sociologies (in the broad sense), and particularly to the question of models, whether mathematical or not. Along with statistical studies, mathematical modelling is the method par excellence of these disciplines.

A model is neither a theory nor a demonstration. It serves at best to relate empirical data, whether this is done mathematically or otherwise. At worst, a model is an imaginary construction simulating a natural process, a simulation that can potentially be computerized. When done properly, such a model can have a certain heuristic value, and hence a certain utility. But this heuristic value can in no case be considered a form of proof.

All modelling is construction. This construction may have an empirical basis, in so far as it is based on data (sometimes statistical) that it seeks to account for, but this does not make the model into experimental science, as the procedure is only analogical and the experimental situation from which it is constructed has only an illustrative value, rather than that of a test.

It is this structural character that permits the generalization of the model. In point of fact, even when the starting point of the construction has been a very specific empirical datum, the structure can be readily exported, by formal analogy, to a whole range of situations. Thus the model constructed for animal society is transferred to human society, or vice versa. The generalization of the model to situations other than that for which it was constructed thus has nothing in common with the generalization that characterizes a theory. *A fortiori*, the mathematicization of a model has nothing in common with the generalization practised within a discipline such as physics.

All this means that, at best, models remain in the empirical domain without themselves being experimental, and that very often they are neither in the one category nor the other, but rather in the category of pure and simple analogy.

Finally, modelling almost invariably involves a resort to ideas that are explicitly acknowledged as 'arbitrary', to the extent that they are not present as empirical data in the situation being modelled, but are simply postulated. In the Nowak/Sigmund model discussed on p. 51 above, the notion of 'brand image' is introduced without any clear idea of what this empirically corresponds to; it is neither measurable nor even observable, simply an interpretation of empirical data.

Very often, models 'function' only by way of arbitrary notions such as these, providing a scaffolding for them. It is thus readily apparent how, with a little skill, it is possible to model anything you like, above all when purely qualitative simulations are concerned; all that is needed is a good choice of notions. Moreover, it is very often these notions that give the models their ideological colouring, a liberty that is permitted the modeller by their arbitrary character.

These points about the limitations of models may be somewhat superfluous, given that biological sociologies are not very careful about epistemological rigour, and it is often quite impossible to apply this standard to judging them. Take for example the Nowak/Sigmund model, which is completely representative of this kind of practice. What is the model actually modelling? Is it a given empirical situation? Not at all. What is modelled by the use of games theory is a proposition of the same order as Darwin's introduction of sympathy, blame or approval (which these authors translate as 'brand image') to explain how altruistic behaviour can be considered to have a selective value.

Darwin's own proposal likewise failed to correspond to any empirical data. It goes without saying that sympathy, approval, blame, brand image, etc., are qualities that can influence social behaviour, and a senseless banality to say this in the case of human society. But does this imply that such factors provide the basis for selecting an altruistic behaviour with a biological and hereditary foundation, and that this selection can explain the advance of morality in human societies (primitive and then civilized, to use Darwin's own terms)? The answer is clearly no. Darwin's claim cannot be considered as describing a well-defined empirical situation. It is not even a vague simulation of this (an informal model), but simply what Novicow called an anthropological fable, a work of pure imagination.

Is game theory's mathematical formulation of this Darwinian proposition (and its computerized simulation) a demonstration of it? Does it give the proposition any meaning? Again, the answer is clearly no. Tacking a few equations onto a hypothesis of this kind, or giving it a computer simulation, is quite insufficient for the task. Saint-Évremond said – of opera – that foolishness decorated with music, scenery and ballet is magnificent foolishness, but foolishness nonetheless. One could say that a fiction embellished with differential equations or matrix calculus is a mathematicized fiction, but fiction nonetheless. Whether mathematicized or not, an anthropological fable is bad literature and certainly not science. The aim of this kind of modelling, moreover, is not to demonstrate such anthropozoological fables and give them significance, but to conceal their intellectual poverty with the aid of a few equations, making them pass for science when they are not.

These modellings generally result from an association of biologists and mathematicians, who each make their distinctive contribution. Without the mathematical apparatus, the poverty of biological ideas would be glaring; indeed, in the example chosen here, it is more or less the same explanation given by Darwin 130 years ago, which no one today would dare to formulate explicitly. But the merit of the mathematical apparatus lies only in its biological application, having little originality in itself. (Game theory, for instance, has been around for half a century.) Taken separately, the biological and mathematical ideas would be devoid of interest. But taken together, they create an illusion. And these remarks apply to the majority of modellings of this type. I am not criticizing here any particular work or any particular authors, but the very principle behind a certain type of approach.

Why does this kind of modelling find such nearly universal favour with biologists, when there is nothing scientific about it? Quite simply because it crudely takes over methods that had the support of geneticists in the first half of the century. To understand it, we must briefly trace the three main stages in the development of genetics.[1]

In the years 1860–1900, the first theories of heredity were physiological. In other words, they sought to explain the mechanism by which characteristics were transmitted from one generation to the next. At that time, it was imagined that living

1 For further details, see A. Pichot, *Histoire de la notion de gène* (Paris: Champs-Flammarion, 1999).

beings were composed of elementary particles (which Weismann termed 'biosphores', and De Vries 'pangenes'). These particles, made up of an assemblage of organic molecules, had the properties of increasing by nourishment and duplicating by way of division. They were of different kinds, and a living being was the way it was by virtue of the particles that composed it.[2] In reproduction, each parent supplied a representative sample of its constitutive particles, and the embryo recomposed itself on the basis of these two mingled samples, thanks to the ability of its particles to take nourishment and divide. This was the basic principle of Weismann's theory of the germ-plasm.

At the start of the twentieth century, these theories lost all credibility, and the methods of genetics completely changed. Given the inability to access the physiological mechanism of heredity, biologists fell back to a 'phenomenist' approach. In other words, they simply studied the apparent characteristics of living beings (their size, shape, eye and hair colour, etc.) – more specifically, by statistical methods, the occurrence of these characteristics within given populations and its variation over generations and crossings. The living being thus had to be dissected into a number of different hereditary characteristics, each of them associated with a 'gene', in the sense of a particle of heredity whose existence had been postulated but whose nature and mode of action were still unknown.

As the decomposition of the organism into distinct characteristics is a rather arbitrary operation on the part of the observer, however – with no assurance that nature acted in that way to assign a gene to each characteristic – the new discovery of mutation was introduced in order to define the characteristics that should be associated with distinct genes. For example, if a particular mutation changed an animal's eye colour (it was at this time that the fruit fly *Drosophila* became the geneticist's favoured experimental subject), it was concluded that a gene 'for eye colour' existed was concluded. This was certainly no more than an approximation, as in general a mutation affects several characteristics simultaneously, and to different degrees. Conversely, a single characteristic can be modified by several different mutations, whether in the same way or differently. Nevertheless, the habit was formed of defining

2 For an exhaustive study of these particle theories of 'living matter', see Y. Delage, *L'Hérédité et les grands problèmes de la biologie générale*. A more summary work is J. Rostand, *L'Atomisme en biologie* (Paris: Gallimard, 1956).

hereditary characteristics by naming them after the mutations affecting them (with *Drosophila*, for example, the characteristics 'white eyes', 'black body', 'vestigial wing', etc.), and associating one gene with each of these defined characteristics, without any real concern for how these genes acted as the bearers of heredity and controlled the formation of the corresponding characteristics.

This manner of explanation was just as hypothetical as that based on the particle theory of living matter, but it functioned the opposite way round. In the first case, the explanation was physiological and deterministic; in the second, it referred to a kind of semiology, in so far as the phenomenon of mutation was interpreted as a sign revealing the existence of a hereditary entity corresponding to the characteristic affected by the mutation. To use anachronistic terms, instead of studying heredity by starting from the genotype and proceeding to the phenotype (which would mean following the direction of physiological determinism), biologists started from the phenotype and assumed that a genotype corresponded to it. The whole physiological dimension of heredity was thus by-passed, with nothing said as to the nature of the gene or the way in which it controlled the heredity of the characteristic in question. And this remained the situation for a full half-century.

The disappearance of the physiological dimension was to encourage the extension of this thinking to all kinds of characteristics. First of all, the interpretation of the mutations affecting eye colour in *Drosophila* as signalling the existence of genes governing this colour led to the assumption that there were 'eye colour genes' in other species, whether related or not (and the same went for other characteristics). In other words, a first extrapolation. Then, the definition of the characteristic in terms of the mutation involved was extended to characteristics for which no mutant form was known but which were assumed to be hereditary. For example, the fact that nose shape is hereditary led to the assumption that there was a gene, or a number of genes, controlling this, even though there was no 'mutant nose' that made it possible to pin down the gene(s) in question. Finally, as psychological characteristics and behaviours were brought under the umbrella of Darwinian theory, inasmuch as they could be ascribed an adaptive value, they were equally associated with one or more genes. So we had genes for intelligence, for musical talent, for stupidity, for crime, etc.

Eventually there were genes for everything, without the need ever being felt to provide a physiological mechanism for the heredity in question. Besides, as mutations are often pathogenic or teratogenic, this invasive heredity took on a rather morbid hue. The proliferation of genes was often a proliferation of 'bad genes', or a degeneration – a very common idea at this time. The state of ignorance as to the physical nature of the gene made it impossible to understand the heredity of any characteristic except the simplest ones, such as eye colour; *a fortiori*, this absolutely prevented the development of any idea, even a hypothetical one, concerning the physiological mechanisms of heredity for a behaviour or psychological characteristic.

This inability did not bother geneticists, since they had by this time developed analytical methods that dispensed with any knowledge of physiology – methods that sociobiology would in turn take up. These methods were purely phenomenalist and mathematical. In other words, they dealt with the apparent characteristics of the living being without concerning themselves with their physiological substrata, and did so by way of mathematical and essentially statistical studies.

Method of this kind had a three-way origin: the rediscovery of Mendel's laws of heredity, De Vries's theory of mutation, and Galton's biometrics (see box below). From this emerged a double path in the study of heredity: population genetics and Morgan's formal genetics.

Phenomenalist and Mathematical Methods in Early Twentieth-Century Genetics

Galton's biometrics (and psychometrics) took over statistical methods developed by Quételet in his statistical anthropology and social statistics, giving these a genetic interpretation. With the rediscovery of Mendel's laws, this interpretation was shown to be erroneous. Statistical methods were still maintained but were corrected with the aid of Mendel's laws and principles, introduced by W. Johannsen, G. Hardy and W. Weinberg among others. This was the origin of population genetics.

This discipline viewed populations as 'gene pools', i.e., a kind of reservoir of genes, and used mathematical models to study how the proportions of various genes in them developed over generations – the way in which certain genes spread and others disappeared.

Population genetics equated the evolution of species with this evolution of different genes within a population. It was thus the main support of Darwinian theory in the first half of the twentieth century.

Morgan's formal genetics developed a different phenomenalist and mathematical approach. Instead of using statistical methods to study the overall genetic composition of populations, it used these to go back to the individual and construct a model of heredity on the basis of the results of various hybridizations (the model having to account for these statistical results).

Morgan thus developed what is usually now called the chromosome theory of heredity, but was actually a chromosome *model*. His method made it possible to localize mutations on the chromosomes and thus produce a mapping – which should not be understood as a mapping of genes, as these were still defined at this time in terms of mutations. The nature and physiology of genes was still unknown, but they were given a degree of concreteness by their localization on a chromosome.

This chromosome model and mapping, attenuating the persistent ignorance about the physiology of heredity, perpetuated the idea of the gene until the early 1950s.

These phenomenalist and mathematical methods had a certain value and interest when used within a very strictly defined framework, but outside these conditions they were worthless. Morgan's method was unusable, and the methods of population genetics could be interpreted any which way.

Biological sociologies and related disciplines were to imitate these phenomenalist and mathematical models (especially those of population genetics), softening them by adapting them to the case of supposedly inherited behaviours. It was this imitation that made them, if not generally accepted by biologists, at least acceptable to certain of them, by recalling patterns with which they were familiar. There is no need for special perspicacity, however, to see that there is a notable difference between a formal genetics that localizes a precise mutation on a chromosome – or population genetics that studies how the occurrence of this mutation evolves within a given population – and a biological sociology that claims to study and model the heredity of characteristics as complex and poorly defined as intelligence, alcoholism, homosexuality or altruism, for which no physiological basis can be shown but for which it is possible nonetheless to imagine one or more genes, without saying anything precise about them.

The methods of biological sociology certainly imitated in their fundamental principles (phenomenalist and mathematical) those of formal genetics and population genetics, but they completely lacked comparable explanatory power. What we have here is a wretched mimicry that certain people, out of naïveté or intellectual laziness, confuse with the original. Formal genetics and population genetics shared the study of heredity in the first half of the twentieth century but were relegated to the background in the early 1950s, with the appearance of molecular genetics. From that point on, a number of studies made it possible to establish the physical nature of the gene, and attempt an explanation of its mode of functioning.

Genetics now came to refocus on the biochemical study of the support for heredity, and the manner in which this heredity was expressed in various characteristics. It thus found itself back on the course of physiological determinism (from the genotype to the phenotype), while the phenomenalist and mathematical methods (from the phenotype towards a supposed genotype) either took second place, as no more than ancillary methods, or disappeared altogether.

This meant a sudden regression for the biological sociologies that knew only phenomenalist and mathematical methods. These now found themselves adrift from the leading edge of genetics (molecular genetics) and came to appear very old-fashioned. Incapable of a physiological approach, all that these disciplines were now able to do was modernize the mathematical apparatus they used, or take advantage of the development of computing to carry out simulations. In both cases, this meant borrowing from disciplines such as economics, which had a similarly imperfect scientific status. Thus it was the rise of molecular genetics, and the eclipse of phenomenalist and mathematical methods, that explains the retreat of biological sociologies after the Second World War (rather than simply the Nazi atrocities with which they were linked – see Parts Two and Three). And it is the problems that molecular genetics confronted that explain their reappearance in the 1970s in the form of sociobiology.

In the mid-1970s questions arose concerning the gene as a linear and continuous segment of DNA. The gene began to lose that simple definition once it was discovered that in the eukaryotes the gene did not possess the structural unity that had been supposed. Biologists were forced to return to a functional definition, characterizing the gene after the protein whose synthesis it controlled. This was, in other words, a return to an approach that was at least partially 'anti-physiological', since instead of starting from the gene to proceed to its product (the protein), the explanation went from the latter back to the gene (just as, in the genetics of the 1930s, the reasoning went from the phenotype characteristic to the gene). One particularly clear symptom of this situation was the return in strength of mutations, especially in human genetics, with the multiplication of 'disease genes' – an offensive term in which a pathogenic mutation serves to characterize a gene, in exactly the manner of Morgan's genetics and its *Drosophila* with 'white eye' or 'vestigial wing' genes.

The biological sociologies, after being relegated to the back of the stage for using only 'antiphysiological' methods, could now raise their heads again and return to being fashionable, given that the principles underlying these methods, if not the methods themselves, were once again current. The 'disease genes' now served as a model for various genes for behaviour (in general, deviant or pathological behaviour: we find once more the morbid aspect characteristic of these approaches), just as the 'white eye', 'vestigial wing' genes and the like had done in the first half of the century. So there was a gene for alcoholism, as a 'behavioural disease', just as there was one for myopia and there had been one for 'white eyes' in *Drosophila*. This is what explains the multiplication of studies that resurrect the problematics and methods of the first half of the century, more or less camouflaged by a modernized mathematical apparatus.

We shall find the same pattern of development in the discourse of heredity in Parts Two and Three. This discourse changes with the methods used by genetics. But the less scientific value these methods can claim, the more they propose applications to society, either in the form of medical applications or directly social ones.

II. GENETICS AND EUGENICS

7

The Origins of Eugenics

The altruistic sociability discussed in Part One provides a general basis on which all kinds of more specific behaviours are superimposed, behaviours that are generally 'negative' from the social point of view (others being included in an undifferentiated fashion in the general explanation of altruism). These negative behaviours (or those reputed to be such) are ascribed a hereditary basis in the same way as is sociability. It is possible, then, to ask that biology, and genetics in particular, resolve all kinds of social ills. This is the origin of modern eugenics, with any medical arguments being no more than pretexts.

Given the rarity of genuinely hereditary diseases, in fact, eugenics has generally focused on psychic or behavioural disturbances, of the kind that pose a problem for society in different degrees. The great majority of individuals sterilized under the eugenic legislation of the first half of the twentieth century displayed disturbances of this kind, rather than genuinely hereditary diseases. Under the 'medical' pretext of eugenics – of which the ostensible goal was to improve the human race or prevent its degeneration – the real intent was the establishment of social order (profitability, the management of 'human resources' after the fashion of livestock).

We should immediately make clear that there is very little originality in the focus on these socially 'negative' behaviours. Here again, we are faced with recurring ideas. Mental diseases, feeble-mindedness, alcoholism, sexual deviance, vagabondage, social maladaptedness, delinquency and other such characteristics (real or imaginary) were the everyday stuff of eugenic theories for the whole of the first half of the century, becoming rarer after the Second World War – only to resurface in the 1970s, at the same time as sociobiology, and to continue to flourish today.

Eugenics is without a doubt one of the last taboos in the history of the twentieth century (see definitions in the box below). As against what is often imagined, this was in no way a marginal phenomenon, the work of a few politicians driven by a Nazi-type ideology. During the first half of the century it was extremely widespread – developed and championed by a very large number of biologists and medical practitioners with a wide range of political and philosophical views, and relayed by different ideologies across a plethora of associations, including eminent institutions such as universities and major philanthropic foundations. The only institutional opposition it encountered came from the Catholic Church and from the Lysenko school in the Soviet Union.

Eugenics gave rise to a great deal of varied legislation, long before Nazism and in completely democratic countries. This legislation led to hundreds of thousands of compulsory sterilizations, most often of people who were in no way affected by a hereditary disease. In Nazi Germany, this sterilization was accompanied by measures of mass extermination, which in a totally deceitful fashion were referred to as 'euthanasia'.

Positive and Negative Eugenics

The word 'eugenics' was invented by Galton in 1883, from the Greek *eugenes*, meaning 'well born'.[1] It claimed to be based on Darwinism and genetics, or, more exactly, the application of these to human society.

Two forms of genetics are generally distinguished: negative and positive.

Negative eugenics seeks to prevent the multiplication of individuals deemed to be 'inferior' from a biological, psychological or intellectual point of view. It postulates that this inferiority is hereditary, and seeks to prohibit the individuals in question from having children – or, more rarely, advises them against doing so. The methods employed are more or less brutal and coercive: the banning of marriage, imprisonment, but above all sterilization (which is particularly the issue here when we speak about eugenics). The harshest version is the pure and simple elimination of so-called inferior individuals, as in Nazi Germany.

Positive eugenics, on the other hand, seeks to improve society by encouraging the reproduction of 'superior' individuals, in the extreme case organizing this either through human 'stud farms', where chosen reproducers are asked to procreate, or with the aid of sperm banks

1 F. Galton, *Inquiries into Human Faculty and its Development* [1883] (London: Dent, 1911), p. 17.

whose repositories have been donated by great men (nowadays, egg banks could also be envisaged).

These two forms of eugenics are distinguished from social Darwinism – a different application of evolutionism to human society – in that they are based on a more or less strong intervention of the state by way of constraining legislation. Social Darwinism, on the contrary, is a laissez-faire doctrine that rejects state intervention. It is an extreme liberalism, even rejecting protective social legislation, with the goal of enabling selection to do its work in society as it supposedly does in nature, i.e., by eliminating the least competitive individuals. Historically, this liberal form was conceived first, with the publication of *The Origin of Species* in 1859, whereas eugenics was only theorized somewhat later, in the 1880s.

Social Darwinism, and both negative and positive eugenics, are related and interacting doctrines. They come in a pessimistic version (when they claim to struggle against the degeneration induced by the disappearance of natural selection in human society) and an optimistic version (when they claim to improve the human species and produce supermen, after the model of the improvement of breeds of domestic animal).

The motivations behind them, whether avowed or implicit, are very diverse. They run from pure and simple economics (damaged individuals cost society dearly) – hence measures such as a charity that would seek to spare unfortunate individuals the wretched life awaiting them in a world for which they are not adapted – through to an idealism that may be racial or aesthetic or espouse a certain virtue (the production of supermen who are beautiful, intelligent, healthy, efficient, etc.). These diverse motivations can be combined in different proportions.

Most history books pass over this phenomenon in silence, not knowing what to make of it, given that it does not fit their customary assumptions. Specialized works are few, and those designed for a broad public often present the question in a very toned-down and biased fashion.

Eugenics nowadays embarrasses all shades of the political spectrum, for the following reasons:

- almost all of them have been compromised by it;
- declarations and practices in this field reached high points of ridiculousness and sinister imbecility;
- the results have been large-scale, both in practical terms (the sterilization and extermination of hundreds of thousands of people) and in ideological terms (what would Nazism have been without eugenics?).

With rare exceptions, it seems that the main concern of general works on the subject has been to minimize eugenics and its consequences, remaining as vague as possible concerning who was responsible. Gobineau, unsurprisingly, serves the usual function of straw target – for example in Jean-Paul Thomas's book *Les Fondements de l'eugénisme* in the popular 'Que sais-je?' series.[2]

Gobineau certainly was a romantic pessimist, an ideologist of decline and degeneration, and it is true that eugenics was bound up with notions of decline and degeneration. But there is degeneration and degeneration; the same word can cover very different ideas, whether we look at the first half of the nineteenth century or the first half of the twentieth, or indeed at different authors from each of these periods.

With Gobineau, as with other romantic thinkers of the same genre, the theme of decline and degeneration derives above all from nostalgia for the ancien régime, rejection of the Enlightenment and the French Revolution. These authors took refuge in the idea of a mythical past in which everything was supposedly magnificent, a marvellous order now destroyed for the benefit of bourgeois industrial society. The Middle Ages were often presented in the manner of a fairy tale, with politics, economics, philosophy, religion, literature, the sciences and arts all peacefully coexisting in peace and clarity. The nineteenth century, by comparison, with its revolutions and wars, was experienced as a period of decadence and decline, even degeneration.

More or less pronounced traces of this current of thought are to be found not only in politics and philosophy but also in literature (with Romanticism and its taste for pseudo-medieval claptrap) and art, in painting and architecture (with the troubadour style in France, the Nazarenes in Germany, the Pre-Raphaelites in England, neo-Gothic more or less everywhere, etc.). This was a constant in the first half of the nineteenth century. The movement continued until 1850, and the revolution of 1848 was certainly one of the factors that spurred Gobineau to write his work on the inequality of races, but gradually a different way of viewing degeneration arose – without there necessarily being a clear dividing line between the two currents.

This first kind of degeneration theory was politically reactionary, often with a religious and Catholic origin; by and large, it came from the traditionalist Catholic right – even the *intégriste* far right. The Catholic Church of the nineteenth century was reactionary and

2 J.-P. Thomas, *Les Fondements de l'eugénisme* (Paris: PUF, 1995).

retrograde in the extreme, doubtless because it felt under attack from the rise of currents of thought and political movements that were deeply foreign or even hostile to it. This was an age when society and thought were becoming more secularized and escaping the church's control, with science staking its claim to replace religion as a criterion of truth. Besides, the existence of the Papal States was threatened by the movement for Italian unification, and in 1870 the pope lost all temporal power. The church was thus weakened on both spiritual and temporal levels, and sought in vain to restore its fortunes.

The apogee of Catholic reaction was reached in 1864 with the publication of Pius IX's encyclical *Quanta Cura* and his *Syllabus of Errors,* which condemned modernity of all kinds, both intellectual and social, denouncing liberalism, socialism, communism, pantheism, materialism, rationalism and scientism, at the same time as they reasserted the primacy of the Catholic Church and the pope. The box below gives some of the theses condemned in the *Syllabus*.

The Syllabus of Errors Condemned by Pius IX

I. 3. Human reason, without any reference whatsoever to God, is the sole arbiter of truth and falsehood, and of good and evil; it is law to itself, and suffices, by its natural force, to secure the welfare of men and of nations ...
III. 18. Protestantism is nothing more than another form of the same true Christian religion, in which form it is given to please God equally as in the Catholic Church ...
VII. 56. Moral laws do not stand in need of the divine sanction, and it is not at all necessary that human laws should be made conformable to the laws of nature and receive their power of binding from God ...
VII. 57. The science of philosophical things and morals and also civil laws may and ought to keep aloof from divine and ecclesiastical authority ...
VII. 58. No other forces are to be recognized except those which reside in matter, and all the rectitude and excellence of morality ought to be placed in the accumulation of riches by every possible means, and the gratification of pleasure ...
VII. 59. Right consists in the material fact. All human duties are an empty word, and all human facts have the force of right ...
X. 80. The Roman Pontiff can, and ought to, reconcile himself, and come to terms with progress, liberalism and modern civilization ...

[www.papalencyclicals.net/Pius09/p9syll.htm]

Gobineau was a good representative of this current of thought: self-professedly aristocratic, Catholic (though with occasional sympathies for Islam), reactionary, anti-democratic and racist (we shall examine his racism in Part Three). Yet to see him, or more generally his ideology, as the origin of eugenics, would be completely fanciful. For Gobineau, sterilization (which, given the state of surgery in his time, would have been scarcely anything other than male castration) would have been a barbarism of the same order as that which he saw at work in bourgeois industrial society – certainly not a remedy for degeneration. There is therefore a confusion here: it was not this kind of degeneration theory that underpinned eugenics.

In the second half of the nineteenth century, at the time when eugenics was theorized, the word 'degeneration' was used to denote something very different from the decline of aristocratic civilization and a romantic nostalgia for the ancien régime. This degeneration was, rather, the other side of progress.

The second half of the nineteenth century saw the triumph of the Industrial Revolution: growing industrialization, proletarianization and urbanization, the concentration in the cities of impoverished populations seeking work. This movement had begun earlier and took place sooner or later, more or less rapidly, in different countries and regions. But by the mid-nineteenth century it had affected almost the whole of Western Europe, often in a very striking fashion.

Its consequences became increasingly noticeable and dramatic. Those most relevant to our present concerns were the multiplication of all the evils inherent in this kind of situation: contagious diseases (tuberculosis, syphilis, cholera), alcoholism, prostitution, mental illness, crime, and so forth. No doubt part of this was the greater visibility of these phenomena because of their concentration in the towns. But the underlying increase was real enough, these towns being dirty and wretched by any standard. Conditions of life were miserable, resulting in a high rate of disease and mortality compared with the countryside. (In the early 1830s, cholera killed between 16,000 and 18,000 victims in Paris, primarily among the lower classes, who lived heaped together in shacks – later removed by Haussmann's redesign of the city.) The abominable working conditions in industry also made their contribution.

Even though reliable statistics for the period are rare, some important indicators are well known – for example, in the work of Adolphe Quételet, *A Treatise on Man and the Development of His Faculties*, first published in 1835, which inaugurated the use of statistical methods in sociology. Quételet's mortality table for Belgium clearly showed that infant mortality for both sexes was higher in the town than in the country. Life expectancy for men was lower at all ages. Women who reached the age of eighteen did marginally better in the town than in the country (their life expectancy was a little longer), but girls suffered a higher mortality than boys during childhood and adolescence.[3] As industrialization and urbanization continued, the situation can only have worsened after 1835, at least until sanitation and the general social situation improved in the latter part of the century.

In the mid-nineteenth century, life expectancy at birth was especially low for the working class, and infant mortality especially high. Child labour had such serious consequences that the French army was concerned for the physical state of conscripts, already worn out and incapable of major effort.[4] In his magnum opus, Quételet compared the size of children of the lower classes in Manchester and Stockport according to whether they worked in factories or not. At nine years old, boys who worked in factories had an average height of 1.22 metres, while that of boys who did not was an average of 1.23 metres – in other words almost identical. But at eighteen years old, those in factories measured on average 1.61 metres, as against 1.78 metres for those who did not – a difference of 16.7 centimetres, which in all likelihood was due to working conditions in the factory. (A significant difference, though less pronounced, was also noted for girls.)[5]

In 1840 a report by Louis-René Villermé described the living and working conditions of French manual labourers. As far as health was concerned, his conclusions were very clear, though he was hardly a socialist agitator. For example:

> In Mulhouse, during the years 1823–4 and at all ages, life was far more certain among some classes of inhabitants than for others. In other

3 A. Quételet, *A Treatise on Man and the Development of His Faculties* [1835] (Gainesville, FL: Scholars' Facsimiles and Reprints, 1969), pp. 30ff.

4 J.-L. Robert, '1841, la première loi sociale en France', *Le Monde*, 20 April 1999.

5 A. Quételet, *A Treatise on Man*, p. 60. These measurements were made with shoes on, and on Sunday; the real height would require subtracting the thickness of shoe soles, and taking account of the slight increase induced by rest.

> words, whether the majority of children reached adulthood or died young depended on the condition or trade to which they belonged, and at all stages of life some had a distinct advantage ... The excessive mortality suffered by the families of workers employed in cotton spinning and weaving in Mulhouse particularly affects the earliest stages of life. In fact, while half of all children born into the class of manufacturers, factory managers or businessmen reach their twenty-ninth year, half of the children of weavers and simple spinning workers had reached the end of their life before the age of two – a fact that is scarcely credible.[6]

Other investigations and reports of similar import were to follow, sometimes spurred by far greater political commitment (Auguste Blanqui, Louis Blanc, Pierre Leroux), leading to the establishment of workers' organizations and the passage of legislation.

For England during the years 1880–2, Vacher de Lapouge presented the table below of deaths by profession per one thousand individuals between the ages of twenty-five and sixty-five, i.e., excluding infant mortality. These figures are quite close to what one might expect – though the high mortality of tailors is rather odd, and that of innkeepers undoubtedly increased by alcohol. That of doctors is perhaps due to their contact with the sick and to the ineffectiveness of treatments of the time.[7]

> Dr Ogle has calculated this general average [mortality] for England during the years 1880–1882. I summarize it here, confining myself to the ages from twenty-five to sixty-five:

Deaths per 1,000 individuals:	
Ministers of religion (all)	22
Farmers and graziers	27
Farm workers	29
Shopkeepers	35
Grocers	32
Builders and casual workers	39
Carpenters and scaffolders	34
Shoemakers	28
Navvies	43

6 L.-R. Villermé, *Tableau de l'état physique et moral des ouvriers employés dans les manufactures de coton, de laine et de soie* [1810], texts selected and presented by Y. Tyl (Paris: Union Générale d'Édition, 1971), pp. 270–1.

7 G. Vacher de Lapouge, *Les Sélections sociales* (Paris: Rivière, 1896), p. 359.

Garment workers	43
Domestic servants	36
Tailors and cutters	43
Bakers	39
Miners	39
Metalworkers	39
Doctors	48
Butchers	47
Innkeepers	60

The few attempts at social legislation in the mid-nineteenth century remained more or less a dead letter. In France, the first law on child labour (that of 22 March 1841) was a result of Villermé's report. It banned the work of children below the age of eight, limited the work of children aged between eight and twelve to ten hours a day, and of those between twelve and sixteen to twelve hours a day, also providing for a weekly rest day. But the employers saw this law as unduly strict, and so it was either not applied or applied only in small part, and not until the 1870s was any real headway made on this front.[8]

For England, one need only read the novels of Dickens to gain an idea of the social situation. This was the homeland of eugenics; in the late eighteenth century, Malthus thundered against the proliferation of the poor, and in 1834 a law was passed that permitted the concentration of the indigent in 'workhouses', where they were kept in exchange for their work, but where men, women and children of the same family were separated, with the express aim of preventing further procreation.[9]

Irrespective of legislation, there was an objective health crisis, a 'degeneration' that had nothing to do with Gobineau's reactionary nostalgia. This social situation, and the wars and revolutions of the time, certainly fuelled Gobineau's pessimism, but the ideology underlying this pessimism cannot be blamed for eugenics.

Social difficulties and the deterioration in public health created a crying need for solutions and remedies. The causes (industrialization, proletarianization and urbanization) are clear enough, and were well understood even at the time; witness the attempts at social legislation, however much these were regularly opposed by those who feared a reduction in their profits. It was

8 J.-L. Robert, '1841, la première loi sociale en France'.
9 J.-L. Robert, 'Enfermer les démunis', *Le Monde*, 16 November 1999.

easier to invoke a 'degeneration' of humanity, and of the poorer classes in particular, as this masked the social causes and relieved industrial civilization of its responsibilities in the matter. Since this civilization was deemed to represent progress, such evils could not be ascribed to it, and a turn was accordingly made towards biology and medicine.

Here, from Vacher de Lapouge, is a belated version of the thesis of the biological inferiority of the poor:

> There exist in my view, among the minus habens of the poor, at least two hereditary elements, distinct in their origins but mixed in these conditions by interbreeding, with the result that degenerates are almost all related to one another, in too complex a fashion to enable them to escape their origins. There are on the one hand the non-eliminated descendants of those unsuited to civilized life, primitive savages with too rudimentary a mentality and lacking any aptitude for sustained work, and on the other hand degenerates, descendants of individuals with an adequate or even superior endowment, who kept their place in their class, or even in a higher one, but whose blood has become spoiled. This results from even the most summary study of the origins of the poor classes.[10]

Otto Ammon offered a different version, more specific and sophisticated, and essentially targeting the subproletariat and the unemployed. For him, these categories were made up of inferior persons unable to keep up with technological progress and the demands of productivity. Since, contrary to Vacher de Lapouge, Ammon believed social selection operated perfectly well as things stood, he saw no need to change anything in the current situation: these individuals were incapable of forming families, so they could not reproduce and would gradually disappear, leading to a general rise in the level of the working class, which would be uplifted by this kind of selection (this is social Darwinism in the pure state, without any eugenic intervention). We should note that this text dates from the 1890s, a period of economic disturbances:

> The more we raise the lower limit of regularly assured existence, the more we condemn individuals to fall into the class of the lumpenproletariat, those who do not always satisfy their appetites by licit means, and who therefore find themselves in frequent conflict with police and justice. What is wrongly

10 G. Vacher de Lapouge, *Race et milieu social* (Paris: Rivière, 1909), p. 230.

known as the 'industrial reserve army', the category of those without work, is only to a lesser extent recruited from real workers who happen to have lost their jobs. The majority of brothers-in-arms who make up this 'reserve' are people of the kind described above, unable to meet the requirements of industrial progress ... Selection in this category is exerted in such a way that it is very hard for the lumpenproletariat to multiply. Their immorality does not drive them to procreate, and they rarely manage to found a family, the indispensable condition for raising children. This is a cruel fatality, but one indispensable for the common good, and it is undeniable that, if we acted from sentiment, we would go against natural selection. Many welfare institutions certainly have a baneful influence, making marriage and family possible for the worst kind of proletarians, and consequently favouring an inverse selection. The same would go for recognition of the right to work, in support of which a good number of reasons are given, but which falls before the consideration that it would amount to multiplying the worst category of individuals and would thus be in fact antisocial ... Every time that a higher output is demanded of the lower class of regular workers, on a regular and ongoing basis, new recruits to vagabondage are provided; and a rise in the minimum sum indispensable for existence supplies new inmates for prisons and penitentiaries. These are sad facts, but they are based on natural laws; they may at best be hidden, they cannot be suppressed.

We have known for a long time that technological advances often lead to an overproduction of goods, along with a simultaneous increase in the number of unemployed. These last are not an industrial reserve army to be recalled in whole or in part when need arises; the majority of them are rather useless rejects condemned to death ... Any flock always contains failed individuals. Stock-breeders get rid of these by sending them to the slaughterhouse. Where people are concerned, a systematic selection of this kind is not possible. We practise humanitarianism by casting these unfortunates out in the world and letting them gradually die off, hunted from place to place, or else we convey them into workhouses or prison. There is a social interest in preventing them from reproducing, and neither legislation nor administration, nor private charity, should act in a direction contrary to this social interest.[11]

Long before these texts of Vacher de Lapouge and Ammon were written, heredity (in particular, 'morbid' heredity) and degeneration had become characteristic themes. The most celebrated among these early illustrations of the new theories included the *Traité*

11 O. Ammon, *L'Ordre social et ses bases naturelles* (Paris: Fontemoing, 1900) pp. 390–2, 493.

philosophique et physiologique de l'hérédité naturelle dans les états de santé et de maladie du système nerveux, avec l'application méthodique des lois de la procréation au traitement général des affections dont elle est le principe, published by Prosper Lucas in 1847–50 and regularly mentioned as one of the forerunners of the eugenic current;[12] followed in 1857 by the *Traité des dégénérescences physiques, intellectuelles et morales de l'espèce humaine et des causes qui produisent ces variétés maladives* by Bénédict Augustin Morel.[13] Some years later saw the appearance of Théodule Ribot's great classic *L'Hérédité, étude psychologique sur ses phénomènes, ses lois, ses causes, ses conséquences.*[14] But all this was only the hors d'oeuvre.

In 1859, Darwinism came along to provide the explanation that this biologico-medical ideology needed: the proliferation of diseases, of behavioural and mental disturbances, was the result not of social conditions but of a biological degeneration caused by the suppression of natural selection in human societies. Far from being a consequence of the harshness of conditions of life, this degeneration was viewed, on the contrary, as the product of undue ease and of too much attention being paid to the weak and incapable (as was said in so many words by C. Royer in the preface to her 1862 translation of *The Origin of Species*; see quote on p. 94 above). Hence the need to restore the law of the jungle in society (the liberal laissez-faire version of social Darwinism) or to substitute a social selection for natural selection (eugenics, the authoritarian as against the liberal version of the same ideology). The eugenic solution came to prevail after the liberal solution had shown its inadequacies, just as unbridled economic liberalism gave way to a more organized capitalism. (On this succession of doctrines, see the quote from A. Béjin, p. 10 above.)

The naturalization of society discussed in Part One certainly played a major role in this process of ascribing biological causes to manifestly social ills. Here we encounter the great anthropological fable of the degeneration of the human race for lack of natural selection. It is a fable that population genetics and biological sociology would later seek to display with their phenomenalist and mathematical models (methods applied at this time in the

12 Éditions Baillière, Paris.
13 Éditions Baillière, Paris.
14 Librairie Philosophique de Ladrange, Paris.

most vague and hazy conditions, hence in a manner quite lacking in value).

Here is the form in which Vacher de Lapouge presents this anthropological fable of the degeneration of humanity. Note how, thanks to science, the terrible fatalism of heredity can be turned round and used to combat the ravages it has caused – indeed, even to improve humans where Christianity, Buddhism and education had failed, unable as they were to modify human biology. (We should remember that Vacher de Lapouge was an absolute pessimist who did not believe in progress, even if he did believe in science.)

> I truly cannot prevent myself from laughing when I see people expecting progress to yield a remedy for their ills. What holy naïveté! When humanity has all eternity before it, it is man that has to be changed. Is such a change possible? Can one remould human nature? This has been tried many times: Christianity and Buddhism are attempts of this kind. In our own day, the positivists have tried to attain this goal by education. Vain efforts! We know why neither Christianity nor Buddhism, nor educators, were able to succeed. Will we find a more powerful lever in science? Is it not possible to employ the formidable power of heredity to combat its own ravages, and oppose a systematic selection to the destructive and uncontrolled selection that has placed humanity in peril? If it is impossible to avoid the final catastrophe, the destruction of the species, can we not hope to defer this demise until intelligence is able to struggle against the powers of destruction? The greater part of social selection, in fact, can be modified by the intervention of human will. The systematic employment of selection can even change this force of regression and death into a principle of grandeur and life.[15]

This is the general context for the appearance of eugenic theories, and it was just at this time that such theories first arose. It is possible of course to find eugenic preoccupations at an earlier date (with Condorcet in the eighteenth century, Schopenhauer and Comte in the nineteenth), and even go back to antiquity with the exposure of ill-formed babies in Sparta, or the choice of parents in Plato's *Republic*. Fantasies of this kind can be found in almost all eras, but to see them as the origin of modern eugenics is rather like rooting modern atomic theory in that of Democritus.

15 G. Vacher de Lapouge, *Les Sélections sociales*, pp. 458–9.

At all events, the context of the appearance of eugenic doctrines in the second half of the nineteenth century has nothing in common with the romantic pessimism of a writer like Gobineau. Politically, it was no longer a cause of the old Catholic and reactionary right but, on the contrary, of secular and progressive circles. By 'progressive', here, I do not mean 'of the left', but rather champions of science and the Industrial Revolution, understood as representing progress and the way forward – and these could be found on both left and right, espousing both socialist and capitalist doctrines.

The ideology in question here is scientism (it was scientists, moreover, who invented modern eugenics and became its propagandists – see below). The religious dimension is secondary, whereas it had been fairly important in the first type of degeneration theory, which referred to a social order supported by the Church and its philosophy. It was a current of thought that was often atheist, sometimes Protestant or Jewish, only far more rarely Catholic.

The Catholic Church did not think much of scientism, still less of Darwinism, and nothing at all of eugenics. It officially condemned the latter in 1930, with Pius XI's encyclical *Casti connubii* (see box below). This undoubtedly explains why Catholic countries generally did not pass eugenic legislation.

Condemnation of Eugenics in the Encyclical Casti Connubii (1930)

Finally, that pernicious practice must be condemned which closely touches upon the natural right of man to enter matrimony but affects also in a real way the welfare of the offspring. For there are some who over solicitous for the cause of eugenics, not only give salutary counsel for more certainly procuring the strength and health of the future child – which, indeed, is not contrary to right reason – but put eugenics before aims of a higher order, and by public authority wish to prevent from marrying all those who, even though naturally fit for marriage, they consider, according to the norms and conjectures of their investigations, would, through hereditary transmission, bring forth defective offspring. And more, they wish to legislate to deprive these of that natural faculty by medical action despite their unwillingness; and this they do not propose as an infliction of grave punishment under the authority of the state for a crime committed, not to prevent future crimes by guilty persons, but against every right and good they wish

> the civil authority to arrogate to itself a power over a faculty which it never had and can never legitimately possess ...
>
> Although often these individuals are to be dissuaded from entering into matrimony, certainly it is wrong to brand men with the stigma of crime because they contract marriage, on the ground that, despite the fact that they are in every respect capable of matrimony, they will give birth to defective children, even though they use all care and diligence.
>
> Public magistrates have no direct power over the bodies of their subjects; therefore, where no crime has taken place and there is no cause present for grave punishment, they can never harm, or tamper with the integrity of the body, either for the reasons of eugenics or for any other reason. St Thomas teaches ... 'No one who is guiltless may be punished by a human tribunal either by flogging to death, or mutilation, or by beating'.
>
> [www.papalencyclicals.net/Pius11/P11CASTI.HTM]

This does not mean there were no Catholic champions of eugenics – consider advocates such as Carrel, who belonged to the *intégriste* wing of French Catholicism, or Hermann Muckermann (1877–1962), a former Jesuit responsible for the eugenics department of the Institute of Anthropology in Berlin – but they were rather rare, and their situation not all that comfortable. Catholic milieux criticized Carrel for his support of eugenics in *L'Homme, cet inconnu*; they were indeed his only critics at that time (those who criticize him today are certainly unaware that they are lining themselves up with the defenders of Pius XI's encyclical).[16] As for Muckermann, his eugenic zeal was cooled by the encyclical; the Nazis found him too moderate, and replaced him in 1933 with Fritz Lenz (1887–1976). Muckermann ended up denouncing the way German law had been abused to sterilize healthy (but poor and delinquent) individuals, and political opponents.[17] As well

16 This is what Carrel's biographer writes, citing an article of the time: 'Catholics took violent issue with the author of L'Homme, cet inconnu on three points: the thesis of the inequality of individuals, eugenic selection, and secular communities as inspirers of public morality. This was inevitable' (R. Soupault, *Alexis Carrel* [Paris: Plon, 1952], p. 275). This author seems to have had a marked ideological affinity with his subject.

17 B. Massin, 'De l'eugénisme à l'opération euthanasia: 1890–1945', *La Recherche*, December 1990, no. 227, pp. 1562–8; Y. Ternon and S. Helman, *Les Médecins allemands et le National-Socialisme, les métamorphoses du darwinisme* (Tournai: Casterman, 1973), p. 172; S. F. Weiss, 'The race hygiene movement in Germany, 1904–1945', in M. B. Adams (ed.), *The Wellborn Science, Eugenics in Germany, France, Brazil and Russia* (Oxford: OUP, 1990), p. 41.

as these few believing Catholics, there were also eugenicists of Catholic origin who had abandoned their religion. That is the case, for instance, with Eugen Fischer (1874–1967), whom we shall discuss below.

In general, the dominant ideological strand among the eugenicists was not just secular but actually anti-religious, more specifically anti-Christian (Christianity, the religion of the fall, was regularly opposed to the religion of progress equated with Darwinism, and particularly to its eugenic component). It was certainly always anti-papist. This is particularly true of Galton, the inventor of modern eugenics, who was anti-papist not simply because he was English (an Anglican and the descendant of Quaker bankers; Karl Pearson, his principal colleague, was also a Quaker), but also and above all because he accused the Catholic Church of being retrograde. It had shown this by its persecution of Protestants, whom Galton viewed as the dynamic fringe in society, bound up with progress and industry. It was in this progressive movement that Galton located eugenics: this was seen as pursuing social and industrial progress on the biological level, by improving the very nature of humanity.[18]

In Germany, the chief proponent of eugenics was Haeckel, whose politically motivated anti-papism we have already discussed and which was just as progressivist as that of Galton,

18 F. Galton, *Hereditary Genius, an Inquiry into Its Laws and Consequences* [1869], (London: Macmillan, 1914), pp. 343–6. Exactly the same kind of reasoning is to be found in a racist theorist such as Madison Grant, who explains very seriously that the 'superstitious and confined' character of present-day Spaniards is due to the fact that the Inquisition burned a large number of heretics in the Middle Ages, this being 'the best method to eliminate lineages of genius in a nation'. The passage continues: 'A similar elimination of intelligence and ability took place in southern Italy, in France, and in the Netherlands, where hundreds of thousands of Huguenots were killed or exiled' (M. Grant, *Le Déclin de la grande race* [Paris: Payot, 1926], p. 78). Vacher de Lapouge gives the same interpretation of Spain (*Les Sélections sociales*, pp. 286–8). We might well believe that the hold of the Catholic Church in Spain or Italy had a negative effect on the development and modernization of these countries, but it is infinitely more likely that this was by way of induced mentalities (through education, thought control, etc.) and political power, rather than by any negative genetic selection that the Inquisition supposedly exercised. Vacher de Lapouge has a still odder theory on the 'anti-eugenic' role of the church: he claims that a number of prominent figures were children of pastors or rabbis, concluding that, without the ecclesiastical celibacy that had prevented them from being born, the children of Catholic priests would certainly have been equally remarkable (*Les Sélections sociales*, pp. 274–7).

but inscribed in the framework of Bismarckian pan-Germanism (see p. 62 above).

In the United States, the main initiators of eugenics were the geneticists Charles Benedict Davenport (1866–1944) and Harry Hamilton Laughlin (1880–1943). Their opinions on the pope and the Catholic Church have not been recorded, but we know that Davenport's father was rather puritanical and devout, being a founder member of the Congregational church in Plymouth (a rather liberal Protestant church), and that he took responsibility himself for the education of his son.[19] As for Laughlin, he was the son of a fundamentalist and reactionary Protestant minister (though his mother was an active feminist in the Women's Christian Temperance Union).[20] On the question of immigration (a typical aspect of eugenics in the United States that bore on regulating the immigration of certain 'races'), Davenport and Laughlin distrusted the Catholic nations (Italy, Poland, Ireland), whose numerous progeny threatened to submerge America's WASP aristocracy.[21] Nothing in their background, in other words, linked either of them to the Catholicism of a Gobineau – who was proud that one of his ancestors had been particularly active in the St Bartholomew's Day massacre.

So much for the religious aspect of the ideology underlying eugenics. As for the political aspect, this likewise had little in common with any aristocratic nostalgia for the ancien régime. Champions of eugenics could be found at all points on the political spectrum. In Germany, for example, while some of them obviously were Nazis (not because eugenics was invented by the Nazis, but because Nazi doctrines derived from eugenic ones), others were socialists, e.g., Karl Kautsky, and others still were biologists and doctors of all persuasions (see below). In the USSR before Lysenko's time, they included Communist biologists such as M. V. Volotsoi and A. S. Serebrovsky (see below on the question of the USSR). In the United States, Hermann Muller was a Communist, Charles Davenport a conservative, and Margaret Sanger a feminist. In Britain, Julian Huxley was a social-democrat, and Karl Pearson a socialist and feminist before developing a sympathy with Nazism. And so on.

19 D. J. Kevles, *In the Name of Eugenics* (London: Harvard University Press, 1995) p. 51.

20 Ibid., p. 103.

21 Ibid., pp. 46–7, 53, 75, 94–5.

Eugenics has been understood in various ways according to political tendency and differing sensibility, from racial idealism, to economic considerations, to charitable solicitude for beings who would have been better off not having been born. It has been construed in its negative form of sterilization or its positive form of the 'human stud farm'; in a voluntary form or an imposed form. All combinations are possible in this field.

The various national laws that provided for the sterilization of a segment of the population were often supplemented by laws authorizing abortion in cases where the infant, for one reason or another, risked not being 'eugenic'.[22] By extension, eugenics has sometimes been understood as an element of birth control: contraception, making it possible to limit the number of births, is complemented by eugenics, making possible a choice of quality. In this capacity it has been championed by feminist and 'family planning' movements, which no doubt explains why Pius XI's encyclical *Casti connubii* combines a condemnation of eugenics, contraception, abortion, and so on.

Eugenics could be envisaged either as a substitute for laws of social protection, or as a complement to these laws, in which case it has sometimes been linked with Marxism. It has also been associated with measures such as family allowances, designed to favour the reproduction of good elements in the working class (good workers are unable to have too many children, given their low wages; bad workers are not constrained by such considerations, breed like rabbits, and are alcoholic into the bargain, hence a degeneration in the working class).[23]

One or another aspect is dominant according to time, place and author, but it is fairly hard to make any logical classification.

Apart from the Catholic Church and the Lysenko school in the Soviet Union, the only current of thought that avoided any compromise with eugenics was extreme liberalism of the Anglo-Saxon type, which rejected the intervention of law and the state into what it saw as pertaining only to private life. Very often,

22 This was the case in Sweden (law of 17 June 1938, supplementing the eugenic legislation of 1934), Denmark (law of October 1939, supplementing the eugenic law of 1929, modified in 1935), Switzerland (the canton of Vaud, where the 1932 law supplemented the eugenic legislation of 1928, and Japan (law of 28 June 1948). See J. Sutter, *L'Eugénique, problèmes, méthodes, résultats* (Paris: PUF, 1950), pp. 153–64.

23 This question is discussed in R. A. Fisher, *The Social Selection of Human Fertility* (Oxford: Clarendon Press, 1932).

however, such liberals appealed to social Darwinism, claiming that the abolition of all socially protective legislation would re-establish natural selection as a sorting mechanism in society; the 'invisible hand' would then be at work in human biology just as in economics. From this point of view, the alcoholic worker might well have a whole tribe of children – often defective – but as he could neither feed nor care for them, they would die off. It was a kind of liberal optimism, opposed to the pessimism of the champions of state interventionism who would impose sterilization on alcoholics.

Apart from the case of Anglo-Saxon extreme liberalism, eugenics was to be found practically everywhere and in all forms, on both right and left, among all those who in one way or another claimed the authority of science, a scientific organization of society, or some such thing. In 1924 a bibliography devoted to eugenics already numbered 7,500 books and articles (a number of them with racist connotations).[24] For a long time eugenics was very fashionable – developing and prospering on the basis of the anthropological fable of the degeneration of humanity, and the appeal to heredity on every topic imaginable.

To conclude this question of the omnipresence of eugenics and its polymorphic character, here is another curiosity worth citing from Vacher de Lapouge. The idea was not his own, however, but was taken rather from an anonymous German work, the author of which explains that it would be possible to attract degenerates and perverts to special settlements, where alcohol would be free, complete with facilities such as hospitable houses of ill-repute and gambling establishments. The free exercise of their vices would keep the degenerates there, and end up destroying and killing them, in the same way as alcohol, gambling, and venereal disease did away with indigenous populations in the colonies.[25]

> It would be very ingenious to achieve the destruction of the degenerate in such an amicable way, by making alcoholism, debauchery and an idle life so easy for them. The author of the Trennungssystem is in no way the inventor of this process: the disappearance of the Red Indians and various peoples of Oceania has been largely facilitated by low-priced alcohol. The

24 S. J. Holmes, *A Bibliography of Eugenics* (Berkeley: University of California Press, 1924).

25 Anon., *Die Aristocratie des Geistes als Lösung der sozialen Frage* (Leipzig: Friedrich, n.d.).

> African Negroes at the present time are being very largely weakened by the grain spirit that they are so greedy for. It is likely that this process will run its full course, using this dangerous passion against them. In Greenland, the Eskimos persist only thanks to measures taken by the Danish government; access to the country is prohibited without special authorization, and the introduction of alcohol proscribed under severe penalties. I can well believe that if there were one town in France in which alcohol was free, alcoholics would congregate there like garden slugs under a buttered cabbage leaf, a succulent and fatal trap. The taste for fatherhood and especially motherhood is very widespread among degenerates of all kind; the sexual act, for them, is simply a matter of pleasure. What an excellent disposition to encourage! ... If the reform of humanity could be achieved in this way by the simple play of human passions, without having to sacrifice people or be in any way violent, it would be likely to proceed at a more rapid pace.[26]

To understand eugenics, we should look not in the direction of Gobineau but much rather towards the Darwinian biology of the late nineteenth century. One need only examine the opinions of the founding fathers of Darwinism and genetics to realize that the majority of them were champions of eugenics, active members of associations promoting it (often along with racism), and drafters of legislation institutionalizing it in a number of countries.

The case of Darwin himself is hard to fathom, since he was in the habit of expressing in his writings quite opposing views on any subject. It is possible to find passages marked with a certain 'humanism', but also passages favourable to Galton. It is the done thing nowadays to whitewash Darwin of any compromise with dubious ideologies, and so it is his least embarrassing texts that are highlighted. Yet Darwin was a Victorian bourgeois, and he had by and large the same kind of philosophy as his peers. He seems to have been on good terms with his cousin Galton, and if he does not mention eugenics in so many words, it is quite likely because the field became theorized only after his death. (Galton became particularly interested in it in the latter part of his career, inventing the word 'eugenics' in 1883, a year after Darwin's death.) Here, in any case, is what Darwin wrote in 1871, in *The Descent of Man*, perfectly clear for the subject that concerns us here (there are other texts in similar vein):

---

26 G. Vacher de Lapouge, *Les Sélections sociales*, p. 486.

> [A]ll ought to refrain from marriage who cannot avoid abject poverty for their children; for poverty is not only a great evil, but tends to its own increase by leading to recklessness in marriage. On the other hand, as Mr Galton has remarked, if the prudent avoid marriage, whilst the reckless marry, the inferior members tend to supplant the better members of society.[27]

There is often a tendency to see Galton himself, Darwin's cousin and the inventor of eugenics, as a mere ideologist. He was, however, a scientist, whatever the value of his theories, and it was he who inaugurated the use of statistical methods in the study of heredity. This biometrics was a regular nursery for eugenicists, starting with Galton's main collaborator, Karl Pearson (1857–1936). Population genetics, which was to take over the statistical methods of biometrics and adapt them to Mendel's laws, was also a breeding ground of eugenics – for example with Wilhelm Weinberg (1862–1937) and Ronald Fisher (1890–1962), to cite only the most famous names.

The quasi-totality of geneticists and evolutionists, in fact, were champions of eugenics, along with a good proportion of other biologists and doctors. If some were not in agreement with it, they did not protest very vigorously. It was only in the 1930s, with Hitler's rise to power, that eugenic discourse experienced some discordant notes. But this was too late; it was during the very period that eugenic sterilization had reached a peak, and not just in Germany, as we see below.

Some of these scientists, including the most eminent, were not content merely with adopting eugenic ideas; they actively propagated them through associations that they often themselves established, and in which they were leading figures. In 1905, Alfred Ploetz (1860–1940) founded the German Society for Racial Hygiene; its honorary committee included Haeckel, Weismann and Galton, all leading biologists. Other German associations and institutions followed, with more or less the same aims.

A Société Eugénique de France was founded in 1912, though with very little success.

The United States had various associations of this kind, the main one being undoubtedly the American Eugenics Society, founded in 1922 by the pioneer of American genetics, Charles Davenport

27 C. Darwin, *The Descent of Man, and Selection in Relation to Sex* (Harmondsworth: Penguin, 2004), p. 688.

(a co-founder being Alexander Graham Bell, inventor of the telephone). As well as this society, there were a large number of associations and institutions pursuing similar goals: the American Breeders' Association, the Eugenics Research Association, the Galton Society, the American Institute of Family Relations, the Race Betterment Foundation, and of course the Eugenics Record Office that Davenport and Laughlin ran from the laboratory at Cold Spring Harbor, an institution we shall discuss below.

In the 1920s the Soviet Union also had eugenic associations and institutions founded by eminent Russian geneticists in Moscow and Petrograd.

In addition to national organizations, there were local societies, including, for instance, the Society for Racial Hygiene in Stuttgart, established by Wilhelm Weinberg, one of the founders of population genetics, as well as a whole range of groups that were more or less scientific and more or less political.

These associations often published periodicals that had both a 'scientific' and a purely propaganda side. There were a large number of these, with very varying success and lifespan. They included *Eugénique*; *Annales Eugéniques*; *Biometrika*; *The Eugenics Review*; *Annals of Eugenics*; *Eugenical News*; *Politisch-anthropologische Revue*; *Archiv für Rassen- und Gesellschaftsbiologie*; *Eugenik*; *Volk und Rasse*; *Zeitschrift für Morphologie und Anthropologie*; *Erb- und Rassenhygiene*; *Das Kommende Geschlecht*; *De Boletin de Eugenica*; *Rassegna di Studi Sessusali e di Eugenica*; *Slovanky Archiv pro Genetiku a Eugeniku*; and *Russkii Evgenichesky Journal*, among others. The more scientific of these publications served as reviews of genetics; for a very long while, moreover, no very clear distinction was made between eugenics and human genetics. Practically all the great geneticists of the first half of the century published articles in them.

Some of these periodicals and associations still exist, but they have changed their names to do away with any reference to eugenics. Thus in 1954 *Annals of Eugenics* became *Annals of Human Genetics*; *Eugenics Quarterly*, the continuation of *Eugenical News*, was rebaptised in 1969 as *Social Biology*; while the American Eugenics Society became the Society for the Study of Social Biology in 1972.

Besides publishing magazines, these associations carried out propaganda and lobbying. They spread their eugenic (and often racist) ideas among the broad public by way of popular pamphlets, lectures, exhibitions, and so on. Their methods were often reminiscent of those of religious groups; thus the American Eugenics Society published a eugenic catechism, and organized competitions for eugenic sermons.[28] This religious and sectarian character of Anglo-Saxon eugenics is related to Galton's statement that eugenics was the religion of the future (of progress as against Christianity, the religion of the fall).[29] All this was clearly perceived at the time; thus, inevitably, we find in Vacher de Lapouge a lucid if caricatured reflection of this current of thought:

> The theorists of recasting [human nature], the selectionists, are zealous and many. For a number of them, in the intelligent classes of America and Britain, selectionism has become a matter of faith or hope, the promised redemption of humanity. It is scarcely possible, I believe, to prevent selectionism from becoming a kind of sect; if something comes of it, the sectarians will be far more responsible than the scientists.[30]

Finally, and above all, these eugenic associations interceded with the public authorities to obtain legislation in conformity with their views (with the scientific backing of the great biologists who belonged to their ranks).

If any more proof were needed of the implication of biology and biologists in eugenics, we could note that there is a good chronological parallel between the development of eugenics and that of genetic and Darwinian theories in the years 1860 to 1960.

The period 1860 to 1900 was the time of the first hesitant efforts, both for genetics and Darwinism on the one hand, and for eugenics on the other.

The period 1900 to 1940 was the time of triumph: Darwinism and genetics found a more or less coherent form, almost universally

---

28 J. Roger, 'L'eugénisme, 1850–1950', in *Pour une histoire des sciences à part entière* (Paris: Albin Michel, 1995), p. 421.

29 Darwinism itself became a kind of new religion, with its devout fundamentalists. This was made fun of by the sociologist Gumplowicz in 1883, though he was himself a fairly hard Darwinist: 'Darwinism, even in its finest details, is still today something sacrosanct for a large section of the world, both scientific and otherwise: its champions recall the ancient adepts of dogmas, even in the blind fanaticism with which they defend their doctrine and condemn, not as heretics, but as "dilettantes for whom the entire kingdom of life remains a closed book", all those who do not swear by it as their alpha and omega' (L. Gumplowicz, *La Lutte des races* [1883] [Paris: Guillaumin, 1893], p. 65.

30 G. Vacher de Lapouge, *Les Sélections sociales*, p. 459.

accepted. At the same time, eugenic legislation proliferated, from 1907 in the United States, then in Europe from the 1920s.

After the Second World War, the Nazi horrors made eugenics suspect. Yet it still had its defenders, and eugenic laws remained in force in the democratic countries that had adopted them. It was only in the 1950s, when conceptions of heredity and evolution had been deeply modified by the development of molecular genetics, that eugenic concerns disappeared (at the same time as phenomenalist and mathematical models gave way to molecular approaches, more physiological and less sensitive to ideology).

Finally, we see the resurfacing of these ideas today, with the theoretical difficulties that molecular genetics is experiencing (following the same pattern as the resurgence of biological sociologies a little earlier).

Let us examine this history in closer detail.

Darwin's *The Origin of Species* dates from 1859, but as we have said, it took nearly fifty years for Darwinism to acquire a finished form, which it achieved only in the years 1900–15, after the rediscovery of Mendel's laws and the beginnings of genetics.

Before 1900, for want of a theory of heredity worthy of the name, as was also the case for a theory of variation (mutation was still considered as a rare and unimportant perturbation), Darwinism as a theory was poorly supported and unformed. The only point on which there was real agreement and a degree of constancy was natural selection. This was the emblem of Darwinism, on which it was entirely based. And this was the moment when eugenics was theorized, also with its focus on the idea of selection.

The scientific foundations of eugenics were non-existent, given the lack of a solid genetic theory. At this time, even Galton put forward an explanation of heredity – known as 'ancestral heredity', refuted in 1900 by the rediscovery of Mendel's laws – in which it would be hard to see sterilization as at all effective except by statistical sleight of hand.

This lack of serious scientific foundation did not prevent eugenics from enjoying success. But, as with that of Darwinism, this success was purely ideological. Eugenics was viewed as the solution to the supposed problem of degeneration, and at the same time it formed a kind of counterpart to Darwinian evolution: if evolution/progress was assured thanks to selection, then the absence of selection must surely lead to degeneration (and if evolution/progress by way of natural selection could not be found in nature, then one resorted to the idea of degeneration for lack of natural selection in society, and thus to eugenics). There was a certain circularity among all these terms, which mutually supported one another without resting on any concrete foundation.

All this, however, remained almost without effect in either social or medical practice at this time. No government passed any real eugenic legislation; there were simply some prohibitions on marriage and a few cases of 'wild sterilization' in the United States and Switzerland in the 1880s and 1890s.

Between 1900 and 1915, genetics gradually took shape (while still using only the phenomenalist and mathematical methods discussed above) and fitted well with Darwinism (which might well be due to its having been constructed in a Darwinian framework). Thanks to this, Darwinism acquired a rather more convincing and scientific form – the form it has retained, which owes very little to Darwin himself, who had already been dead for some time.

Eugenics then entered a phase of practical application, though without any new theories appearing on the subject. Eugenic associations were now established to campaign for the adoption of eugenic measures and legislation. The first eugenic law was passed in Indiana in 1907, followed by other American states: Washington, Connecticut and California in 1909, and so on. Europe remained behind; there were very active campaigns, especially in Britain and Germany, but no legislation.

We can assume that, by giving Darwinism a more convincing form, genetics, and especially population genetics, made eugenics credible as a substitute for natural selection in human societies, and brought it into practical application. The problem is that the same genetics showed, at the same time, that eugenics could have no significant effect – a fact that no doubt explains why eugenics remained fixed in its late nineteenth-century form, without experiencing a theoretical development parallel to that of genetics. Geneticists today like to claim that their predecessors at this time did not support eugenics, apart from some ideologically driven marginal figures. It is, however, an established fact that the majority of geneticists at this time did indeed support eugenics, against the implications of their own discipline – in the same way as Galton invented this eugenics in the context of a theory of heredity in which it could scarcely have any effect, other than by statistical juggling.

It was no mere whim that led governments to pass eugenic legislation, especially in the democratic countries; this was rather at the instigation of biologists and doctors. These champions of eugenics were active in the United States at the start of the century, and supported similar legislation in Europe in the late 1920s and early 1930s. Politicians followed suit, no doubt because these ideas ran in a direction that they liked, but also because they believed in the eugenic assertions of biologists, without realizing that these assertions actually contradicted the view of reality that genetic theories offered.

It was only after 1933, when they realized what Hitler was doing with the theories they had put forward, that some eugenicist biologists started to retreat, though without completely retracting their ideas. In 1935, Muller criticized negative eugenics as practised in the United States and Germany (the sterilization of 'defective' individuals), but recommended to Stalin a positive eugenics (encouraging 'superior' individuals to reproduce).[31] J. B. S. Haldane was far more critical in 1938, but he too continued to champion a vague positive eugenics, certainly more presentable in humanitarian terms but just as ineffective as negative eugenics.[32] These sporadic criticisms, moreover, were totally without effect, as it was at this very time that sterilization reached a peak in the United States, and that European countries also adopted eugenic legislation.

The case of Wilhelm Weinberg is particularly instructive on the question of support for eugenics against the findings of genetics. The Hardy-Weinberg law that he discovered in 1907, in tandem with the mathematician Hardy, demonstrates the ineffectiveness of eugenics. Yet Weinberg was a militant eugenicist, and even a founder of Stuttgart's Society for Racial Hygiene.[33] The implications of the Hardy-Weinberg law were certainly not as clear to its discoverers as they are to geneticists today. Genetics in general was then at its beginnings, not what it has since become. But the problem is more complex, as other geneticists understood perfectly well the ineffectiveness of eugenics, yet were nonetheless eugenicist themselves.

---

31 H. J. Muller, *Out of the Night: A Biologist's View of the Future* (London: Gollancz, 1936).

32 J. B. S. Haldane, *Heredity and Politics* (London: Allen & Unwin, 1938). In the 1960s, Haldane supported Muller and Graham's project of the Repository for Germinal Choice; we see here the constancy of eugenic ideas in this generation of geneticists (D. J. Kevles, *In the Name of Eugenics*, p. 262).

33 B. Massin, 'De l'eugénisme à l'opération euthanasia', p. 1565; Massin, preface to P. Weindling, *L'Hygiène de la race*, vol. 1, *Hygiène raciale et eugénisme médical en Allemagne, 1870–1933* (Paris: Odile Jacob, 1989), p. 34.

A text by Jean Rostand captures this situation wonderfully well. Certainly Rostand was not a major scientist, but he was a well-informed writer who always presented things from a moderate humanist point of view, and his opinion is often interesting. To appreciate his text, we should understand that in 1930, the year it was written, eugenic legislation in the United States went back to 1907, and had been followed by Switzerland and Canada in 1928, Denmark in 1929, Sweden in 1935, etc. The Fischer referred to here is probably Eugen Fischer, a German anthropologist and geneticist, and the author, along with Baur and Lenz, of what at the time was the bible of human genetics,[34] a bible which supposedly served Hitler as a model for his political biology;[35] Fischer was a eugenicist, and compromised by Nazism. Here is Rostand's text:

> All kinds of objections, both practical and theoretical, were raised against the application of eugenic ideas. In particular against the sterilization of severely defective individuals, in which connection the discouraging slowness of the possible effects was adduced. Fischer calculated that it would take twenty-two generations to reduce the proportion of mentally ill from 1 in 100 to 1 in 1,000, and ninety generations to reduce it to 1 in 10,000. The more serious objections, however, were of a moral order; to many consciences, any restriction on the right to procreate was seen as an injury to individual liberty and the dignity of the human person. The advantage that might be obtained would in no way compensate for the moral prejudice; are there not intangible spiritual achievements that take precedence over any progress of the species?
>
> Without getting into this debate, it is possible right away to ask whether humanity, too threatened by decadence, will not see itself compelled one day to adopt the eugenic programme, despite the repugnance it feels towards it; this would be a tragic conflict, with the instinct of legitimate defence on one side and the concern to persist in its moral being on other, between biological necessities and the imperatives of sentiment.[36]

---

34 E. Baur, E. Fischer and F. Lenz, *Menschliche Erblehre und Rassenhygiene* (Munich: Lehmans, 1921).

35 See, among others, J. Sutter, *L'Eugénique, problèmes, méthodes, résultats*, p. 236; B. Müller-Hill, *Science nazie, science de mort, L'extermination des Juifs, Tziganes et des maladies mentaux de 1933 à 1945* (Paris: Odile Jacob, 1989), p. 194. On the level of Hitler's possible understanding of the Baur, Fischer & Lenz book, see P. Weindling, *L'Hygiène de la race*, vol. 1, p. 225.

36 J. Rostand, *De la mouche à l'homme* (Paris: Fasquelle, 1930), pp. 210–1.

To put it another way, according to Rostand eugenics has no noticeable effect, but despite being ineffective, we will in any case be forced to apply it when the human race is too threatened by degeneration. But for what gain, if it is ineffective? And what is the basis of this idea of degeneration? This is a perfect example of the attitude of the geneticists of that time. And the same is more or less true until the 1950s.

The question arises, therefore, whether these biologists really believed in the eugenic theories they put forward, against the findings of their own science. In my view, they believed in them because they had a need to, in a kind of self-intoxication, despite being well aware that what they believed had no serious scientific basis.

More generally, eugenics has to be placed in the context of the anthropological fable of human degeneration for want of natural selection. This fable may well be contradicted by genetics, but it has a sufficiently strong purchase that it rises above these objections. We should not forget that, in its first half-century, Darwinism was a fairly shapeless doctrine, maintained only by way of various anthropological and zoological fables in which selection held centre stage as the pivot on which the whole theory revolved. An idea hammered in for fifty years is not so easily abandoned, no matter how absurd it might be. There was thus a kind of schizophrenia split between Darwinism and genetics, despite their close linkage.

On the one hand, Darwinism postulated an 'evolution/progress through natural selection', and for fifty years, despite the lack of evidence, justified this by an argument from 'degeneration for lack of natural selection', which imposed the need for eugenic measures. On the other hand, genetics maintained the ineffectiveness of these measures, while seeking to give a more scientific and convincing form to evolution through natural selection, albeit by phenomenalist and mathematical methods whose scope was very limited, and which could be made to say whatever was wanted. The nature and physiology of the gene began to be known only as of the 1950s, and at that very point eugenics went into decline.

This schizoid attitude is related to a similar phenomenon that we mentioned in the Introduction to this book, i.e., the coexistence between racial biology and a discourse of the non-existence of races (see pp. xiv–xv above). Eugenics justified the work of geneticists, but their work did not justify eugenics. According to whether they chose one or the other of these alternatives, geneticists could maintain either position. The former enabled them to maintain a eugenic discourse, which often served to make up for the lack of a genuinely scientific framework, a lack from which their work suffered. The latter prevented them from giving a scientific basis to this discourse, which served purely as an ideological framework. They thus oscillated from one position to the other according to the needs of the day, or even sought to reconcile the two by sophistic ruses. The geneticists of today, of course, maintain only the second alternative, which permits them to claim that genetics was never eugenic and that only ideologically driven marginal figures could have defended such a doctrine.

8

The Extension of Heredity to Disorderly Behaviour

Let us turn now to those hereditary degenerative conditions that eugenics was supposed to remedy.

The first of these were naturally those few genuinely hereditary diseases that were known at this time, such as the haemophilia of inbred royal families. With the work of Archibald Garrod on alkaptonuria, between 1902 and 1909, a better understanding of these diseases was achieved.[1] They were now viewed as inborn errors of metabolism, comparable therefore with hereditary physical deformations that had long been familiar (for example, six-fingeredness, studied in the eighteenth century by Maupertuis and Bonnet). Other, similar hereditary conditions were identified in this way, yet they remained very rare, each affecting only an extremely small number of individuals. There was nothing here to match the apocalyptic predictions of a degeneration of humanity for lack of natural selection.

Heredity was then extended to all kinds of disturbance, in particular mental illness and behavioural problems (vagabondage, alcoholism, sexual deviance, etc.), and even criminality. As we have seen, heredity was an obsession of the time. This is what Vacher de Lapouge wrote on the subject in 1896, forcing the point, yet representative of contemporary opinion:

> The extent of heredity is ... as universal, and its force as irresistible, as that of gravity. When exceptions to it seem to appear, this no more means the disappearance of heredity than the rise of a balloon in the air, or of a cork to the surface of the water, means the disappearance of gravity.
>
> Contemporary science has discovered the manner in which two formidable forces respectively act, heredity and electricity, capable in each

1 A. E. Garrod, *Inborn Errors of Metabolism* (London: Hodder & Stoughton, 1909).

> case of transforming material and social life in a radical fashion. Of these two discoveries, the most important is heredity. Electricity scarcely rivals it in its possible application. Institutions and ideas did not hold back the progress of the dynamo, but the same is not true of selection.[2]

At a time when genetics was still far from acquiring the scientific character of physics, anything and everything could be maintained on the question of hereditary characteristics. And this was indeed the case. Biologists and ideologists could embroider to their hearts' content on the basis provided by the general sentiment concerning degeneration, essentially due to the social conditions already mentioned.

We can survey very rapidly the crude genealogical studies that were used as evidence for the heredity of intellectual characteristics. The first systematic works of this kind were those of Galton, who listed a whole series of genealogies of judges, statesmen, writers, scientists, musicians, painters, etc., in his 1869 book *Hereditary Genius*.[3] This approach had a long life. Seventy years later, Otmar von Verschuer (1896–1969), a recognized scientist who became a Nazi eugenicist, could still write the following:

> The heredity of mathematical gifts is particularly pronounced. There is the case of the Bernoulli family, who produced eight remarkable mathematicians in the course of three generations ... The gift for technical inventions can be followed over three generations in the Krupp family. In the Siemens family, out of fourteen siblings, three (or four) brothers were notable inventors and successful entrepreneurs. Eminent scientific gifts, particularly for the natural sciences, arose frequently in the Darwin and Galton families. Charles Darwin and Francis Galton were cousins. Their common grandfather Erasmus Darwin conceived the fundamental idea of the theory of evolution before Lamarck. Four of Darwin's sons were celebrated men of science.[4]

It is clear that what holds for the genealogy of genius holds equally for the genealogy of debility, perversion and crime. For characteristics of this kind, we likewise find the inevitable twin studies, which it is pointless to discuss again here, as they have

2 G. Vacher de Lapouge, *Les Sélections sociales*, pp. 41, 477.

3 F. Galton, *Hereditary Genius, an Inquiry into its Laws and Consequences* (London: Macmillan 1914), passim.

4 O. von Verschuer, *Manuel d'eugénique et hérédité humaine* (Paris: Masson, 1943), pp. 72–3.

served to prove anything and everything. On p. 75 of Verschuer's book, the author compares educational results, but makes similar comparisons elsewhere for tuberculosis and cancer, crime and suicide – Verschuer was a past master of this, as we shall see.

Coming to more properly scientific approaches, we can take two examples in which, thanks to a bit of rhetoric and the phenomenalist and mathematical methods previously mentioned (and in the absence of any physiological explanation), phenomena that had nothing hereditary about them were artificially 'hereditized' – one of these being an ill-defined behaviour, the other the infectious disease tuberculosis.

Charles Davenport, the pioneer of genetics in the United States, conducted a number of studies on the heredity of behaviour, particularly in 1915 on the heredity of what he called 'nomadism'.[5] The fashion in which he proceeded is very interesting, since, no matter how ridiculous a caricature, this is typical of the methods by which behaviour was hereditized.

For Davenport, nomadism meant a certain tendency to wander, expressed in a large number of forms ranging from the migrations of the Gypsies to the tendency of certain children to run away, as well as including the taste for travel on the part of sailors or missionaries, and even the drifting of beggars and the homeless. Nomadism, in fact, covered more or less everyone who moved, whether voluntarily, compulsively or forced by social and economic conditions. Despite being expressed in such different ways, all these behaviours, normal or pathological, were grouped together under the name of nomadism, in the same way as, in Part One, we saw how all behaviour impossible to classify in the framework of a war of all against all was grouped under the name of altruism. This is the nomadism that Davenport sought to hereditize, just as certain scientists hereditized altruism.

He starts off by noting that there is a nomadic instinct in all kinds of animals, particularly migratory birds. Then he moves on to the anthropoid apes, which form wandering bands in the forests. The next stage is that of 'primitive' human societies such as the Bushmen or Hottentots, who move across the savannah, and finally the Gypsies, who lead a nomadic existence in Europe

5 C. Davenport, *The Feebly Inhibited: Nomadism, or the Wandering Impulse, with Special Reference to Heredity; Inheritance of Temperament, with Special Reference to Twins and Suicides* (Washington: Carnegie Institution, 1915).

(and are not only 'the terror of farmers and the despair of the legislator' but 'have no ethical principles and do not recognize the obligations of the ten commandments'; without even taking into account an 'extreme moral laxity in the relations of the two sexes').[6]

Davenport concludes from this that the sedentary pattern of the majority of modern peoples is a recent acquisition, due to the inhibition of an ancestral nomadic instinct. The reappearance of nomadism among certain individuals (all those with itchy feet, whether runaway children or travelling salesmen) is due to a failure of the inhibition of this ancestral nomadism, a failure expressed in a certain behavioural deviance in relation to the social norm. Hence the title of his book, *The Feebly Inhibited*, after the model of 'the feeble-minded'. It is the heredity of this failure of inhibition that Davenport sets out to study.

For this project, he took a hundred families that presented cases of nomadism, as recorded in the files of the Eugenics Record Office at Cold Spring Harbor. We shall see in due course what this organization actually was, but suffice it to say here that it possessed an enormous database of the diverse characteristics of all kinds of American families. Davenport classified his hundred families according to whether the father, mother, sons, daughters, grandfathers, etc., were or were not nomads, with a view to establishing a hereditary pattern for this characteristic. A further factor was its relationship to sex – men being more nomadic than women. He also studied the relationship between this nomadism and other characteristics such as alcoholism, criminality, epilepsy, eccentricity, feeble-mindedness, madness, neurosis, erotomania, etc., according to whether one or several members of these families were affected or not.

Here are his conclusions from the study:

(1) The wandering instinct is a fundamental human instinct, which is, however, typically inhibited in intelligent adults of civilized peoples.
(2) Nomadism is probably a sex-linked recessive monohybrid trait.
(3) Sons are nomadic only when their mothers belong to nomadic stock.
(4) Daughters are nomadic only when the mother belongs to such stock and the father is actually nomadic.
(5) When both parents are nomadic expectation is that all children will be.

---

6 Ibid., p. 11.

(6) The nomadic impulse frequently occurs in families showing various kinds of periodic behavior, such as depression, migraine, epilepsy, and hysteria. It is concluded that these periodic states are not the true cause of nomadism, but rather that, for the better inhibited part of the community, the nomadic tendency is released in the periodic state which paralyses the inhibitions. The feeble-minded and demented may wander without going into a periodic state. The periodic psychoses are frequent concomitants, but not the fundamental cause, of nomadic impulses. They merely permit the nomadic impulses to appear.[7]

The same book contains a second study in the same style, devoted, in the words of its title, to the 'inheritance of temperament, with special reference to twins and suicide'.

We should note in passing, before going on to discuss this in more detail, that these studies were financed by two American billionaires: Mrs E. H. Harriman, the leading patron of the Eugenics Record Office at this time, and the rather better-known John D. Rockefeller.

Studies of this kind have absolutely no scientific value. We should recall that in 1915 population genetics was still in its infancy. Davenport's tables describing a few families were followed by more elaborate statistical studies, and much later by various mathematical models. Davenport's work was in fact the ancestor of the study of altruism by Nowak and Sigmund discussed above. The mathematical apparatus of this study was a considerable advance on that of Davenport, but the biological discourse remained similar; and, as we saw in Part One, it is not the sophistication of the mathematical apparatus that makes for scientific value, inasmuch as all the mathematics one likes can be applied to a weak anthropological fable.

In any case, Davenport's studies, along with those of other scientists working on additional 'hereditary' traits such as alcoholism, crime and genius, were one of the foundations of eugenics, to the extent that they extended heredity to a whole series of behaviours or psychological characteristics, thus bringing them into the orbit of laws providing for sterilization. It is clear that in the case of nomadism and its connections with various psychological characteristics, a particular deviant category of the population was targeted, i.e., the migrants and travellers of the time.

7 Ibid., p. 26.

It is a commonplace today to claim that these eugenicists abused genetics and Darwinism, failing to understand its principles, that they were vulgar ideologists and certainly not scientists. The truth, as we have seen here, is that it was geneticists and evolutionists themselves who invented and popularized these eugenic ideas by extending heredity in all directions. Davenport's studies are only one example among many.[8]

The reach of such genetic Darwinian theories was not confined to psychological and behavioural characteristics. Thanks to the notion of hereditary predisposition, infectious diseases (the origin of which had been discovered by Pasteur and his microbiology a few years earlier) would also be brought within the field of heredity, and sometimes, indeed, into that of eugenics. The method was very easy, requiring no more than a bit of rhetorical savoir-faire.

Tuberculosis, for example, had long been considered hereditary. After the bacillus was discovered in 1882 by Robert Koch, it was believed, like syphilis, to be both infectious and hereditary. The nature of infectious diseases now being better understood, a turn was made towards the 'terrain' on which they developed, and in this way the role of heredity was preserved. Tuberculosis and all kinds of other diseases were now studied in twins, with a view to establishing a hereditary predisposition. This was the particular speciality of Verschuer, whom we have just mentioned in relation to genealogies, and whom we shall discuss below in connection with the work he conducted with his celebrated disciple, Dr Mengele.

In 1939, vaccination against the tuberculosis bacillus was already well established (dating from 1921), but Jean Rostand could still write:

> The study of twins tends to show that resistance to tuberculosis depends to a great extent on genetic factors. According to Diehl and Verschuer, identical twins react identically to tuberculosis infection in 70 per cent of cases, whereas this similar reaction is shown in only 25 per cent of cases for non-identical twins.
>
> According to H. Edmund Stoddart, sensitivity to septicaemia is hereditary; he cites a family in which sixteen individuals died of septicaemia over

8 To stay with the same author, see the thick volume C. Davenport and M. Steggerda, *Race Crossing in Jamaica* (Washington: Carnegie Institution, 1929), and also C. Davenport, *The Genetical Factor in Endemic Goiter* (Washington: Carnegie Institution, 1932).

> four generations, while four individuals showed an excessive sensitivity to infections of the blood. This sensitivity is transmitted as a dominant characteristic, not linked to sex.[9]

This is not completely false. There may well be a hereditary predisposition to certain diseases, due to a particular sensitivity to certain germs, and this sensitivity may even be more pronounced in certain 'races' than others. It remains, nonetheless, that considerations of this kind, which are resurfacing today with what is called 'predictive' medicine, do not mean much. Theories of this kind are based on an old adage of the geneticist August Weismann, who claimed that every acquired characteristic is only acquired because there is a hereditary predisposition to acquire it.[10] This is an admirable paralogism, which opens the door wide to a total genetic determinism. It is then possible to find a hereditary predisposition for anything and everything, and particularly for all diseases, according to the principle that the simple fact of living implies a certain predisposition to die.

Public health clearly gains no benefit from this, though the object of these kinds of theories is not really medical, but ideological. The point is to make the greatest possible number of things hereditary, in order to bring human life into a particular philosophical framework and to justify social applications – and also for geneticists, of course, to justify their own studies.

The struggle against tuberculosis made headway because the pathogen was discovered by Robert Koch, vaccination subsequently developed (BCG was invented by A. Calmette and C. Guérin), and finally antibiotics became widely available. Preoccupation with the hereditary terrain made no contribution to this at all. It was even somewhat damaging. For example, this preoccupation led Rostand to write, in the same work cited above: 'Far from vaccination increasing the inborn resistance of the race,

---

9 J. Rostand, *Biologie et médicine* (Paris: Gallimard, 1939), p. 102. We may note that in 1939, Verschuer was still seen as a completely acceptable international scientific authority, despite having long since compromised with Nazism. The studies of Diehl and Verschuer cited by Rostand were published in K. Diehl and O. von Verschuer, Zwillingstuberkulose. Zwillingsforschung und erbliche Tuberkulosedisposition (Jena: Fischer, 1933). This work also claims the role of hereditary dispositions in all kinds of disease, infectious and otherwise, including polio, mumps, pneumonia, cancers, allergies, etc., as well as mental illness.

10 A. Weismann, *Essays on Heredity and Kindred Biological Problems* (Oxford: Clarendon Press, 1899), p. 429.

it can only weaken this, to the extent that, by permitting the survival of genetically vulnerable individuals, it counteracts the effects of natural selection.'[11]

We should not believe that this is an isolated or insignificant remark. It is inscribed in a long litany of observations in the same style. Here, for example, is how eugenics was justified in 1894 by the progress of therapies for infectious diseases; selection was supposedly needed to eliminate those individuals no longer eliminated by microbes:

> The microbes and other selective agencies have been improving the race, or at any rate have in the past been preventing its deterioration, but it by no means follows that this action is to be permitted to them in the future ... But if the selecting microbe is to disappear, we have to replace it by something else ... But if we remove selective influences without replacing them by others, then racial decay is certainly and inevitably upon us. At the present time people with strong strains of insanity or phthisis marry freely ... A man may be summoned for neglecting to send his son to school, but at present there is no strong public feeling against knowingly begetting a son who all his life may suffer from weak lungs or brain, and hence obvious disease is no bar in the marriage market ... With the judicious selection of parents to be the race-producers, we need have no fear as to the care that modern civilization and preventive medicine bestow upon the individuals. If the community undertakes its own selection we can dispense with the selecting influence of the micro-organism or whooping-cough, scarlet fever, or tubercle.[12]

Ammon, likewise, reflects on the differences in sensitivity to tuberculosis, but this time between different social classes. Tuberculosis is less common in the upper classes 'because they have been refined and selected over many generations'. This disease, in fact, 'is less dependent on housing and standard of living' than on innate predisposition, and 'the poor, if they lived in palaces, would still be as decimated by tuberculosis as they are in their wretched shacks'.[13] Such is the force of heredity, even in infectious diseases.

Genetic and Darwinian theories, applied in this way, displaced Pasteur's concern for public health in favour of eugenic (and

11 J. Rostand, *Biologie et médecine*, pp. 101–1.

12 J. B. Haycraft, *Darwinism and Race Progress* (London: Sonnenschein, 1894), pp. 81–7, 170, cited in G. Vacher de Lapouge, *Les Sélections sociales*, pp. 466–7, 468–9).

13 O. Ammon, *L'Ordre social et ses bases naturelles*, pp. 206–7.

incidentally racist) doctrines. In other words, for purely ideological reasons the attempt was made to replace public health and medical techniques that had already proved themselves with ineffective politico-biological principles. This was no innocent process, not simply a scientific rivalry between the schools of Pasteur and Darwin, but rather an attempt to privilege the innate over the acquired with a view to concealing social factors to the benefit of biological terrain, thus showing the uselessness or even ineffectiveness of social measures, including those of public health underpinned by Pasteur's principles.

The results of such doctrines in the case of tuberculosis are sadly familiar. In the case of Dr Kurt Heissmeyer, for example, of whom E. Klee writes:

> Heissmeyer quite simply aimed to take the place of Robert Koch in the public consciousness ... He based himself on the studies conducted on twins by Karl Diehl and Otmar von Verschuer, according to which the essential element in tuberculosis was not the bacillus but rather a congenital sickly constitution.
>
> Robert Koch had obtained his results by experimenting on animals. But for Heissmeyer, animal experiments were devoid of scientific value: 'Since the creation of the ideological concept of Volk and race, it is impossible to establish on the basis of animal experiments the way in which human tuberculosis makes its appearance. This means either denying the role of physical constitution in the appearance of human tuberculosis, or else placing the constitution of animal and human on the same level.'[14]

Strong in these principles, which were not yet called 'predictive medicine',[15] Heissmeyer went on to conduct various contamination experiments on children taken from Auschwitz, who were subsequently hanged and autopsied. He continued his career as a pneumologist after the war, until he was arrested in 1963 and condemned to life imprisonment.

The objection can perhaps be made that there is no strict relation of cause and effect between 'constitutional' (or predictive) medicine and the misdeeds of someone like Heissmeyer, that

---

14 E. Klee, *La Médecine nazie et ses victims* (Paris: Solin-Actes Sud, 1999), p. 127.

15 The phrase 'constitutional medicine' was often used, since, as this text indicates, it is the individual's genetic constitution rather than the pathogen that is highlighted – whereas the expression 'constitutional disease' served rather as an equivalent for hereditary disease.

different conclusions on the field of heredity could equally have been drawn from such considerations. Perhaps. But the fact is that these were the conclusions, and not any others. I even believe it is not very surprising that this was the case. The importance of terrain has certainly been appreciated in medicine ever since Hippocrates, and no one has ever denied that this should be taken into account, but the very idea of constitutional or predictive medicine involves a certain perversion of the medical ideal (see below).

Here at all events are two cases – one bearing on the 'deviant' behaviour of nomadism, the other on an infectious disease – that perfectly illuminate the way in which geneticists 'hereditized' everything in order to justify the idea of human degeneration, and hence the eugenic doctrines that were supposed to remedy this degeneration. We shall now return to our chronology, and follow what took place in a number of countries, principally the United States, France, the Soviet Union and Germany.

9

Eugenics in the United States and Europe

The origin of modern eugenics lay in Britain, with Galton, Pearson, biometrics, the Eugenics Education Society, and so forth. But Britain never passed any eugenic legislation, properly speaking. The propaganda and lobbying of associations and biologists came up against the English democratic tradition, and the political authorities opposed the demands of these scientists. Several bills for sterilization measures were rejected by the House of Commons,[1] particularly in 1931 and 1934, the time when eugenic legislation crossed the Atlantic from the United States, where it had been pioneered.

It was in the United States that eugenic propaganda found its first legislative responses. Indiana adopted a law on eugenic sterilization in 1907, followed by some thirty other states. Before discussing this legislation in greater detail, we shall focus on the support this received from scientists.

The main institution of scientific eugenics in the United States was the Station for Experimental Evolution at Cold Spring Harbor (Long Island, New York), which still exists and is today, under another name, a very fashionable centre of molecular biology, frequented by the elite of this discipline (until recently under the direction of James Watson, who received a Nobel Prize in 1962 together with Francis Crick for their discovery of the structure of DNA).

This station was established in 1904 by the American geneticist Charles Davenport (initially a champion of Galton and Pearson's biometrics who then converted to Mendelism), with the financial backing of the Carnegie Institution of Washington. Davenport was the author of the studies of nomadism discussed in the

---

1 Y. Ternon and S. Helman, *Les Médecins allemands et le National-Socialisme*, p. 159.

previous section, which were carried out at Cold Spring Harbor and published by the Carnegie Institution.

In 1910, thanks to an extremely rich philanthropist, Mrs E. H. Harriman,[2] and with the laudable object of public health, an annex was established for eugenic research, the Eugenics Record Office (ERO). This annex, which in the 1920s was placed along with the entire station under the wing of the Carnegie Institution, more particularly studied human genetics, the rest of the station focusing on animal and vegetable genetics and evolution.

The ERO comprised, among other things, a vast data bank recording various characteristics of American families, which counted some half a million files in 1918 and nearly a million by 1935 (including those of families presenting cases of nomadism, epilepsy, alcoholism, erotomania, and so on). It also conducted eugenic propaganda, particularly in youth organizations, and what is today called 'genetic counselling' for people intending to marry. This eugenic annex of the Cold Spring Harbor Laboratory was directed by Davenport's assistant, Harry H. Laughlin, who reigned there for some thirty years.

Cold Spring Harbor became a kind of world reference point for questions of eugenics. It was from here that 'scientific' eugenics spread across the United States and thence to Europe. As for Laughlin, he became an authority on eugenic legislation, as well as on the (racist) regulation of immigration into the United States, serving as eugenics expert for the immigration committee of the House of Representatives.

Laughlin was a rather odd character. He was himself epileptic, which would have made him eligible for sterilization under the laws he himself championed. He was rather mediocre as a scientist and is thought to have falsified his results in a way that served his ideology. He was linked with the American racist theorist Madison Grant, and in 1936 received from the Nazis an honorary doctorate at Heidelberg University. (The German law of 1933 was modelled on what had long been done in the United States on Laughlin's initiative.) Laughlin did not conceal even his Nazi sympathies, becoming an active propagandist for Nazism in the United States; in particular, he tried to distribute the Nazi film *Erbkrank* ('Hereditary Disease') with the financial support of the

---

2 Mary Williamson Harriman, widow since 1909 of the railroad magnate Edward Henry Harriman.

Pioneer Fund that he had helped set up along with men such as the textile billionaire Wickliffe Draper. (After the war, the Pioneer Fund continued to provide financial support for research in Anglo-Saxon countries on subjects such as the heredity of intelligence, race and intelligence, etc., – for example, the work of Arthur Jensen and Hans Eysenck. The protégés of the Pioneer Fund also included William Schockley, co recipient of the 1956 Nobel Prize for physics, and the only person to have acknowledged having donated his sperm to the repository founded on the initiative of Muller and Graham.)[3]

It was not until 1935 that the Carnegie Institution expressed concern about this situation; in 1939 it pressed Laughlin to retire, and early the following year – when the Nazis had begun gassing the mentally ill – it put an end to eugenic activity at Cold Spring Harbor;[4] Davenport had already retired in 1934.[5]

After this brief presentation of this centre of 'scientific' research (the studies of nomadism give a good idea of its activity), let us return to the subject of legislation.

At the end of the nineteenth century, the United States started to forbid certain categories of the population from marrying (the mentally backward, the abnormal, alcoholics, those suffering from venereal disease). In 1896, Connecticut passed a law to this effect; its eugenic bearing was evident in so far as it was applicable only if the woman was under the age of forty-five, and thus capable of bearing children. This was followed in 1905 by Indiana, and then rapidly by several other states, some thirty by 1914.[6] There were also laws banning interracial marriage, sometimes likewise including eugenic considerations.

The laws on sterilization, starting with the Indiana legislation of

---

3 S. Kühl, *The Nazi Connection: Eugenics, American Racism, and German National Socialism* (Oxford: OUP, 1994), pp. 5–10, 48–9.

4 D. J. Kevles, *In the Name of Eugenics*, p. 199.

5 For the history of the Eugenics Record Office and the character of Laughlin, see G. E. Allen, 'The Eugenics Record Office at Cold Spring Harbor, 1990–1940, an essay in institutional history', *Osiris*, 2nd series, 1986, 2, pp. 225–64.

6 D. J. Kevles, *In the Name of Eugenics*, p. 100. According to Vacher de Lapouge, writing in 1899, the Connecticut legislation not only prohibited marriage for epileptics, imbeciles and the feeble-minded, but sexual activity in general (on penalty of a minimum of three years' imprisonment); in Massachusetts, epileptics, alcoholics and syphilitics were affected; in Pennsylvania, a proposed law (not yet adopted in 1899) targeted individuals suffering from syphilis, blennorrhoea, epilepsy, dipsomania, tuberculosis and mental alienation (L'Aryen, *son rôle social* [Paris: Fontemoing, 1899], pp. 504–5.

1907, thus supervened on existing measures that simply banned marriage. The Indiana example was followed by other states: Washington, Connecticut and California in 1909; Nevada and Iowa in 1911; Kansas, Wisconsin and North Dakota in 1913; etc. By 1930, thirty-three states possessed laws of this kind.[7] They varied from one to another, they were not always well accepted, and certain were attacked as unconstitutional – a fact that explains why they were also rather unevenly enforced.

Like all eugenic legislation, these laws had no serious scientific foundation. They did, however, reflect quite well the fantasies of American society as represented by Laughlin and others of his ilk. These laws were an odd concoction. On the one hand, like the majority of eugenic texts, they only seem ridiculous to any normally constituted individual. On the other hand, when we consider that there were people who could formulate them and apply them to other human beings, laughter seems the wrong response.

D. J. Kevles's book *In the Name of Eugenics*, a standard reference for eugenics in the Anglo-Saxon countries, is rather too restrained, as he does not explicitly describe this legislation or give the number of people who fell victim to it. I shall thus refer here to the study by Jean Sutter, already rather old and with less of a 'complex'. Sutter, incidentally, was one of Carrel's colleagues in his final years, and was himself rather well-disposed towards eugenics, displaying the unconcerned approach characteristic of eugenicists in the first half of the twentieth century.

The conditions that were cause for compulsory sterilization, by vasectomy for men (in rare instances, by castration) and salpingectomy in women, were first and foremost mental illness, intellectual deficiency and behavioural disturbances. Such diseases were not always hereditary – they were not always even diseases – and the legislation was extremely vague. Sutter writes:

> The fact of heredity did not often interest the legislators. Mention of 'disease' or 'hereditary forms' only appeared in a few states such as California, Mississippi, Utah and the two Virginias; in Nebraska it was heavily stressed. The expression 'future reproduction of the feeble-minded' was used in Maine and New Hampshire; in Idaho, the phrase was just 'hereditary tendency', in Michigan the legislation was intended 'to prevent

7 J. Sutter, *L'Eugénique, problème, méthodes, résultats*, p. 125.

> the increase in the number of idiots, the feeble-minded, etc.' We should note that in certain states sterilization was viewed as a therapeutic means for certain forms of mental illness: in North Carolina, the operation could be practised 'if it may lead to the mental, moral or physical improvement of the patient'. In Vermont, the 'well-being of the patient' was also mentioned.[8]

Apart from 'mental illnesses' and other forms of intellectual backwardness, crime was particularly targeted, and sexual crime above all – as if the intent was to punish the criminal in the place where he sinned. To quote Sutter again:

> In half the cases, sexual delinquents were indicated by the following terms: 'sexual perverts who may become a danger to society' (Idaho, Michigan), 'habitual criminal sexual tendencies' (Oregon, Utah, the two Virginias), 'authors of sexual offences' (California); 'bestial or sexually perverse habits' were cited in the state of Washington; sodomy in North Dakota and Oregon; Nevada and California specified that 'sterilization can be performed on someone who has committed a carnal act or rape on a girl under ten years of age'; rape alone was cited in New Jersey and Idaho.[9]

It should be said that the United States had a certain tradition in this area, as laws providing for castration in the case of sexual offences already existed in the nineteenth century. In Kansas, for example, an 1855 law condemned black and mixed-race people to castration for the rape of a white woman.[10] According to Vacher de Lapouge (1899), abduction and rape were also punished by castration in Ohio.[11] Laws on the sterilization of sexual criminals were the extension of these practices under the pretext of eugenics. We should note that castration remained authorized in some states in the twentieth century, particularly Utah; in 1930, 175 men were castrated under these laws in the United States.[12]

More general crimes under common law could also lead to sterilization. In the states of Washington and California, for example, this was the case for criminals with three convictions. The most curious case is that of Michigan, where those condemned

8 Ibid., p. 124.
9 Ibid., p. 123.
10 Y. Ternon and S. Helman, *Les Médecins allemands et le National-Socialisme*, p. 180.
11 G. Vacher de Lapouge, *L'Aryen, son rôle social*, p. 505.
12 Y. Ternon and S. Helman, *Les Médecins allemands et le National-Socialisme*, p. 180.

for life could be sterilized; in the absence of mixed prisons, it is hard to see how these poor wretches could have reproduced. This shows how far into the realm of fantasy such legislation extended; moreover, if 'nomadic' behaviour is viewed as hereditary, it is easy enough to develop a genetics of criminality. Yet these laws were actually applied, however haphazardly, and claimed tens of thousands of victims.

The United States did not invent this kind of theory. The inheritance of criminality, and the sterilization of delinquents, has a long history. Here, for example, is what Vacher de Lapouge wrote in 1896, simply reflecting a widespread opinion that had already been championed by many writers, European as well as American:

> It is not enough, from the social point of view, for the criminal to be punished. This does not even matter all that much. Ancient ideas of chastisement and rehabilitation raise a smile. In the present, what is needed is to put him in a position where he can do no more harm; in the future, it is to suppress his posterity. Every descendant of a criminal, even if he himself is the most honest man in the world, bears within him the germ of criminality. A stroke of atavism, an unfortunate match, can awaken it in any generation. Disfavour and distrust are quite legitimate, both towards those descendants who exist at the moment of the offence, and also to the family as a whole. For future posterity, the important thing, if the death penalty cannot be applied, is to place the individual in a position where he can no longer sully with his descendants the social body of which he forms a part.[13]

Among other strange legislation in America – again cited by Sutter, who is fairly uncritical of it – we find sterilization in Idaho on the grounds of syphilis (it is hard to see the prophylactic effect of such a measure), as well as, in certain states, on grounds of epilepsy, Down's syndrome, toxicomania, and the genuinely hereditary disease of Huntington's chorea.

By comparison, Sutter cites the case of Denmark, where between 1929 and 1945, as well as 400 castrations of 'abnormal' adult men and 'sexual criminals', 3,608 sterilizations were carried out, 2,803 of these being for mental debility. As for the 805 others:

---

13 G. Vacher de Lapouge, *Les Sélections sociales*, p. 321. O. Ammon is not far from ideas of this kind: *L'Ordre social et ses bases naturelles*, pp. 80–81.

> The causes most commonly cited were the following: slow-wittedness, epilepsy, psychopathy, schizophrenia, depressive mania, other mental illnesses, psychoses and nervous troubles of pregnancy, malformation of various kinds, blindness, deaf-mutism, and serious illnesses including haemophilia, diabetes, muscular dystrophy, Friedreich's disease, etc.[14]

Sterilization on the grounds of haemophilia is also hard to understand, given that surgical intervention is generally avoided wherever possible for sufferers from this disease.

To return to the United States: according to Sutter, the number of individuals sterilized from 1907 (the first Indiana legislation) to the beginning of 1949 was 50,193 – 20,308 men and 29,885 women.[15] I am inclined to believe that this figure is not that reliable, being simply the count made at the time. It is rather a minimum, and there were certainly cases that escaped enumeration.

It is fairly hard to get any more detailed figures, as studies have been rare, difficult and not exactly encouraged. (We can imagine all the lawsuits and demands for compensation that survivors could make if they were brought together in the context of such studies.) Ternon and Helman give the following figures over time: 2,250 sterilizations between 1909 and 1933; 2,000 from 1933 to 1935; 3,000 in 1935; and a total of 50,193 by 1 January 1949.[16] Kühl gives 3,233 sterilizations between 1907 and the start of 1920; 2,689 from 1921 to 1924; then between 200 and 600 per year until 1930; and between 2,000 and 4,000 per year in the 1930s.[17] The figures given thus vary somewhat between different authors, but the general import is the same.

At all events, it was only in the 1930s – at the same time as Europe, and especially Germany, started to adopt similar legislation – that sterilization really proliferated in the United States. Meanwhile, certain geneticists expressed concern at the turn that eugenics was taking in Nazi Germany. Yet this was also the time that the consequences of the 1929 economic crisis were making themselves felt. And from the scientific point of view, it was when the paradigm of Morgan's genetics started to go into decline. Chromosomes could now be mapped – and those of *Drosophila* largely had been, thanks to more than four hundred

14 J. Sutter, *L'Eugénique, problème, méthodes, résultats*, pp. 136–7.
15 Ibid., p. 134.
16 Y. Ternon and S. Helman, *Les Médecins allemands et le National-Socialisme*, p. 157.
17 S. Kühl, *The Nazi Connection*, p. 24.

mutations – but this did not cast any light on the physical basis of heredity or its physiological mechanism, which remained unknown. It was also at this time that Ronald Fisher published *The Genetical Theory of Natural Selection*, in which he applied the statistical methods of population genetics to the study of natural selection.[18] This work was one of the origins of the synthesis that brought genetics and evolutionism together. It expresses a certain reorientation of genetics towards Darwinism, as an advance on the Morgan paradigm. It included a number of eugenic arguments (which took up about a third of the book) and is dedicated to Major Leonard Darwin, son of Charles and a major figure in the international eugenic movement.

Whatever the causes may have been, it was certainly in the 1930s that eugenic sterilization became most widespread in the United States. It is notable here that women were more often sterilized than men, even though the operation is more difficult in the female sex. This was the case in all democratic countries, except in the state of California, already at the forefront of science and progress. It was certainly California eugenicists who had the closest connections with their Nazi counterparts.[19] By January 1949, California had conducted 19,042 sterilizations – almost 40 per cent of the American total, despite representing no more than 8 per cent of the US population. These broke down into 9,845 male and 9,197 female, i.e., almost half and half. In Germany, too, if we can trust the incomplete figures given by Sutter as being at least representative samples, sterilization was also practised more or less equally on both sexes.

In the democratic countries, this large number of women is certainly explained by the abuse of eugenic laws for the purpose of contraceptive sterilization: feeble-minded girls were sterilized, less to prevent the transmission of their mental backwardness than to avoid an unwelcome pregnancy. One curious abuse is that in Switzerland in the 1920s, before the passage of eugenic legislation, women in good health were sometimes sterilized if they were unfortunate enough to have a psychopathic husband.[20]

Sterilization continued in the United States after the Second World War (2,322 cases in 1948) but gradually slowed down.

18 R. A. Fisher, *The Genetical Theory of Natural Selection* (Oxford: OUP, 1930); rev. ed. (New York: Dover, 1958).

19 On the connections between American and Nazi eugenics, see S. Kühl, *The Nazi Connection*.

20 M.-T. Nisot, 'La sterilisation des anormaux', *Mercure de France*, 1929, 209, pp. 595–603.

According to Ternon and Helman,[21] the total had reached 60,166 by 1960, which would mean fewer than 10,000 after 1949. After 1960, there was in all likelihood a sharp decline.

We can compare this with the situation in Sweden. There, the first eugenic law was passed in 1934, but this led to only a small number of sterilizations. A second law was passed in 1941, which considerably increased the number. According to Sutter, between 1935 and February 1949, some 15,486 individuals were sterilized: 12,108 women and 3,378 men, i.e., a rate of sterilization far higher than in the United States. The general estimate today is that around 60,000 sterilizations were performed in Sweden between 1935 and 1976,[22] three-quarters of these apparently after the Second World War.

To sum up, the American project, quite contrary in this respect to the situation in Nazi Germany, was never systematic, being a general purge of society rather than a specifically ethnic one – even though, given the manner in which these laws were applied, it was better to be white and rich than black and poor. It was a purge of everything considered undesirable, whether from a biological point of view (hereditary diseases, or those believed to be so), a psychological one (mental illness, intellectual deficiency, behavioural disturbance), or a social one (alcoholism, delinquency, criminality). Geneticists had tried to make all these things hereditary, and eugenics sought to establish a social order free from this kind of problem.

The quintessence of American eugenics is found in the model legislation concocted by Laughlin, which was never put into effect but is revealing to the point of caricature. Here is the list of 'socially inadequate persons' liable to sterilization (should we laugh or cry?):

> 'A socially inadequate person is one who by his or her own effort, regardless of etiology or prognosis, fails chronically in comparison with normal persons, to maintain himself or herself as a useful member of the organized social life of the state …
>
> 'The socially inadequate classes, regardless of etiology or prognosis, are the following: (1) Feeble-minded; (2) Insane (including the psychopathic);

---

21 Y. Ternon and S. Helman, *Les Médecins allemands et le National-Socialisme*, p. 157.

22 B. Massin, 'La science nazie et l'extermination des marginaux', *L'Histoire*, January 1998, 217, pp. 52–9.

> (3) Criminalistic (including the delinquent and wayward); (4) Epileptic; (5) Inebriate (including drug habitués); (6) Diseased (including the tuberculous, the syphilitic, the leprous, and others with chronic, infectious and legally segregable diseases; (7) Blind (including those with seriously impaired vision); (8) Deaf (including those with seriously impaired hearing); (9) Deformed (including the crippled); and (10) Dependent (including orphans, ne'er-do-wells, the homeless, tramps, and paupers.'[23]

To escape the wrath of Laughlin, you had to be a good American citizen, preferably Anglo-Saxon, white, Protestant, healthy, moral and productive. Laughlin's reign lasted from 1910 to 1940, but he rarely appears in books of social or political history and is little known except to historians of biology. At any rate, we can see the purpose and the result of studies such as Davenport's on nomadism, and of the files on American families held at the Eugenics Record Office.

Over to Europe

The transfer of American legislation to Europe was not an easy matter. All the propaganda of the eugenic associations and major geneticists of the time did not immediately persuade the European countries to follow the United States on this path. Sterilization seems for a long time to have been repugnant to Europeans, reminiscent of castration or at least a form of mutilation. They preferred positive measures of public health and social policy.

One point to mention in this context is the existence of measures designed for especially gifted children. A German, Joseph Petzoldt, had called for special schools for such children as early as 1905.[24] In 1917 such an establishment was founded in Berlin. Other cities followed suit: Hamburg, Göttingen, Leipzig and Hanover, with the selection made on the basis of tests devised by psychologists. Measures in this direction were also taken in Austria, Belgium and Switzerland. The main concern was to sort children with a view to giving them the education that suited them. These measures were justified in the following terms:

---

23 Cited by J. B. S. Haldane, *Heredity and Politics*, pp. 16–17. (Haldane notes that Milton would have fallen under category 7, Beethoven under category 8, while Jesus would have been classed under category 10 as both homeless, a tramp and a pauper.)

24 J. Petzoldt, *Sonderschulen für hervorragend Befähigte* (Leipzig, 1905).

> More than ever, in the present economic conditions, each nation has an interest in drawing the greatest benefit from the citizens that compose it, not wasting the higher intelligences by failing to provide them with the culture they need, or wasting time and money by giving higher culture to those unable to profit from it.[25]

It would be interesting to know the outcome of this experiment, and what eventually came of these schools for the specially gifted. The advice that Otto Ammon offered his contemporaries in 1894 should also be noted, and will ring a bell with many teachers today:

> It is of public interest that study programmes and examinations should take into account as far as possible the conditions of practical life. There is a great deal here that needs reform; for example, the exclusive or predominant use of dead languages as a criterion of aptitudes. At a time of such advance in industry and technology as our own, this criterion can often yield erroneous results, given that the aptitude for dead languages and the aptitude for pure and applied natural sciences are not necessarily combined, and often even mutually exclusive.[26]

As is so often shown, there is nothing new under the sun. At all events, social concerns of this kind were evidently more pleasing to Europeans than was eugenic sterilization. More precisely, we should say that they pleased the majority of Europeans, while there were those who had no such appreciation for education, considering it a vain attempt to improve the masses – Vacher de Lapouge, for instance, who had an opinion on more or less everything. For him, the most to be expected from instruction was that individuals would develop their hereditary potentials; it would not change them in any way:

> It is one of the most widespread prejudices of our time to see the extension of education as a panacea. Moralists, criminologists and economists have long been preaching this doctrine, and in the last twenty years it has been the basis of a regular political campaign in France. This false idea has produced beneficial results in bringing forward the time when each civilized individual will be provided with the knowledge needed for him to attain his

25 É. Claparède, *Comment diagnostiquer les aptitudes chez les écoliers* (Paris: Flammarion, 1929), pp. 20–1.
26 O. Ammon, *L'Ordre social et ses bases naturelles*, p. 78.

true worth ... This is not the first occasion that an error of principle will have produced good results in practice, but from the scientific point of view it is still a crude error.

Prejudice as to the effectiveness of education for the development of nations and humanity stems from a certain confusion. As regards the individual, the idea is correct. It is clear that any man will benefit from education in all respects. However inferior he may be by nature, and however limited the education that his faculties allow him to receive, there is always a significant difference between his condition before and after his exposure to culture. It is advantageous therefore for him and for society that he should receive the maximum education that can usefully be applied to him ...

The power of education is not unlimited. All culture requires a foundation. Wheat cannot be grown on bare rock ... Similarly, there are minds whose primitive coarseness will be changed very little by education. Others, on the contrary, can acquire a prodigious sum of knowledge and habits, but this receptiveness always has a limit, and these fortunate minds are infinitely rarer, not just than the mediocre ones, but also than the inferior ones. We may even say that our intense civilization transforms into an ever deeper abyss the gap that previously existed between men in terms of this receptiveness ... For necessary reasons, even the most assiduous cultivation can do no more than provide the individual with the maximum value that his organization renders him capable of. Observation shows that it cannot modify this organization itself, which is correlated with that of the brain ...

To improve the mass of people by instruction and education is thus a utopia. Of all the changes of environment, the least effective is change in the intellectual environment. This only serves the individual according to his own nature, and yields nothing that can be transmitted by heredity.[27]

At all events, it was not until the late 1920s and early 1930s that Europe followed the United States in eugenic legislation. Switzerland rather than Germany was the first in this respect, in particular the canton of Vaud in 1928.[28] (The same year, the Canadian province of Alberta adopted a similar law.) This was followed by Denmark in 1929, Germany in 1933, and then a number of other countries: Sweden, Norway, Finland and Estonia.

The decisive factor here was perhaps the economic problems of the interwar period, in particular the economic crisis of 1929.

---

27 G. Vacher de Lapouge, *Les Sélections sociales*, pp. 101–3, 125.

28 The history of this Swiss legislation, with contemporary commentaries, can be found in M.-T. Nisot, 'La sterilisation des anormaux'.

Undoubtedly, though, the general climate of the time played a part, as well as the effect of accumulated propaganda over many long years. We should note, however, that if in 1907 the virtual non-existence of scientific genetics could excuse Indiana, by 1930 genetics was well established, and it was a proven fact that eugenic sterilization was ineffective (see the quotation from Rostand on p. 135 above, which dates precisely from that year). All these laws, however, were adopted with the support of biologists and doctors, if not at their instigation. This is hard to explain, though we have tried to do so above (p. 136).

One factor that undoubtedly had a certain importance was the intervention of the Rockefeller Foundation (based on oil wealth), which, if it did not directly support eugenics in Europe, at least backed the work of some of the leading eugenicists in France, Sweden, Germany, and elsewhere. It generously supplemented local funding, where this existed, or else replaced it where it was lacking for want of resources or political will. In this respect, it played a similar role for Europe as the Carnegie Institution did for the United States itself.

American foundations more generally played a large part in the development of scientific research in both the United States and Europe, where the Rockefeller Foundation was particularly active. They had considerable financial resources and could thus make up for the inadequacies of government-backed research. By virtue of their philanthropic vocation, moreover, they were less focused on work that could be expected to have immediate and profitable applications.

These financial resources, along with the detachment from immediate utility, led these foundations to develop regular science policies with a more or less long-term scope. The very notion of 'science policy' – taking into account the present state of a discipline, its perspectives, possible utility, etc. – is more or less of their making. But 'policy' means also the possibility of misguided policy, and 'science policy' was sometimes understood as the science of politics rather than as the politics of science.

The involvement of the Rockefeller Foundation in biological research was such that modern biology is sometimes said to be practically its creation. Given the place that eugenics held in the biology of the first half of the twentieth century, it was thus inevitable that both the Rockefeller and other major foundations should be caught up in it.

In the United States, John D. Rockefeller, Jr had helped to finance the American Eugenics Society[29] as well as the Eugenics Records Office at Cold Spring Harbor (together with Mrs Harriman and the Carnegie Institution); we also saw on p. 142 how he financed Davenport's work on nomadism. In Europe, he was not personally involved, but his foundation was very much so.

France

In France, among a large number of other projects, the Rockefeller Foundation financed the Société Française d'Eugénisme – a fairly inoffensive body, and not very successful.[30]

France, as we have said, never had any specifically eugenic legislation, nor even a very strong eugenic movement. What is sometimes called eugenics in France is more properly described as public health policy. To speak of eugenics in this case is a play on words: etymologically, 'eugenics' simply means the science of good births, and in France these good births were seen as resulting from the health of the pregnant woman, conditions of childbirth and breast-feeding,[31] rather than from selectionist measures to sterilize individuals deemed genetically incorrect. This particular aspect of French 'eugenics' was chiefly due to the influence of Pasteur and Lamarck, and no doubt also to Catholicism. France was strongly attached to the work of Pasteur, a national hero, and long remained reticent towards Darwinism, preferring the work of Lamarck. It is good form nowadays to claim that this held back the development of biology and genetics in our country (which remains to be proved), but it at least had the benefit of sparing us eugenic folly.

There remains the case of Carrel, which has been so much discussed, and on which anything and everything has been said. Though Carrel was French by nationality, his entire career was spent at the Rockefeller Institute in New York. His position in eugenics, accordingly, fitted completely with long-established practice in the United States, as well as with the politics of his employers.

Under the Vichy regime, however, Carrel did establish the Fondation Française pour l'Étude des Problèmes Humains, which

29 D. J. Kevles, *In the Name of Eugenics*, pp. 55, 60.
30 J. Roger, 'L'eugénisme, 1850–1950', p. 416.
31 See A. Carol, *Histoire de l'eugénisme en France* (Paris: Le Seuil, 1995).

was effectively a Pétainist enterprise. A number of remarks on this are needed here. First of all, Carrel was, so to speak, Pétainist even before the Vichy regime. He and his wife were politically reactionary, anti-parliamentary and indeed anti-democratic (Mme Carrel was active in Colonel La Rocque's Croix-de-Feu movement from the start, and Carrel himself a sympathizer),[32] and they both cultivated a fairly *intégriste* Catholicism. Carrel had also turned into a rather disagreeable character by this time, soured by the social and political turn of his time.

A second point to note is that this foundation was a project that Carrel had already sought to establish in the United States before the war.[33] He was already elderly, having received the Nobel Prize back in 1912, and his work on tissue culture was no longer pioneering. This is undoubtedly the reason why his thoughts turned towards the social applications of science. The international success of his book *Man, the Unknown*, translated into multiple languages, could only have encouraged him.

Carrel saw the Pétain regime, which matched his own political ideas, as providing the opportunity to realize the foundation he had been unable to set up in America. Its name was very likely borrowed from the Centre d'Étude des Problèmes Humains that certain intellectuals had established before the war, with the object of promoting the application of natural science to resolving social problems (this is a point we shall return to below). Besides Carrel, this centre had also attracted the likes of Alfred Sauvy, Teilhard de Chardin, André Siegfried and Henri Focillon, as well as Paul Desjardins, René Capitant, Le Corbusier and Aldous Huxley.[34]

It is not clear whether the Vichy government attached much importance to Carrel's foundation, given that it failed to provide the government with any premises. It was in fact housed by the Rockefeller Foundation in its Paris building at 20 rue de la Baume.[35]

This Fondation Française pour l'Étude des Problèmes Humains existed from 14 January 1942 to 21 August 1944, on which date Carrel was dismissed from his post by the new government (he died on 5 November that year). The present Institut National

32 R. Soupault, *Alexis Carrel*, p. 192.
33 Ibid., pp. 181ff.
34 F.-G. Dreyfus, *Histoire de Vichy* (Paris: Perrin, 1990), p. 24.
35 R. Soupault, *Alexis Carrel*, pp. 238–9.

d'Études Démographiques is its legal successor (by a decree of 24 October 1945, with A. Sauvy replacing Carrel).[36] The only general study of eugenics conducted in the immediate post-war period, moreover, was published by this institute in 1950. It was the work of Jean Sutter, as cited above, who was one of Carrel's last colleagues.

The statutes of this foundation are given in the box on this page. It had neither administrative nor legislative power. Undeniably it was a Pétainist body, but it was not involved in any eugenic crimes.

Official Texts Concerning the Fondation Française pour l'Étude des Problèmes Humains, Cited in Robert Soupault, *Alexis Carrel 1873–1944* (Paris: Librairie Plon, 1952), pp. 235–8

Law of 17 November 1941 establishing the Fondation Française pour l'Étude des Problèmes Humains (*Journal Officiel, 5 December 1941*)

We, Marshal of France, head of the État Français, in the presence of the Council of Ministers, decree:

Article 1. – A public body of the state is established, with legal status and financial autonomy, under the name of the Fondation Française pour l'Étude des Problèmes Humains. This body, with its seat in Paris, is directed by a regent.
Article 2. – Dr Alexis Carrel is appointed to fulfil the functions of the regent of the Foundation.
Article 3. – The Foundation will receive, by way of initial endowment, an extraordinary grant, its amount to be fixed by a subsequent law which will also approve the statutes of the body.
A grant to cover the running expenses of the Foundation will be included each year in the budget of the Secretariat of State for Family and Health.
Article 4. – The present decree will be published in the Journal Officiel and put into practice as a law of the state.

Vichy, 17 November 1941

Marshal of France, head of the État Français

36 F.-G. Dreyfus, *Histoire de Vichy*, p. 221.

P. Pétain
Admiral of the Fleet, Vice-President of the Council of Ministers
A. Darlan
Secretary of State for the Interior
Pierre Pucheu
Secretary of State for Family and Health
Serge Huard
Secretary of State for National Economy and Finance
Yves Bouthillier

[This law was later supplemented by the law of 14 January 1942, approving the foundation's statutes (as below) and setting its initial endowment at forty million francs (*Journal Officiel*, 1 February 1942).]

Statutes of the Fondation Française pour l'Étude des Problèmes Humains

Article 1. – The Fondation Française pour l'Étude des Problèmes Humains has as its object the study, in all its aspects, of the measures most suited to safeguarding, improving and developing the French population in all its activities.

It is particularly charged with carrying out investigations both in France and abroad, establishing statistics, constituting a documentation on all human problems, equipping laboratories, seeking all practical solutions and proceeding to all demonstrations with a view to improving the physiological, mental and social state of the population.

Article 2. – For this purpose:

First, it may receive public and private funding, gifts and legacies in money or in kind.

Second, it will assure the financing of investigations, scientific or practical studies, and centres of demonstration, as needed for the accomplishment of its mission.

Third, it may make grants in money or in kind, and loans without interest, to various associations and works of social utility whose aim is in conformity with its own, as well as to individuals whose activity runs parallel with the realization of its object. It will offer encouragements and rewards.

Fourth, it may in each country draw upon the assistance of delegates, correspondents, and information centres.

[Various articles follow concerning the functions of the regent, supervisory committee and so on.]

The case of mental patients who died of hunger in French asylums has sometimes been raised against Carrel. It is certainly true that there were deaths of this kind; there are even photos in which the degree of malnutrition is reminiscent of that seen among prisoners in German concentration camps. Nevertheless, if we believe the doctoral thesis of Max Lafont (which, although certainly not perfect, is the only study of this question to date),[37] these patients, a total of some thirty or forty thousand,[38] were not victims of any systematic policy of extermination decided on by the Vichy government (as against that which emptied German mental hospitals on the orders of the Nazi government). Their deaths had two main causes. On the one hand, the general food rationing always had even more severe consequences in closed environments, where insufficient rations could not be supplemented by the black market or other informal arrangements. On the other hand, there was a lack of care and a general indifference on the part of political leaders, the administration and the hospital staff – to which one might add the possible redirection of food.[39] Mortality varied from one asylum to another, by size (the larger being more affected than the smaller), situation (in town or country,

37 M. Lafont, *L'Extermination douce, La mort de 40 000 malades mentaux dans les hôpitaux psychiatriques en France sous le régime de Vichy* (Ligné: Éditions de l'AREFPPI, 1987), pp. 165–7.

38 Lafont writes that these are the figures generally put forward by various writers (ibid., pp. 60–1), but he does not make clear which authors these are. According to him, the figures need to be verified. He says that he attempted to do this, in so far as was possible given the archives made available to him, but that he never obtained a reply from the competent authorities. An anonymous article in his bibliography, 'L'extermination des maladies mentaux sous le régime national-socialiste' (La Raison, 1951, 2, pp. 15–45) offers the figure of 40,000 victims in French psychiatric hospitals. This article is also unclear about its sources; it seems to be based on a letter from a certain Dr Poitrot (perhaps Robert Poltrot?), who also wrote on the subject of the extermination of the mentally ill in Germany. At all events, this article states: 'In fact, certain psychiatric hospitals had nothing to envy the Nazi concentration camps. 40,000 French mental patients died of hunger and cold in France during the occupation. The annual rate of mortality reached 47 per cent of the population in one establishment; rates above 30 per cent were not uncommon, especially in the Paris region.' This is likely the origin of the figure of 40,000 that is often given, though no one seems to know exactly on which studies this was based.

39 According to P. Weindling, during the First World War malnutrition led to many deaths in German psychiatric establishments (up to a third of their patients dying due to rationing and the diversion of food to 'useful mouths'). B. Massin, in the same volume, gives the figure of 45,000 deaths from hunger in Prussia alone at this time (P. Weindling, *L'Hygiène de la race*, vol. 1, p. 222).

according to region and the zone of occupation), and according to the abilities of those in charge to make ends meet.

It is by no means impossible that this incompetence corresponded to an unavowed policy decision, though to my knowledge this has never been established. Apparently, however, there is no sign here of any involvement on Carrel's part; he seems simply to have been used as a handy target to mask a far more widespread and indefinite responsibility. It would certainly be more justifiable to question the way in which the administration and the doctors in charge treated the mentally ill before the war as well as during it.[40] Deeper studies would be needed to settle the matter more precisely, but historians of the Second World War and the Vichy regime have not exactly trampled each other in their rush to study this kind of question, the death rate of mental patients being the least of their concerns – just as it was for the administration of the time.

The Soviet Union

The case of the Soviet Union is particularly interesting, as it clearly shows that eugenics was by no means exclusively linked with far-right politics, as is often believed. It is true that the Soviet Union had no eugenic legislation, and also that during the Lysenko era it stood, along with the Catholic Church, as the only institutional opposition to this doctrine. In this regard we can repeat the conclusion of Sutter (citing A. Sauvy), who wrote that 'Marxists are thus opposed to the capitalists who try to plan men, though they refuse to plan the production of goods'.[41]

This does not mean that the Soviet Union did not have a eugenic movement or that there were not Communist eugenicists. The history of eugenics, however, is rather complicated in this case, and quite closely linked with the question of Lysenko and his doctrine. I shall try to summarize its broad lines, following two of the rare studies of this question.[42]

During the tsarist period, Russia more or less followed Western movements in this field. In the late nineteenth century, it produced

40 On the subject of psychiatry between the wars, and the way in which mental patients were treated, a good source is the reportage of Albert Londres: 'Chez les fous' (1925), in *Oeuvres complètes* (Paris: Arléa, 1992), pp. 185–243.

41 J. Sutter, *L'Eugénique, problèmes, méthodes, résultats*, p. 245.

42 L. R. Graham, 'Science and values: the eugenics movement in Germany and Russia in the 1920s', *The American Historical Review*, 1977, 82, pp. 1133–64; M. B. Adams, 'Eugenics in Russia', op. cit., pp. 153–216.

the same kinds of works on degeneration as were produced in Western Europe. An example would be *Usovershenstvovani i vyrozhdenie cheloveschеskago roda* ('The Improvement and Degeneration of the Human Species') by F. V. M. and V. N. Florinskii (St Petersburg, 1866). There had also been translations of Western authors, for example in 1913 Charles Davenport's *Heredity in Relation to Eugenics* (1911). But the eugenics movement in the full sense got under way only after the 1917 revolution and the civil war.

Two biologists were at the origin of Soviet eugenics, Nikolai Konstantinovich Koltsov (1872–1940) and Yuri Aleksandrovich Filipchenko (1882–1930), both of whom had worked in Western Europe before the 1917 revolution – in Naples and Munich respectively. In the summer of 1920, Koltsov established a eugenics section at the biology institute he directed in Moscow, followed, in the autumn of the same year, by the founding of the Russian Eugenics Society, which published the *Russkii Evgenitchesky Journal*. In February 1921, Filipchenko and Koltsov jointly established a eugenics bureau in Petrograd, under the aegis of the Russian Academy of Sciences, that would publish its own bulletin. (Theodosius Dobzhansky, who after his emigration to the United States became a founding father of the new synthesis, was a student of Filipchenko's at this time, and began his career at this eugenics bureau.)[43]

The discourse underlying the foundation of these eugenic organizations was more or less the same as that prevalent in the West at this time. Research programmes were generally in the same style: the study of genealogies (especially those of 'geniuses'); the study of hereditary diseases and malformations, or at least those supposed to be hereditary; the study of mental and behavioural disturbances; and demographic studies. And the same kind of people could be found in both cases: biologists (especially geneticists), psychiatrists and psychologists, anthropologists and demographers. The only difference is that there was a certain attempt made to accommodate eugenic discourse to the Marxist discourse prevalent in Soviet Russia, with a view to producing a 'Bolshevik eugenics'.

The claims of the eugenicists were also similar to those met with in the West. In 1923 one of Koltsov's disciples, M. V. Volotsoi (whom Adams terms a 'Marxist activist') proposed

43 L. R. Graham, ibid., p. 1145.

a sterilization programme comparable to that undertaken in the United States. Filipchenko opposed this on both moral and scientific grounds. One reason this programme met with rejection on all sides is that the Soviet Union suffered from a severely depressed birth rate, and there could be no question of rushing into sterilizations.

This was followed by an attempt at a Lamarckian eugenics, viewed as more compatible with Marxism. But this attempt, too, was soon abandoned – first, because it did not run along the same lines as the genetics of the time (in 1927, Muller discovered the mutagenic effect of X-rays, i.e., an explanation of variations that had nothing to do with the inheritance of acquired characteristics), and second, because Filipchenko untied the link between Marxism and Lamarck by maintaining that, if there was an inheritance of characteristics acquired under the influence of the environment, then the proletariat would not be the vanguard of society but instead a biologically inferior class, made inferior by the wretched conditions in which it had lived for generations. Lysenko's doctrine was not yet on the agenda in the USSR, but it was at this time, and in reaction to eugenics, that Mendelian genetics began to be dismissed as 'bourgeois', and a certain 'Lamarckian' tendency became prominent.[44] At all events, eugenic ideas tended to decline. At the Academy of Sciences, the *Bulletin of the Bureau of Eugenics* became in 1925 the *Bulletin of the Bureau of Genetics and Eugenics*, and in 1928 simply the *Bulletin of the Bureau of Genetics*.

In 1929 Aleksandr Sergeivich Serebrovsky (another 'Bolshevik activist', and a protégé of Koltsov's) proposed the solution of a positive eugenics: artificially inseminating women chosen for their qualities with the sperm of men similarly chosen. The solution seemed more in harmony with the Marxist spirit than negative eugenics by sterilization:

44 I use the term 'Lamarckian' here in the common meaning of 'inheritance of acquired characteristics', but we should remember that this notion was as much Darwin's as Lamarck's – even more so, since Darwin, as opposed to Lamarck, proposed a theory of it. It was only Weismann, well after Darwin's death, who rejected the inheritance of acquired characteristics. Contrary to a common opinion, Lysenko's doctrine did not oppose Lamarck to Darwin, but rather Darwin to Weismann and Morgan. Lysenko himself, moreover, wrote that he did not see himself in the line of Lamarck, but rather that of Darwin (T. Lysenko, Agrobiologie [Moscow: Éditions en Langues Étrangères, 1953]), pp. 178, 314, 534–5.

> [W]ith the current present state of artificial insemination technology (now widely used in horse and cattle breeding), one talented and valuable producer could have up to 1,000 children ... In these conditions, human selection would make gigantic leaps forward. And various women and whole communes would then be proud ... of their success and achievements in this undoubtedly most astonishing field – the production of new forms of human beings.[45]

This project, which would be taken up by Muller a few years later but which already went back to Vacher de Lapouge,[46] remained a dead letter, since in 1930 eugenics was declared a 'bourgeois deviation' of genetics. The eugenics association, its journal and its institutes were all closed. Stalin had by then taken full power, and eugenics evidently did not tally with his ideas.

After some upheaval, the leading biologists found new positions and continued their careers, with the exception of Filipchenko, who died suddenly of meningitis in May 1930 and was replaced in his laboratory by the geneticist Nikolai Ivanovich Vavilov (1887–1943).

Gradually a network formed that brought together the senior Russian eugenicists around an institute devoted to what was now known as 'medical genetics'. The geneticist Solomon Grigorevich Levit (1884–1938?) had profited in 1930–1 from an award from the Rockefeller Foundation, which enabled him,

---

45 A. S. Serebrovsky, 'Antropogenetika i evgenika v sotsialistischekom obshchestvye', 1929, cited in M. B. Adams (ed.), *The Wellborn Science*, p. 181.

46 This is what Vacher de Lapouge, always the caricature prophet, wrote in 1896: 'This delay [in the improvement of humanity] could be considerably reduced by the employment of artificial fertilization. This would mean the substitution of animal and spontaneous reproduction by zootechnical and scientific reproduction, and the definitive dissociation of three things that are already in the process of separation: love, sexual pleasure, and fertilization. By operating in determined conditions, a very small number of male individuals of absolute perfection would be sufficient to fertilize all women worthy of continuing the race, and the generation thus produced would have a value proportionate to the most rigorous choice of male reproducers. Sperm, in fact, can be diluted in various alkaline liquids without losing its properties. A one-thousandth part dilution in an appropriate vehicle remains effective with a dose of two cubic centimetres injected into the uterus. With Minerva replacing Eros, one single reproducer in a good state of health would thus suffice to ensure two hundred thousand annual births. The sperm can also be transported; in one of the experiments that Darwin recommends, I obtained a fertilization in Montpellier with sperm sent from Béziers by post, and repeated this without the protection of an incubator' (*Les Sélections sociales*, pp. 472–3).

along with Izrail Iosifovich Agol, to go and work in Muller's laboratory in Texas. On his return to Moscow in 1931, Levit established an institute for the study of biological, pathological and psychological questions from a genetic point of view – very broadly, the same focus as eugenic studies had had prior to 1930. It was to this institute, renamed in 1935 the Maxim Gorky Medical Genetics Institute, that the Russian eugenicists gravitated. Only the name had changed – and the concern, to conform more closely with Bolshevik ideology.

In 1933, after witnessing Hitler's rise to power in Germany, Muller came to work in the USSR at Vavilov's laboratory; he was also in contact with Koltsov and Levit. In 1935, while still living in the Soviet Union, he published in the United States his book *Out of the Night*, a kind of treatise on political biology designed for the USSR and, more broadly, for socialism.[47] Its final chapter includes a programme of positive eugenics perhaps inspired by that of Serebrovsky – the programme which had been condemned by Stalin in 1930.

In May 1936, Muller sent his book to Stalin, along with a letter in which he wrote:

> [I]t is quite possible, by means of the technique of artificial insemination, which has been developed in this country, to use for such purposes the reproductive material of the most transcendently superior individuals, of the one in 50,000, or one in 100,000, since this technique makes possible a multiplication of more than 50,000 times.
>
> A very considerable step can be made even within a single generation. And the character of this step would in fact begin to be evident after only a few years, for by that time many children have already developed enough to be distinctly recognizable as backward or advanced. After twenty years, there should already be very noteworthy results accruing to the benefit of the nation. And if at that time capitalism still exists beyond our borders, this vital wealth in our youthful cadres, already strong through social and environmental means, but then supplemented even by the means of genetics, could not fail to be of very considerable advantage for our side ... We hope that you will wish to take this view under favourable consideration and will eventually find it feasible to have it put, in some measure at least, to a preliminary test of practice.[48]

47 H. J. Muller, *Out of the Night*.

48 H. J. Muller, letter to Stalin, 1936, in Muller Papers, Lilly Library. Cited in M. B. Adams (ed.), *The Wellborn Science*, p. 195.

At this time, therefore, just five months after Himmler had set up the *Lebensborn*, similarly based on positive eugenics, Muller proposed that Stalin should put into practice Serebrovsky's programme. With a view to this, Muller linked genetics with eugenics, thus drawing Stalin's attention to the geneticists, especially the group gravitating around Levit's 'medical genetics'.

Muller was also one of the organizers of the Seventh International Congress of Genetics, slated to be held in Moscow in August 1937. The Soviet authorities wanted to avoid the mention of human questions at this congress, and especially eugenics, so as not to offer a platform for Fascist biologists. The situation was all the more tense because a group of rather critical American geneticists had written to Levit, who served as general secretary for the congress, asking him for a section to be reserved for discussion of the scientific foundations of racist theories and eugenics. The Nazis, for their part, had countered by organizing a boycott of the congress.[49]

In November 1936, Levit was accused of championing Nazi doctrines. The next month, despite the warnings of Vavilov and Serebrovsky, Muller openly attacked the supporters of Lysenko at a session of the Academy of Agricultural Sciences, just after it was announced that Agol had been arrested as an 'enemy of the people'. In spring 1937, Vavilov succeeded in getting Muller to leave for Spain (through the International Brigades). Agol was killed the following day. The international congress was cancelled.[50] Levit was arrested in January 1938, and probably murdered in May of the same year; Koltsov died of a heart attack in 1940 (his wife committed suicide the next day); Vavilov was arrested in 1940 and died three years later in a camp. And this is only what happened to the people mentioned above; a number of other geneticists of the same school were also arrested and in some cases killed.

49 S. Kühl, *The Nazi Connection*, p. 78.

50 This congress was eventually held in Edinburgh in 1939. A manifesto published on that occasion, the *Genetico Manifesto*, condemned racism but accepted eugenics. The text of the manifesto was published in the *Journal of Heredity*, 1939, 30, pp. 371–3. Its initial signatories were F. A. E. Crew, J. B. S. Haldane, S. C. Harland, L. T. Hogben, J. S. Huxley, H. J. Muller and J. Needham. These were subsequently joined by G. P. Child, P. R. David, G. Dahlberg. T. Dobzhansky, R. A. Emerson, C. Gordon, J. Hammond, C. L. Huskins, W. Landauer, H. H. Plough, E. Price, J. Schultz, A. G. Steinberg and C. H. Waddington.

Lysenko triumphed in the Soviet Union against Western genetics, which was decisively associated with eugenics, racism and Nazism. As for Muller, he received the Nobel Prize in 1946, and in the early 1960s, with the backing of the billionaire. Graham, relaunched the programme that Stalin had rejected – this time in the form of the Repository for Germinal Choice, a sperm bank for the emissions of great men, which was finally realized only after his death.

Biologists and certain historians are in the habit of seeing Lysenkoism as a kind of ideological deviation in science. They are not wrong in this, and yet they should understand the full story, and be clear that Western genetics, for its part, was in no way spotless and transparent. It is understandable in any case why, on the subject of Lysenko, they generally avoid speaking of eugenics and Muller's campaigns. We should add that it is very likely that the organization of an international genetics congress in Moscow in 1937 had the object of bringing together a large number of geneticists to maintain, in chorus, and to be heard by Stalin, that Lysenko's genetics was worthless and Muller's correct.

It is customary to accuse Lysenko of three things. First of all, that he sent various geneticists to the Gulag, particularly Vavilov. Second, that he ruined Russian agriculture with his agronomic principles. And third, that he considerably delayed the development of modern genetics in the USSR. The first charge is certainly well founded: Lysenko did indeed play a role in the arrest of Russian geneticists, including Vavilov. The second is much exaggerated: Stalin's policy was already sufficient to ruin Russian agriculture, and Lysenko's agronomics did not make much of a contribution to this. As for the backwardness of Russian genetics, no one is really interested in this today, including those who blame Lysenko for it. We should note, however, that in the late 1930s the paradigm of Morganian genetics was nearly exhausted, no longer productive of new findings, and that it was only after the war that molecular approaches relaunched genetics in the West, on a completely different path than that followed by Morgan, who had neglected the physico-chemical dimension of heredity. It was at this point that Lysenkoism certainly acted as a brake.

That said, it is true that Lysenko was an aberration from the scientific point of view, that he was a forger and a villain. Muller was undoubtedly an incomparably greater scientist, but in human terms he was scarcely any better – a dubious character in several respects. He may well have criticized the negative eugenics prevalent in the United States and Germany, but his programme of political biology was scarcely better than that of a raving Nazi geneticist. Among other fantasies, he proposed that animals might be used as 'birth mothers' for human foetuses, a kind of human parthenogenesis; it is understandable that this shocked a former seminarian such as Stalin.[51]

History is not Manichaean, or is only so after the event, and rarely consists in a pure confrontation between good and evil. More often it is one villain against another, and the common people pay the price. If Lysenko, whatever his motives, spared the Russian people a eugenic policy in addition to Stalinism, we can be glad of his victory over Muller. Never mind the champions of political correctness and scientific progress.

51 We should not think that Muller was exceptional in this respect; there were many strange theories going round. Here is one taken from the inexhaustible Vacher de Lapouge: 'Mme Clémence Royer [the translator of Darwin] and M. Meunier have tried to resolve the formidable problem of manual labour in the perfect society by the domestication of anthropoid apes. To my mind, this does not seem very serious ... A very long-term process of selection would be needed even to realize this programme imperfectly. As for hybridization, I do not see it as impossible a priori to obtain hybrids, both between two species of anthropoid ape and even with man. The gap is less than between the macaques on the one hand, the baboons and guenons on the other, and these monkeys of different families have frequently produced hybrids. With artificial fertilization one could manage to create semi-humans, but it would perhaps be premature to count on the use of hypothetical hybrids of which no example has yet been produced' (G. Vacher de Lapouge, *Les Sélections sociales*, pp. 483–4).

10

German Eugenics Before and Under Nazism

We turn finally to Germany. Before 1933 and Hitler's coming to power, the situation there was not essentially different from that in other countries. The usual concerns about the degeneration of the human race were being sung in chorus in their different variants by biologists, anthropologists, demographers and sociologists, as well as by ideologists of all political colours.

Industrialists were also present, more or less philanthropic, more or less preoccupied with social questions, and seeking to promote eugenics, or more broadly the application of biological principles to society. Of particular note was Friedrich A. Krupp, son of the industrialist and arms manufacturer Alfred Krupp, and himself an amateur biologist, who organized in 1900, on Haeckel's suggestion, a competition on the theme: 'What can the theory of biological evolution teach us about domestic political development and state legislation?' The results were published in 1903, the winner being Wilhelm Schallmayer (1857–1919), for his *Vererbung und Auslese im Lebenslauf der Völker* ('Heredity and Selection in the Life Process of Peoples'). The title speaks for itself, an application of Darwinian theory to society.[1]

It was this competition that launched the fashion for eugenics in Germany. In 1904, it was followed by the establishment, by Alfred Ploetz, of the periodical *Archiv für Rassen- und Gesellschaftsbiologie* (Archive for Racial and Social Biology) and a year later, again by Ploetz, of the *Deutsche Gesellschaft für Rassenhygiene* (German Society for Racial Hygiene). This was the first German eugenics association.

'Racial hygiene' was the more common phrase in German, rather than 'eugenics', but this did not necessarily have a more

1 A presentation with comments on the main entries in this competition is given in G. Vacher de Lapouge, *Race et milieu social*, pp. 309–24.

pronounced racist connotation. As in other countries, certain eugenicists were more or less racist, more or less champions of theories of the 'Nordic race', more or less anti-Semitic (Ploetz, Lenz, Fischer), while other were not, or were only slightly so (Schallmayer, Ostermann, Muckermann).

The main scientific centres of German eugenics were Berlin and Munich. The Dahlem suburb of Berlin was the seat of the Kaiser-Wilhelm-Institut für Anthropologie, menschliche Erblehre und Eugenik (Kaiser Wilhelm Institute for Anthropology, Human Genetics and Eugenics),[2] which hosted, simultaneously or successively, a large number of well-known geneticists. Its governing body was headed by Eugen Fischer (who also directed its anthropology department), also including Alfred Grotjahn (who died 1931) and Erwin Baur (died 1933). Other major figures here included Otmar von Verschuer, who headed the department of human heredity (he subsequently left this post, and was replaced by Fischer and Lenz, before returning to direct the institute when Fischer retired in 1942); the former Jesuit Herman Muckermann, responsible for the eugenics department and dismissed by the Nazis in 1933 when he was replaced by Fritz Lenz; Karl Diehl, whose research focused on tuberculosis; Wolfgang Abel, in charge of racial studies; Hans Nachtsheim, responsible for experimental studies of genetic pathology; and Wouter Ströer for embryology. This Berlin centre received funding from the Rockefeller Foundation, and Davenport, as president of the International Federation of Eugenic Organizations, attended the inauguration of the new building in 1927.[3] Likewise in Berlin, and prior to the creation of this institute, there was the Kaiser Wilhelm Institute for Biology, which also housed eugenic research. This institute was headed by Richard Goldschmidt, who was Jewish; dismissed in 1933 by the Nazis, he emigrated to the United States.

The other major scientific centre of German eugenics, in Munich, was the Kaiser Wilhelm Institut für Genealogie und Demographie der Deutschen Forschungsanstalt für Psychiatrie (Kaiser-Wilhelm-Institute for Genealogy and Demography of the German Institute for Research in Psychiatry). The Munich institute was founded

2 P. Weindling, 'Weimar eugenics: The Kaiser Wilhelm Institute for Anthropology, Human Heredity and Eugenics in social context', *Annals of Science*, 1985, 42, pp. 303–18. H. Friedlander, *The Origins of Nazi Genocide, from Euthanasia to the Final Solution* (Chapel Hill: University of North Carolina Press, 1995), pp. 13–14.

3 S. Kühl, *The Nazi Connection*, pp. 20–1.

in 1918 by Emil Kraepelin (1856–1926), with the support of the Rockefeller Foundation. After 1924 it was sponsored by the Kaiser-Wilhelm-Gesellschaft, the main German scientific organization (somewhat equivalent to the Centre National de la Recherche Scientifique today in France). From 1931 the institute's department of genealogy and demography was headed by Ernst Rüdin, and its research focused in particular on the genetic and neurological bases of mental disease and criminality. The institute received substantial backing from the Rockefeller Foundation for the construction of buildings as well as various research grants.[4]

After institutions, men. Here is a small bouquet of statements by prominent German scientists of the time, culled from a 1934 doctoral thesis by Jean Giraud on the Nazi eugenic legislation:

- In September 1932 (when the Nazis were not yet in power), Arthur Ostermann wrote: 'Science will now be able to establish with such confidence the hereditary prognosis in particular cases that it will be possible for doctors to take measures to prevent the unfit having progeny'.
- In July 1933 (i.e., after the Nazis took power), Eugen Fischer wrote: 'Today we know enough about human heredity, and with sufficient certainty, to be able to take the most important decisions for the future bearing on particular individuals, families, and lineages – with full awareness and responsibility', and he demanded that a detailed inventory be made of all hereditary defects existing in Germany, something that he was already in the process of doing with funding from the Rockefeller Foundation.
- In August 1933, Otmar von Verschuer wrote: 'The scientific foundations are already sufficiently assured for the application of practical eugenic measures'. Rüdin was to approve this a month later.[5]

We are very far here from the doubts that other geneticists might have had as to the effectiveness and propriety of eugenic measures. The majority of these scientists did indeed have Nazi sympathies. Indeed, even in the thesis from which these quotes

4 P. Weindling, 'The Rockefeller Foundation and German biomedical sciences, 1920–1940: from educational philanthropy to international science policy', in N. A. Rupke (ed.), *Science, Politics and the Public Good: Essays in Honour of Margaret Gowing* (London: Macmillan 1988), pp. 119–40; S. Kühl, *The Nazi Connection*, p. 20.

5 'J. Girard,' *Considérations sur le loi eugénique allemande de 14 juillet 1933*, thesis in medicine (Strasbourg, 1934), pp. 22–25, 28.

are taken, though Girard made certain criticisms of the German eugenic legislation, he could himself write:

> Whatever opinion we might have, we should not lose sight of the fact that the German medical corporation only decided to act in presence of a serious danger for the race; more or less throughout the literature, statistics are found that demonstrate the significant and continuous growth in the number of abnormals in relation to healthy individuals. German writers have made a particular study of the comparative fertility of defective families and healthy families ... all their findings concur: in these last few years, the defective families showed a markedly higher fertility than those with a healthy heredity.[6]

This is the argument previously put forward by J. Rostand (quotation on p. 135). It is necessary merely to refer to a number of problematic families, and the degeneration of the human race can be demonstrated.

There were movements in Germany for eugenic legislation long before the Nazis took power, indeed from as far back as 1907, when legislation of this kind was adopted in Indiana. In 1923–4, at the instigation of a doctor inspired by the American experience, Gerhard Boeters, the Saxon parliament drafted a law providing for the sterilization of individuals affected with dementia praecox, severe forms of manic depression, essential epilepsy, hereditary alcoholism with mental disturbance, congenital feeble-mindedness and Huntington's chorea. Although this project initially seems to have received a good deal of support, the law was never put into effect, as a study on the question found that sterilization was not actually practised on a wide scale in the United States. Only in the 1930s did it really take off.[7]

In 1927, the Social Democrats in the Prussian parliament called for a new study on the question of sterilization.[8] A legislative proposal was almost passed in Prussia in July 1932. Among the spokesmen and experts on this project were Muckermann, the above-mentioned former Jesuit who was dismissed by the Nazis, and the geneticist Richard Goldschmidt, soon forced to emigrate to the United States on account of his Jewish origin. The draft law never came into force, because when the Nazis

---

6 Ibid., p. 58.

7 Ibid., p. 21. See also S. Kühl, *The Nazi Connection*, pp. 23–4.

8 S. Kühl, *The Nazi Connection*, p. 24.

took power in 1933, they promulgated their own law, directly inspired by the Prussian draft but with the addition of a paragraph on alcoholism (see below).

This 'Nazi' law was enacted on 14 July 1933, and came into force on 1 January 1934. Here is its first article, which defined which individuals were subject to sterilization; subsequent articles specify how such sterilization is to be requested, decided and carried out:

> 1. Any person affected by a hereditary disease may be sterilized by way of a surgical operation, if, according to the experience of medical science, it is believed with high probability that the person's descendants will be affected by severe hereditary disorders, either mental or physical.
>
> Persons suffering from the following diseases are considered as affected with a hereditary disease in the sense of this law:
>
> 1) Congenital feeble-mindedness
> 2) Schizophrenia
> 3) Bipolar disorder [manic depression]
> 4) Hereditary epilepsy
> 5) Hereditary St-Guy's dance [Huntington's chorea]
> 6) Hereditary blindness
> 7) Hereditary deafness
> 8) Severe hereditary bodily malformations.
>
> Persons subject to severe crises of alcoholism may also be sterilized.[9]

This law of 14 July 1933 is taken as a Nazi law, yet it was based on existing practice in the United States. Laughlin was soon to be appointed *doctor honoris causa* at Heidelberg University, and welcomed the convergence of views between the United States and Germany on this subject.[10] It was California in particular that served as a model, and there were many connections between Californian and German eugenicists.

Moreover, before 1933 the American model had already been exported to Switzerland, Denmark and Canada. Weimar Germany, as we have just seen, already saw such legislation being

---

9 J. Girard, *Considérations sur le loi eugénique allemande*, p. 33. The text of the law is also given in Y. Ternon and S. Helman, *Les Médecins allemands et le National-Socialisme*, pp. 160ff.

10 Laughlin had already published in German in 1929 an article on legal eugenic sterilization in the USA: H. H. Laughlin, 'Die Entwicklung der gesetzlichen rassenhygienischen Sterilisierung in den Vereinigten Staaten', Archiv für Rassen- und Gesellschaftsbiologie, 1929, 21, pp. 253–62.

proposed. A number of other democracies followed, in particular the Scandinavian countries. It is probable, therefore, that even without the Nazis, Germany would at some point or other have adopted and put into effect legislation of this kind. Besides, it was only the Catholic Church that made any institutional protest, particularly in the person of the bishop of Münster, Clemens August Graf von Galen – whom we shall meet again later on, and who condemned eugenic sterilization in a pastoral declaration of 29 January 1934.[11]

It is notable, in fact, that the German law was less harsh than the majority of its American counterparts, bearing only on hereditary diseases (or those that were supposedly such), and not at all on criminality; this was so that sterilization, decided on by a special tribunal and based on medical advice, should not appear as a punishment but rather as a medical measure.[12] Besides, contrary to some of the American legislation, it only authorized sterilization (vasectomy or salpingectomy), and not castration. We should note, however, that a different law, that of 24 November 1933, did permit the castration of sexual offenders; by 1936, there had already been 1,116 castrations of this kind.[13] The eugenic sterilization of the diseased (or those supposedly such), however, remained quite distinct from the sterilization or castration of criminals, contrary to the American situation. On this subject, there seems to have been a kind of reciprocal fascination between German and American eugenicists; as of 1907, the former had envied the latter for managing to obtain legislation on sterilization, while after 1933, it was the latter who envied the former, because the German legislation was clearly conceived and applied on a more properly medical basis than the American.

We may well consider that, even with these nuances, this kind of law was indeed typically Nazi, in so far as Nazism was marked by its pretension to base politics on biology – this being the way in which it was different from other totalitarianisms such as Italian fascism. In this case, however, the equivalent laws in the United States, Switzerland, Canada, Denmark, Sweden, etc. would have all to be categorized as Nazi laws, with German Nazism being distinct only in so far as it was 'Hitlerite'. We could then say that,

11 Y. Ternon and S. Helman, *Les Médecins allemands et le National-Socialisme*, p. 183.

12 J. Girard, *Considérations sur le loi eugénique allemande*, p. 61.

13 Y. Ternon and S. Helman, *Les Médecins allemands et le National-Socialisme*, p. 181.

if the German eugenic legislation was indeed Nazi in its essence, it was not specifically Hitlerite; indeed, in Germany as elsewhere, eugenics had the backing of many Jewish biologists, doctors and intellectuals, a point that we shall return to below.

At its inception, the law of July 1933 did not have clear racist connotations. This was not the case, however, with a further eugenic measure taken by the Nazis even before their seizure of power. This was the marriage code for the SS, which had been developed in 1932 according to the biological ideas of the agronomist Richard Walther Darré, adviser to Himmler and future minister of agriculture.[14] To be allowed to marry, an SS man had to provide certificates establishing that his fiancée's parents had no physical or mental disease, that she was not sterile, and that there were neither Jews nor Slavs in her genealogy as far back as 1750. In this case, the eugenic component was explicitly combined with a racist component, anticipating the racial laws of 1935.

A further Nazi eugenic initiative had a very marked racist aspect, that known as *Lebensborn* (Source of Life). This organization, established by Heinrich Himmler on 12 December 1935, applied a positive eugenics (like that which Muller recommended to Stalin five months later), rather than a negative one. This consisted in organizing the production of 'Aryan' infants on the part of men and women with particularly pronounced characteristics of this kind, as well as the rearing and education of such children who were already born. It had at its disposal eight childbirth facilities and six homes for children who met the racial criteria. It is estimated that a minimum of 92,000 children passed through this system, of whom 80,000 were taken from their families and 12,000 were born to order.[15]

This institution would seem to have been purely German and Nazi. There was nothing comparable in other countries. We should, however, note this passage from Vacher de Lapouge, writing in 1896 on the subject of the American slave states:

> I am not acquainted with any more fruitful attempts than those of certain rearers of black flesh, who set up regular stud farms in the Southern states: this practice goes back to Cato in antiquity, and contributed to the creation

14 Y. Ternon and S. Helman, *Histoire de la médecine SS, ou le mythe du racisme biologique* (Tournai: Casterman, 1969), pp. 45–6.

15 Ibid., p. 47. M. Hillel, *Au nom de la race* (Paris: Fayard, 1975).

of the superb creole Negro race. By comparison with the African Negro, that in the United States is certainly a selected being.[16]

The further Nazi eugenic legislation that remains to be mentioned is the law of 18 October 1935, which supplemented the sterilization law of 14 July 1933 by prohibitions on marriage. Here is its first article:

> I. – No marriage can be concluded:
> a) When one of the parties suffers from a contagious disease, leading to the fear of significant damage to the health of the spouse or the descendants;
> b) When one of the parties is prohibited from marriage or under temporary wardship;
> c) When one of the parties, without being prohibited from marriage, suffers from a mental disorder, which leads the marriage to be undesirable for the racial community;
> d) When one of the parties suffers from a hereditary mental disorder, as defined by the Law for the Prevention of Offspring with Hereditary Diseases [the law of 14 July 1933].
> The intent of section I.d is not opposed to marriage if the other party is sterile.[17]

The last sentence indicates very well the eugenic intent of this law. The contagious diseases specified in section I.a were essentially venereal diseases (syphilis in particular) and tuberculosis.

In Germany as in France, the eugenic movement enjoyed the generosity of the Rockefeller Foundation. If the Société Française d'Eugénique was never very dangerous, and if even a Pétainist such as Carrel did not slip into criminal eugenic practices, what happened in Germany was quite different – and far more embarrassing for the future image of genetics.

As we saw above, the Rockefeller Foundation played a certain role in the establishment and financing of the two major scientific institutes devoted to eugenics, in Berlin and Munich respectively. More specifically, it contributed, from the 1920s on, to financing the work of various German eugenicists. As Paul Julian Weindling writes:

16 G. Vacher de Lapouge, *Les Sélections sociales*, p. 460.
17 Cited in O. von Verschuer, *Manuel d'eugénique et hérédité humaine*, p. 123.

> The Rockefeller Foundation facilitated the extension of German genetics to areas of social interest. Its 'human biology programme' was set up to tackle what those in charge of the Foundation perceived as a social crisis of poverty, criminality, and hereditary diseases. Poll, a geneticist who specialized in human heredity and was active in the League for Regeneration, was one of the eugenicists who benefited from the Rockefeller funds; this was also the case with Bluhm for her research on heredity and alcoholism, Nachtsheim for his Mendelian research, the statistician Siegfried Koller, and Fetscher and Grotjahn for their work on social hygiene, while substantial funds were also allotted to Rüdin and Eugen Fischer for psychiatric and anthropological studies. The Rockefeller Foundation was interested in genetic and neurological research on such mental conditions as crime and mental illness, and in 1925 agreed to allocate $2.5 million to the German Institute of Psychiatry, reputed to be the best centre in Europe for psychiatric research.[18]

The wording here is careful and diplomatic. If we work through the various names cited, all were eugenicists, but they varied greatly in their political shading and attitude towards Nazism.

Agnes Bluhm (1862–1943) was the first woman eugenicist in Germany. She specialized in the question of alcoholic degeneration and was also interested in the decline of breast-feeding.[19] For a while she was involved with Ploetz; I do not know her attitude towards Nazism.

Rainer Fetscher (1895–1945) was a socialist eugenicist. He was killed by the SS in 1945 while trying to make contact with the Red Army.[20]

Siegfried Koller (1908–1998) focused on the statistics of heredity, and was one of the leading German specialists in this field. He seems to have been sympathetic to Nazism.

Alfred Grotjahn (1869–1931) was another socialist eugenicist. He criticized the Aryan ideology, and died before the Nazis came to power.[21]

Heinrich Poll was, according to Weindling, the key person here, as secretary to the committee that the Rockefeller Foundation set

18 P. Weindling, *L'Hygiène de la race*, vol. 1, p. 254.

19 Ibid.

20 P. Weingart, J. Kroll and K. Bayertz, Rasse, Blut und Gene, *Geschichte der Eugenik und Rassenhygiene in Deutschland (Frankfurt: Suhrkamp*, 1988), pp. 538–9.

21 S. F. Weiss, 'The race hygiene movement in Germany, 1904–1945', in M. B. Adams (ed.), *The Wellborn Science*, pp. 9–10.

up to organize the funding it provided to German scientists.[22] He was of Jewish origin, a eugenicist, and had to leave Germany after the Nazis took power, emigrating to Sweden where he later committed suicide.

We shall study the case of Hans Nachtsheim in Part Three. It need only be noted here that he headed the department of experimental genetic pathology at the Kaiser Wilhelm Institute for Anthropology, Human Genetics and Eugenics, in Berlin.

Eugen Fischer and Ernst Rüdin (1874–1952), who also received funds from the Rockefeller Foundation, were unequivocal supporters of Nazism. If they did not yet belong to the Nazi party, they had already shared its ideas for some time. Rüdin, the founder of genetic psychiatry,[23] had long maintained notions concerning the superiority of the Aryan race. He had been involved in the establishment of Ploetz's German Society for Racial Hygiene in 1905. In November 1933, this society was placed under the supervision of the Reichskommissariat for public health, and Rüdin appointed Reichskommissar; he became, in other words, the representative here of Nazi power. Rüdin was one of the experts who drew up the official commentary on the eugenic law of July 1933, though he did not join the Nazi party until 1937.[24] He headed the Munich institute, where he chiefly studied the genetic and neurological foundations of mental illness and criminality.

As for Eugen Fischer, the Rockefeller Foundation contributed financially to an anthropological study he directed, with the aim of determining the hereditary characteristics of the German populations, a study that Rüdin and Verschuer were also involved in, as well as the racial theorist Hans F.K. Günther, whom we shall discuss later. This funding lasted until 1935, well after the Nazis were already in power.[25] Fischer, who headed this project, was one of the nine leading German scientists who took a solemn pledge of allegiance to Hitler on 12 November 1933.[26]

---

22 P. Weindling, 'The Rockefeller Foundation and German biomedical sciences', p. 125.

23 This genetic psychiatry has nothing in common with Piaget's theory that goes under the same name. It was a biologistic and hereditarian interpretation of mental disorders.

24 S. F. Weiss, 'The race hygiene movement in Germany, 1904–1945', in M. B. Adams (ed.), *The Wellborn Science*, pp. 41, 44, 48.

25 P. Weindling, 'The Rockefeller Foundation and German biomedical sciences', pp. 132–3.

26 Y. Ternon and S. Helman, *Les Médecins allemands et le National-Socialisme*, p. 102.

We should also note that Otmar von Verschuer, whose work on twins (heredity of mental characteristics as well as of crime, cancer and tuberculosis – on the last of these, see the quotation from Rostand on pp. 143–4) was financed by the Rockefeller Foundation, though his political views were perfectly clear.[27] Verschuer was one of the leading Nazi eugenicists, the teacher and superior of the infamous Dr Mengele, and he pursued his studies on twins with 'material' that his student sent from Auschwitz. There is testimony from Miklos Nyiszli, a Hungarian Jewish doctor and a forced collaborator in Mengele's studies, on the relations that existed between Auschwitz and the Institute of Anthropology, Human Genetics and Eugenics headed by Verschuer,[28] and this is what Verschuer himself wrote to the director of the Frankfurt paediatric clinic, B. de Rudder, on 4 October 1944:

> We have taken plasma substrata from more than two hundred individuals of different race, from pairs of twins and from a number of families ... The object of our concurrent efforts is no longer to establish that genetic influence plays a major role in certain infectious diseases, but to determine the way that it does so.[29]

Here is a testimony on Mengele's work at Auschwitz, which gives some idea on the way in which 'we' gathered these plasma samples:

> Dr Mengele had selected a dozen or so twins. He wanted to check whether the humoral and organic constitution was identical in each twin pair. It was necessary therefore to obtain readings for all the elements of blood and urine, as well as conducting serological tests. Finally, to conclude his examination, he did not hesitate to kill the children, so as to be able to carry out autopsies and study the question in depth.[30]

We can compare this with the quotations from Rostand and Heissmeyer on pp. 143–4 and 146 above. Verschuer was not a mad scientist, but an acknowledged expert whom Rostand cited without hesitation in 1939. After the war, he continued his career as a specialist in human genetics at various German universities.

27 P. Weindling, *Hygiène de la race*, p. 259.
28 E. Klee, La Médecine nazie et ses victims, p. 350.
29 Ibid., p. 354.
30 A. Lettich, *Trente-quatre mois dans les camps de concentration* (Tours: Imprimerie Corporative, 1946), p. 36 (cited in Y. Ternon and S. Helman, *Histoire de la médicine SS, ou le Mythe du racisme biologique*, p. 19).

Weindling emphasizes that the Rockefeller Foundation's funding of German eugenicists fell sharply after the Nazis came to power, and in 1937 its president insisted that the foundation should not be involved in studies of a racial character. In 1938 it rejected various projects of Rüdin's, in particular on the heredity of homosexuality and of goitre.[31] In June 1939 it took the decision not to make any further major grants in Germany.[32]

Despite these reservations, we can hardly avoid noting that, by supporting 'the extension of German genetics to areas of social interest' (as Weindling puts it), the Rockefeller Foundation had contributed very generously to the financing of what was, or would shortly become, the *fine fleur* of Nazi biology.

Does this mean that the foundation specifically financed Nazism? Certainly not, as eugenic ideas were the common property at this time of all political colours (see the above list of German eugenicists whose research was supported), and major foundations, including the Rockefeller, supported a wide range of projects. It remains true however, that these major foundations supported projects that corresponded to their own political, philosophical and scientific views; there was never any question of their supporting anti-eugenic Catholic movements, or Communists who demanded social reforms rather than measures of political biology. We are forced to remark that eugenic projects were broadly supported by fairly characteristic individuals and groups: Krupp (steel and armaments), Harriman (railways), Carnegie (steel), Rockefeller (petrol), Wickliffe Draper (textiles), to list only the names already encountered. The mildest comment would be that these fairy godparents who watched over the cradle of eugenics made a mistake in their philanthropic aim, and that the Rockefeller Foundation's funding of Nazi biology is at the very least the sign of a certain blindness – comparable with that of the Carnegie Institution, which did not put an end to the eugenic activity at Cold Spring Harbor until 1940, when the laboratory was drifting into becoming a centre of Nazi propaganda.

---

31 The genetics of goitre had already been studied by Davenport, who found two genes involved in it: a dominant one linked to sex, and a dominant autosomal one: C. Davenport, *The Genetical Factor in Endemic Goiter* (Washington: Carnegie Institution, 1932).

32 P. Weindling, 'The Rockefeller Foundation and German biomedical sciences', pp. 134–5.

What would the future of eugenics have been without the millions of dollars thus placed at the service of geneticists who promoted it in the United States and Europe? Without these dollars, what would have been the intellectual credit of these geneticists? Why did Carnegie in the United States, and Rockefeller in Europe, finance projects of this kind, when genetics itself – if not these particular geneticists – maintained that eugenics had no scientific foundation, was ineffective and useless?

In the quotation on p. 182 above, Weindling offers an explanation of the actions of the Rockefeller Foundation in Germany: it had to tackle 'a social crisis of poverty, criminality, and hereditary diseases'. A biologizing interpretation of the German social crisis of the 1920s and 1930s became useful, and an attempt to remedy it by science, more particularly by Darwinian biology. The Rockefeller Foundation thus took up an old theme: the interpretation of social problems in biological terms, and the attempt to resolve them by biological rather than social measures, the whole thing underpinned by a deranged hereditarianism and wrapped up in a broader and vaguer concern – control of human evolution.

In the 1920s, the American psychologist Edwin R. Embree was in charge of European awards at the Rockefeller Foundation. This is what he wrote in 1930, in the introduction to a collective volume whose twenty-eight contributors included, to mention only the most famous, A. Carrel, W. B. Cannon, C. S. Sherrington, C. Davenport, R. A. Millikan and J. Dewey:

> With these tentative findings in our possession in physics, medicine, biology, psychology and the social sciences ... the question arises as to whether it may not now be possible to make another great push forward in human evolution ... It is beside the point to dispute as to the relative importance of inheritance and education, of nature vs. nurture; for any great advance must include attention to both the biological and the social. We must, for instance, find some way to avoid wars if the race is not to destroy itself with its ever-increasing knowledge of physics and chemistry which may be used for mutual benefit or equally for world destruction ... A fundamental question of the future is: Can we to some extent control the direction of the evolution of the race?[33]

33 E. Embtree, in E. V. Cowdrey (ed.), *Human Biology and Racial Welfare* (New York: Hoeber, 1930), pp. ix–x.

And here is what Warren Weaver, a mathematician who headed the natural sciences division at the Rockefeller Foundation from 1932 to 1955, wrote in 1934:

> There is a strong and growing belief, held by many thoughtful scientists – even by many of the ablest specialists in the physical sciences – that the past fifty or one hundred years have seen the supremacy of physics and chemistry, but that hope for the future of mankind depends in a basic way upon the development during the next fifty years of a new biology and a new psychology. As one views the present state of the world, with its terrific tension, its paradoxical confusion of abundance, and its almost uncontrollable mechanical expertness, one is tempted to charge the physical sciences with having helped to produce a situation that man has neither the wits to manage nor the nerves to endure ...
>
> The challenge of this situation is obvious. Can man gain an intelligent control of his own power? Can we develop so sound and extensive a genetics that we can hope to breed, in the future, superior men? Can we obtain enough knowledge of the physiology and psychobiology of sex so that man can bring this pervasive, highly important, and dangerous aspect of life under rational control? Can we unravel the tangled problem of the endocrine glands, and develop, before it is too late, a therapy for the whole hideous range of mental and physical disorders which result from glandular disturbances? Can we solve the mysteries of the various vitamins ... ? Can we release psychology from its present confusion and ineffectiveness and shape it into a tool which every man can use every day? Can man acquire enough knowledge of his own vital processes so that we can hope to rationalize human behaviour? Can we, in short, create a new science of Man?
>
> This point of view has recently been realized by various scientists, philosophers and statesmen; many of the techniques are at hand; but direction, stimulation, support and leadership are for the most part lacking. The foundation has a unique chance to correlate and direct existing forces and to stimulate the creation of new forces for a coherent and strategic attack. The proposed program recognizes here one of the most inspiring opportunities with which science has ever been faced.[34]

To understand the details of this text, it is necessary to understand that the Rockefeller Foundation had developed a programme of

---

34 W. Weaver, 'Progress report, the NS', 14 February 1934, pp. 1–3 (R.F.915.1.7), cited in R. E. Kohler, 'The management of science: the experience of Warren Weaver and the Rockefeller Foundation programme in molecular biology', *Minerva*, vol. 14, 1976, pp. 279–306 (quotation on p. 291).

psychobiology, which undoubtedly explains its financing of Rüdin and his genetic psychiatry, and that hormones and vitamins were the latest thing at this time. Beyond these details, it is clear that the main concern was to resolve the social problems of the 1930s by science, and more specifically by biology and psychology (reduced to neurobiology), disciplines that were seen as the sciences of the future. Added to this was a dose of American Puritanism (the dangerous character of sex, which had to be rationalized), which comes back to what we said previously about American eugenic legislation in relation to sexual offences.

To sum up: in order to resolve the social problems of the 1930s, science had to be utilized, especially biology, in order to rationalize humanity and manage human resources, along the lines of the model that Frederick Winslow Taylor (1856–1915) had developed for the rationalization of industrial work.

All this is familiar scientistic rhetoric, such as we have already encountered in Part One on the naturalization of society by Darwinism. It is also a rhetoric that enjoys a revival today: the idea that biology is the science of the future is voiced every day in the media, with exactly the same constituents as in Weaver – genetics and neurobiology (to which psychology is reduced).

There is a sense in which the work of the German biologists discussed above fitted perfectly into this programme of the Rockefeller Foundation. And to the extent that this work involved reputable scientists rather than mere fantasists, there was no reason why Rockefeller should not fund it. After all, a journal as prestigious as *Nature* published in 1936 an article signed E. W. M. (perhaps E. W. MacBride), which proposed to resolve social problems by way of compulsory sterilization, with a view to punishing people who appealed to state aid for raising their children.[35]

Warren Weaver, during his tenure at the Rockefeller Foundation, reoriented it towards molecular biology, a discipline he himself helped to create, even inventing the term. When molecular biology really did take off in the 1950s, it relegated phenomenalist and mathematical approaches to genetics into the background – and eugenic doctrines along with them (something that Weaver did not envisage). Weaver was also the co-author, together with Claude Shannon, of the mathematical theory of

35 E. W. M., 'Cultivation of the unfit', *Nature*, 1936, vol. 137, 3454, pp. 44–5.

communication, which led to computing. In other words, though little known to the public at large, he played a major role in the origins of what has come to be known as scientific modernity. His opinion on these questions is therefore interesting, and still crops up today in many places. But we must first return to German eugenics.

A particular point is striking in the German case in a way that it is not in other countries: the presence of Jewish biologists and doctors among other eugenicists, sometimes even in leading positions (which is all that we discuss here). But this is surprising only to the extent that we have got into the misguided habit of considering eugenics as a Nazi speciality, and more particularly as an extension of the extermination of Jews and certain other categories of people. In reality, there were a number of Jewish eugenicists in all countries, as we have already noted: Muller was a German Jew by origin. In the USSR, Levit was a Baltic Jew, and Agol, given his first name of Izrail, was also most likely of Jewish extraction. In Germany, Weinberg, Goldschmidt and Poll, to cite only names already mentioned, were likewise of Jewish extraction.

These were by no means isolated or exceptional cases. According to B. Massin,[36] a fairly large number of German-Jewish writers championed eugenics and wrote in such publications as Ploetz's *Archiv für Rassen- und Gesellschaftsbiologie* or Woltmann's *Politisch-Anthropologisch Revue*. On the one hand, eugenicists were recruited primarily in the biomedical professions, where Jews were disproportionately represented. (According to Ternon and Helman, in 1932 some 13 per cent of German doctors were Jewish – rising to 60 per cent in Berlin – though Jews made up barely more than 1 per cent of the German population.)[37] On the other hand, contrary to Catholicism, Judaism did not condemn eugenics, any more than did Protestantism; Verschuer was a pious Protestant – in an anti-Nazi church – and regularly attended Sunday services.[38]

In the USSR, Levit and Agol were among the first scientists arrested. Perhaps this was because they were Jewish as well as being eugenicists; given their connections with Muller and the Rockefeller Foundation, they could have been accused of being involved in a 'Judeo-capitalist plot' aiming to establish eugenics in

36 B. Massin, preface to P. Weindling, *L'Hygiène de la race*, vol. 1, pp. 52–3.

37 Y. Ternon and S. Helman, *Les Médecins allemands et le National-Socialisme*, p. 70.

38 B. Müller-Hill, *Science nazie, science de mort*, p. 121.

the Soviet Union. In Germany, Weinberg, who belonged to the first generation of eugenicists, died in 1937. Poll, as we said, emigrated to Sweden and committed suicide. Goldschmidt emigrated to the United States, where he continued his career.

The most curious case is that of the Jewish doctors and biologists who were forced to flee Germany at the time of the anti-Semitic laws of 1935 but who continued to support eugenics, sometimes finding themselves in American eugenic associations where they rubbed shoulders with sympathizers of Nazism such as Laughlin. Goldschmidt fits this particular profile, though it is true that he was also someone who preferred Hitler to Bolshevism. Müller-Hill, who considered him one of the fathers of the eugenic legislation of 1933 (modelled on the Prussian draft law of 1932 which Goldschmidt had helped to write), explains his reasoning as follows: it was not Goldschmidt's eugenics that should be condemned, but what the Nazis made of it (a customary argument among post-war eugenicists, as can be seen in the quotation from Sutter on pp. 217–8 below). In Goldschmidt's words,

> In this way the Nazis took over this project en bloc, but without ever mentioning its origin. In their application of our draft law, however, which was humane and based on a great awareness of our responsibility, they employed methods most unworthy of humanity and the most condemnable that can be imagined.[39]

The case of a psychiatrist such as Franz Kallmann is still more revealing about the capacity for blindness on the part of these eugenicists. In Germany, he had worked with Rüdin and demanded the sterilization of 10 per cent of the German people who allegedly carried the gene for schizophrenia. On emigrating to the United States, he joined the American Eugenics Society, continued to support German eugenics, and called for the sterilization of brothers and sisters of schizophrenics. He was appointed to the chair of psychiatry at Columbia University, and research director at New York State Psychiatric Institute. After the war, he helped to clear Rüdin in the de-Nazification procedures.[40]

---

39 R. Goldschmidt, *Im Wandel des Bleibenden: Mein Lebensweg* (Hamburg/Berlin, 1963), p. 264; cited in B. Müller-Hill, *Science nazie, science de mort*, p. 23. See also S. Kühl, *The Nazi Connection*, p. 52.

40 B. Müller-Hill, *Science nazie, science de mort*, pp. 23–4 and 136–7. B. Massin, preface to P. Weindling, *L'Hygiène de la race*, vol. 1, pp. 18, 57.

The kind of Manichaeanism characteristic of contemporary judgements on this period is often far removed from actual historical reality. The situation was terribly confused, and no one seemed to see very clearly the connections and consequences of the theories in vogue. Jewish biologists, doctors and intellectuals acted more or less like their non-Jewish colleagues. They approved eugenic measures, even though it was clear these were drifting towards racist applications. This drift took concrete shape in 1937 with the sterilization of mixed-race offspring of the French-African forces in the Ruhr and Rhineland, but it was evident enough earlier on, as such sterilization had already been envisaged in 1927 under the Weimar republic.[41] These Jewish biologists do not seem to have conceived that such eugenic measures could be turned against themselves as soon as the Nazis categorized Jews as an 'inferior race'.

At all events, this shows that there was no necessary connection between eugenics and anti-Semitism, even if certain eugenicists, and not just German ones, professed a theory of the superiority of the Nordic race. (Indeed, the superiority of the white race in general, Jews included, was accepted by more or less all of them.) Weindling even mentions that the finances of certain eugenic associations in Germany were managed by Jewish banks,[42] and that the Munich Institute was partly financed by the American Jewish banker James Loeb. According to Kühl,[43] this financing lasted until 1940.

Eugenic sterilization in Germany was particularly common and systematic. It was completely legal, performed under the supervision of medicine and law (1,700 special tribunals), even if there was of course a political animus. It is accepted today that between 350,000 and 400,000 sterilizations were carried out in Germany between 1934 and 1945.[44] Before the war, estimates of the number of people who should be sterilized varied according to different authors. Kallmann gave as high a figure as 10 per cent of the population; Lenz even foresaw 20 per cent, before the law was passed. Hütt, Rüdin and Ruttke, the official commentators on the law, put forward the more modest number of 1.2 million, or about 2 per cent of the population. Verschuer opted for

41 P. Weindling, *L'Hygiène de la race*, vol. 1, p. 214. Weindling speaks of some 500 to 800 mixed-race offspring.

42 Ibid., p. 233.

43 S. Kühl, *The Nazi Connection*, p. 20.

44 B. Müller-Hill, *Science nazie, science de mort*, p. 29.

400,000.[45] A 'neutral' French observer, Girard, likewise gave a figure of between 300,000 and 400,000 in his 1934 thesis.[46] There is thus a similarity in the order of magnitude between this 'neutral' estimate before the event and the *a posteriori* figure given today.

The majority of individuals affected were sterilized on grounds of feeble-mindedness or mental illness (understood in a very broad sense); others suffered from various diseases or disorders that were claimed to be hereditary, just as in other countries that passed this kind of legislation. Certainly in Germany, more than elsewhere, the legislation tended to be used for political and racist applications, but the precise extent of this is hard to assess. Sterilizations also seem to have often been badly performed. According to Ternon and Helman, some 1 per cent of men and 4 per cent of women died from the operation – a very high figure for such simple procedures.[47] By way of comparison, in 1934 Girard gave the figure of three deaths out of 2,500 sterilizations performed in the United States by Dr Gosney.[48]

45 Y. Ternon and S. Helman, *Les Médecins allemands et le National-Socialisme*, p. 178.

46 J. Girard, *Considérations sur le loi eugénique allemande*, pp. 39, 50.

47 Y. Ternon and S. Helman, *Les Médecins allemands et le National-Socialisme*, p. 177.

48 J. Girard, *Considérations sur le loi eugénique allemande*, p. 52.

11

The Extermination of the Mentally Ill

In Germany, the sterilization of the ill, the mad, the handicapped and the 'abnormal' was followed by extermination. This is what we must turn to now, and its relationship with eugenics in the proper sense of the term. (This extermination should also be differentiated from that practised on prisoners in concentration camps when they were sick or incapable of working.) It was often termed 'euthanasia', a term that we shall also discuss later on.

The history of 'euthanasia' in this sense has been traced by Yves Ternon and Socrate Helman, as well as by Willi Dressen.[1] As soon as the Nazis came to power in 1933, funds for psychiatric hospitals were reduced, affecting wages and salaries, medicine, the upkeep of buildings, and the feeding of patients. In August 1939, shortly before the outbreak of war, Hitler decided to proceed to a planned extermination. The pretext was provided by a couple who had requested authorization to put an end to the life of their incurably malformed infant. This authorization was granted, and a circular of 18 August 1939 ordered doctors to register the birth of malformed children, who would then be eliminated. Soon after, the decision was taken, by verbal order, to exterminate mental patients in the same fashion.

In October 1939, Hitler signed a decree, backdated to 1 September, 'authorizing' the extermination of the mentally ill (in other words, ordering this extermination, even if it was left to the responsibility of the doctors themselves). The operation was given the name 'Operation T4', after the administrative centred located in Berlin at 4 Tiergartenstrasse.

1 Y. Ternon and S. Helman, *Le Massacre des aliénés, des théoriciens Nazis aux practiciens SS* (Tournai: Casterman, 1971); W. Dressen, 'Euthanasia', in E. Kogon, H. Langbein and A. Ruckerl, *Nazi Mass Murder* (New Haven: Yale University Press, 1993).

Gas chambers were constructed at a number of sites, where the victims were taken according to their region of origin: the castle of Grafeneck (replaced by Hadamar in January 1941), the prison at Brandenburg an der Havel (transferred to Bernburg an der Saale in November 1940), the castle of Hartheim, and a clinic at Sonnenstein an der Pirna. These gas chambers were put into operation early in 1940. Jewish patients were systematically eliminated, non-Jewish ones according to their degree of infirmity. In certain districts the extermination was extended to people suffering from senility, alcoholics, the disabled, the bedridden, and various 'antisocials' (indigents, vagabonds, prostitutes and others).

The operation was to be kept secret, but it was impossible to empty mental hospitals in such a massive way without distressing the patients' families, who all learned at the same time of the death of their relatives. There were protests by the families as well as by a number of doctors (including Kurt Schneider, Karl Bonhoeffer, K. Klesit, H. Berger and Kurt Klare) who sought to restrain the process. Above all, however, protests from both Catholic and Protestant churches: Pastor Fritz von Bodelschwingh, of Bethel, who actually managed to save his patients from extermination; as well as Pastor Braune of the German Evangelical Church; the archbishop of Freiburg, Conrad Gröber; Cardinal Bertram; Bishop Theophil Wurm of Wurttemberg; Pastor Schlaich; Cardinal Faulhaber; and others.

The protest that became most famous, undoubtedly because it was public and led to the closing of these gas chambers, was that of the bishop of Münster, Clemens August von Galen, who in August 1941 denounced this extermination *ex cathedra* and accused those responsible of murder (extracts in the box below).

Sermon of Clemens August von Galen, Bishop of Münster (Westphalia) at the Church of St Lambert, 3 August 1941[2]

For the past several months it has been reported that, on instructions from Berlin, patients who have been suffering for a long time from apparently incurable diseases have been forcibly removed from homes and clinics. Their relatives are later informed that the patient has died,

2 Cited in Rev. H. Portmann, *Cardinal von Galen* (Norwich: Jarrold, 1957), pp. 239ff.

that the body has been cremated and that the ashes may be claimed. There is little doubt that these numerous cases of unexpected death in the case of the insane are not natural, but often deliberately caused, and result from the belief that it is lawful to take away life which is unworthy of being lived.

This ghastly doctrine tries to justify the murder of blameless men and would seek to give legal sanction to the forcible killing of invalids, cripples, the incurable and the incapacitated ...

I am assured that at the Ministry of the Interior and at the Ministry of Health, no attempt is made to hide the fact that a great number of the insane have already been deliberately killed and that many more will follow ...

When I was informed of the intention to remove patients from Marienthal for the purpose of putting them to death I addressed the following registered letter on July 29th to the Public Prosecutor, the Tribunal of Münster, as well as to the Head of the Münster Police ...

I have received no news up till now of any steps taken by these authorities. On July 26th I had already written and dispatched a strongly worded protest to the Provincial Administration of Westphalia which is responsible for the clinics to which these patients have been entrusted for care and treatment. My efforts were of no avail. The first batch of innocent folk have left Marienthal under sentence of death, and I am informed that no less than eight hundred cases from the institution of Waestein have now gone. And so we must await the news that these wretched defenceless patients will sooner or later lose their lives. Why? ... It is simply because according to some doctor, or because of the decision of some committee, they have no longer a right to live because they are 'unproductive citizens'. The opinion is that since they can no longer make money, they are obsolete machines, comparable with some old cow that can no longer give milk or some horse that has gone lame. What is the lot of unproductive machines and cattle? They are destroyed. I have no intention of stretching this comparison further. The case here is not one of machines or cattle which exist to serve men and furnish them with plenty. They may be legitimately done away with when they can no longer fulfill their function. Here we are dealing with human beings, with our neighbours, brothers and sisters, the poor and invalids ... unproductive – perhaps! But have they, therefore, lost the right to live? Have you or I the right to exist only because we are 'productive'? If the principle is established that unproductive human beings may be killed, then God help all those invalids who, in order to produce wealth, have given their all and sacrificed their strength of body. If all unproductive people may thus be violently eliminated, then woe betide our brave soldiers who return home, wounded, maimed or sick.

Once admit the right to kill unproductive persons ... then none of us can be sure of his life. We shall be at the mercy of any committee that can put a man on the list of unproductives. There will be no police protection, no court to avenge the murder and inflict punishment upon the murderer. Who can have confidence in any doctor? He has but to certify his patients as unproductive and he receives the command to kill. If this dreadful doctrine is permitted and practised it is impossible to conjure up the degradation to which it will lead. Suspicion and distrust will be sown within the family itself. A curse on men and on the German people if we break the holy commandment 'Thou shalt not kill' which was given us by God on Mount Sinai with thunder and lightning, and which God our Maker imprinted on the human conscience from the beginning of time! Woe to us German people if we not only license this heinous offence but allow it to be committed with impunity!

According to J. Rovan, the Nazis did not dare to arrest Mgr von Galen, but dozens of priests who distributed his sermon were indeed sent to Dachau.[3] F. Bayle, who gives a French translation of von Galen's sermon, says that Goebbels suggested hanging the bishop, but Hitler's immediate entourage rejected this, because Hitler had not made public the extermination decision or promulgated any law for it: 'the bishop was arrested, but demonstrations in Westphalia were so forceful that he had to be released'.[4] According to H. Friedlander, Hitler rejected the arrest of von Galen so as not to make him a martyr or create outrage among Catholics; he goes on to say that no leader of either the Protestant or the Catholic Church was victimized for their opposition to this extermination of mental patients, except one who had also protested against the persecution of the Jews.[5] Ternon and Helman maintain that several leading Nazis, including Bormann, demanded the hanging of von Galen, but Goebbels opposed this, for fear of disturbances in Westphalia, where the extermination of mental patients had already led to tensions (according to them, a number of priests died at Dachau for having opposed this 'euthanasia'). The same authors also offer an alternative version, which they believe apocryphal,

3 J. Rovan, *Histoire d'Allemagne* (Paris: Point-Seuil, 1998), p. 691.

4 F. Bayle, *Croix gammée contre caducée, les expériences humaines en Allemagne pendant la Deuxième Guerre mondiale* (Neustadt: no imprint, 1950), p. 75.

5 H. Friedlander, *The Origins of Nazi Genocide*, pp. 115–7.

claiming that the bishop donned his robes and mitre, and took up his crook, in preparation for the arrival of the Gestapo, who then did not dare arrest him.[6] Different historians give rather differing versions, and though it is a matter of an objective event, not a matter of subjective interpretation, clearly it is one that is not well known. At all events, von Galen survived and became a cardinal.

The protest had a certain effect: Operation T4 was suspended on 24 August 1941, and the gas centres closed; the elimination of malformed infants, which was a separate matter, continued.

This was only a temporary reprieve, however, and the extermination of the mentally ill subsequently took other forms: gassing in mobile installations, injection with various toxic substances, and deprivation of food (particularly for infants). Historians customarily refer to this as 'wild euthanasia', as against euthanasia at fixed centres, which was more systematically planned and organized. We shall return below to this term 'euthanasia' and its relation to eugenics. Here we need only note that it began just before the Second World War and that Operation T4 was simply a more organized form of it.

The number of people exterminated in this way is hard to assess. In light of statistics established by the Nazis themselves in 1941, when the operation was suspended, it would seem that the six gassing centres claimed more than 70,000 victims in the period from January 1940 to August 1941. This is the figure cited by Dressen, Hilberg, Müller-Hill and Friedlander.[7]

As for the wild euthanasia, which is not recorded in the archives, B. Massin tried to give an estimate of this, based on the known number of patients still alive in certain asylums in 1945 compared with the number counted in 1938, a ratio he applied to the number of psychiatric beds in Germany as a whole. The resulting total, including both the gassings of 1940–1 and the later, wild euthanasia is between 187,000 and 235,000 victims (of whom between 117,000 and 165,000 would have been subjected to wild euthanasia).[8] Müller-Hill's estimate is that there could well have been 100,000 deaths from malnutrition in asylums between 1942

6 Y. Ternon and S. Helman, *Le Massacre des aliénés*, pp. 125–7.

7 W. Dressen, 'Euthanasia', in E. Kogon et al., *Nazi Mass Murder*, p. 57; R. Hilberg, *The Destruction of the European Jews*, (New Haven: Yale University Press, 2003), p. 930; B. Müller-Hill, *Science nazie, science de mort*, p. 204; H. Friedlander, *The Origins of Nazi Genocide*, p. 109.

8 B. Massin, 'L'Euthanasie psychiatrique sous le IIIe Reich, La question de l'eugénisme', *L'Information psychiatrique*, 8, October 1966, pp. 811–22.

and 1945; according to him, only 40,000 patients remained in these asylums at the end of the war.[9]

None of these four authorities, however, refer to a report drawn up by Dr Theo Lang, who in January 1941 met Professor H. Goering, cousin of the Nazi leader and director of the German Institute for Psychological Research and Psychotherapy in Berlin, supposedly with a view to organizing resistance to the elimination of mental patients. According to this report, written in December 1941, a few months after the closing of the gassing centres, 200,000 mental patients had already been exterminated, as well as a further 75,000 old people (see the box below).

Testimony of Theo Lang
on the Extermination of Mental Patients

Report addressed to the International Commission for the Investigation of War Crimes on 10 May 1945 by Dr Theo Lang, medical head of an institution in Herisau, Switzerland

I wish to state that all German doctors, and especially those at psychiatric clinics and mental asylums, were aware at the latest by the end of 1940 of the extermination by gas of patients suffering from mental and nervous disorders, as well as of the execution of Germans and Poles who were perfectly well but who did not match racial criteria. This was confirmed to me in particular by Professor H. Goering, cousin of the marshal and director of the Berlin Institute of Psychotherapy. I had a discussion with him at his institute on 20 January 1941, with a view to taking action against this.

Naturally, though in my presence he expressed his opposition to these exterminations by gas, he refused to sign a declaration I had prepared on the subject, which we would then have circulated in medical circles and addressed to the government.

Annex B to Dr Theo Lang's report

1. The sources of the extermination can be found in a secret law which became known in summer 1940.
2. As well as the minister for public health, Dr Conti, Reichsführer SS Himmler, interior minister Frick and others, the following individuals played a part in this secret law:

9 B. Müller-Hill, *Science nazie, science de mort*, p. 70.

a) ministerial counsellor Dr Linden;
b) Dr Stähle of Württemberg;
c) medical counsellor Pfannmueller, director of the Eglfing-Haar asylum;
d) Profesor Werner Heyde, directory of the psychiatric and neurological clinic of Würzberg.

3. As I declared, the number of patients exterminated, in a careful calculation, came to 200,000: the mentally ill, imbeciles, neurological cases, those chronically ill (not only incurable cases), and at least 75,000 elderly people.

4. The main exterminations were carried out at Münsingen in Württemberg and Linz on the Danube [these were the castles of Grafeneck and Hartheim respectively]. Several gas chambers and crematoria were constructed there.

5. Transport from the establishments to the gas chambers was carried out by SS commandos, working with a Berlin transport company.

Patients from small and medium-size establishments were exterminated almost without exception. The larger establishments kept a number of their patients; for example 500 out of 2,500 at Berlin-Buch; 160 out of 600 in Thuringia, 200 out of 2,000 at Kaufbeuren in Bavaria. Among larger establishments, the following have already been closed now for some time:

Illenau in the duchy of Baden: 800 patients;
Berlin Hezsberg: 2,500 patients;
Kreuzburg in Upper Silesia: 1,500 patients;
Sonnenstein in Saxony: 800 patients;
Werneck in Lower Franconia, Bavaria: 1,111 patients;
Steinhof in Vienna: 3,000 patients;
as well as others, very likely Schleswig with 1,000 beds, Günzburg with 400 beds, etc.

The following procedure was used with old people still in perfect health and living at home: a political leader summoned them; then a doctor, generally from the SS, confirmed that they were mentally deficient. He suggested placing them in care and sending them to an institution; from there these old people went to the gas chambers.

Written in December 1941
Dr Theo Lang

(Cited in F. Bayle, *Croix gammée contre caducée, les expériences humaines en Allemagne pendant la Deuxième Guerre mondiale* [Neustadt, 1950], pp. 764–5)

This report is cited by M. Lafont in his thesis, who evidently became aware of it via the anonymous article 'L'extermination des maladies mentaux sous le régime national-socialiste' published in *La Raison* in 1951. This article is a compilation of extracts from Bayle's book, the only summary (more than 1,500 pages) of the 1946 doctors' trial at Nuremberg, which is therefore the primary source. Being quite a rarity,[10] it may be that the authors cited above were unaware of it. (No doubt Theo Lang's report rests somewhere in the archives of the trial.)

Bayle attended this trial as a member of the French scientific commission on war crimes. Despite his tendency to adopt an emphatic and literary tone, there is no reason to believe that he invented this report, or to ignore it. According to Bayle, Theo Lang's report was presented at the Nuremberg trial, and its figures recorded by the Tribunal, as against the declarations of two accused with major responsibility for the extermination, K. Brandt and V. Brack, who acknowledged only between 50,000 and 60,000 victims:

> If Karl Brandt and Viktor Brack estimated that at most sixty thousand people were exterminated in the course of the euthanasia programme, many indications lead us to believe that the true figure was four or five times higher; the verdict of the International Military Tribunal declared that at least 275,000 were killed in this way (pages 16,916 and 16,917).[11]

Lang's report thus seems to have been taken seriously at the time, and it is rather odd that the figure of 70,000 victims should be cited today – a figure close to that admitted by the Nazis, rather than the 275,000 that the Nuremberg Tribunal accepted as a minimum figure.

Ternon and Helman[12] cite Bayle's book in their bibliography, but they do not refer to Theo Lang's report. They write that the Nuremberg Tribunal's figure of 275,000 is excessive, since the Nazi statistics which showed somewhat over 70,000 victims were not yet available. They do not say, however, that this figure of 275,000 includes 75,000 old people, and only 200,000 mental patients. Moreover, they do see the figure of 70,000 as too low,

10 This book of over 1,500 pages appeared without a publisher's imprint, and no doubt in a small print run. Besides, it was written in French but printed in Germany. (See note 13, p. 201.)

11 F. Bayle, *Croix gammée contre caducée*, p. 727.

12 Y. Ternon and S. Helman, *Le Massacre des aliénés*, pp. 140–1.

given that a statistician from Grafeneck gave 10,654 as the number exterminated at this centre, whereas the true figure is seen as closer to 12,000 or even 15,000. (Bayle cites a study by Dr Poitrot, from immediately after the war, which describes the extermination at Grafeneck Castle and estimates the figure as at least 15,000, whereas Dressen gives a figure of 9,839 for this centre on the basis of the archives discovered.)[13]

Instead of 70,000, Ternon and Helman put forward the figure of 100,000 victims for Operation T4 alone (that is, the gassings of 1940–1 but not the so-called wild euthanasia), which they see as corresponding to about a third of the 300,000 to 350,000 German mental patients from before the war.[14] A few lines further on, however, they write that almost all the small and medium-sized establishments were emptied of their patients, as well as certain of the larger ones, and they give some examples of large asylums that remained open: Berlin-Buch in 1941 had 500 patients for some 2,500 beds; Stadtroda, 150 for 600 beds; Kaufbeuren, 200 for 1,000 beds. This information is clearly taken largely from Theo Lang's report (see box on pp. 198–9), and corresponds to an extermination of well over a third of all patients. At this remove, it is no longer possible to establish accurate figures.

How credible is the Lang report, as cited by Bayle? Müller-Hill does not mention this report, but he mentions Lang in the account he published of his interview with Professor Edith Zerbin-Rüdin, the daughter of Ernst Rüdin,[15] who, we recall, headed the German Institute for Research in Psychiatry at Munich. According to this interview, Theo Lang was a colleague of Rüdin's, and administrative head of the Munich institute until 1941, the year of his report;[16] he was very well placed, therefore, to know what was happening in the German mental asylums. This gives credibility both to his meeting with Professor Goering and to the

13 F. Bayle, Croix gammée contre caducée, p. 766; W. Dressen, 'Euthanasia', in E. Kogon et al., *Nazi Mass Murder*, p. 55. This Dr Poitrot (who appears more correctly as Robert Poltrot in Ternon and Helman's book, and is presented by them as a French psychiatrist accompanying the occupation forces) is perhaps the same who first gave the figure of 40,000 dead in the French asylums that is cited in the article from La Raison cited above (see note 38 on p. 165 above).

14 According to them, Germany in 1931 had more than 30,000 private charitable institutions, both religious and secular, which housed and cared for nearly half a million individuals, including many mentally ill (Y. Ternon and S. Helman, Le Massacre des aliénés, p. 31).

15 B. Müller-Hill, *Science nazie, science de mort*, p. 135.

16 Ibid., p. 66.

way in which, in Appendix B of his report, he seems to have been able to count the victims in each asylum. He had joined the Nazi party, then left it quite early on, but kept sufficient connections to obtain the money he needed to continue his research on goitre in Switzerland, even before the end of the war. This is perhaps linked to the refusal of the Rockefeller Foundation to fund the work on goitre that Rüdin proposed (see p. 185 above); Lang's move to Switzerland may also attest to a desire on his part to leave Germany after his reaction to the extermination of sick people. He was then appointed head of a Swiss establishment, at Herisau. Finally, still during the war, he informed the British secret service of Rüdin's involvement in studies of mass sterilization by means of X-rays, which may indicate, if not the initial destination of his report on the extermination of mental patients, at least the means by which this reached the Nuremberg Tribunal. Lang later committed suicide, following various problems.

There is thus a fairly coherent pattern, and I cannot understand why Theo Lang's report is practically never mentioned, even to be criticized or rejected. The difference between the figures in his report and those of other estimates may of course arise from an exaggeration on Lang's part (intended to whitewash himself by blaming former colleagues) or, on the other hand, to his inclusion of those eliminated in a 'wild euthanasia' operation parallel to the 'official' Operation T4.

According to Bayle, Theo Lang wrote his report in December 1941. It has nothing to say, therefore, on what might have happened after the closure of the initial gassing centres, i.e., in the wild euthanasia.

Bayle's book cites the figure of 476 people killed at Hadamar on or about 5 July 1944; likewise 600, including children, at the Wernigerode asylum between 1943 and 1944; 2,000 children in Vienna between July 1942 and April 1945; as well as some other cases for which the dates are not clear.[17] Ternon and Helman also cite a number of cases pertaining to the wild euthanasia. The Eichberg asylum, for example, with 1,500 beds, had forty deaths in 1936; between January and August 1941, 800 patients were sent off to Hadamar to be gassed (i.e., more than half, rather than a third); there were a further 470 deaths in 1941, 737 in 1942, 753 in 1943, and 583 in 1944. In other words, for an asylum of 1,500 beds, a total of 3,343 deaths between 1941 and 1944. Clearly, this

17 F. Bayle, *Croix gammée contre caducée*, pp. 769, 770, 775 and passim.

asylum was repopulated several times after being emptied, either with persons from other asylums that were finally closed down or with undesirables 'collected' here and there: Lang's report explains the procedure practised with the elderly, but cases are also known of runaway children who were exterminated after being arrested by the police.

Still according to Ternon and Herman, the Hadamar files count 4,159 deaths between 15 August 1942 and 31 March 1945, six hundred deaths at Wiesloch between 1941 and 1945, and so on.[18] There is no need to list all the asylums; it is enough to mention that besides deaths actively produced by various injections, a large number of patients died of hunger as a premeditated act (and not a result of 'incompetence', as in the French asylums).

Added to this is the extermination of children, which was not part of Operation T4 and so was not suspended in August 1941.

The psychiatric hospitals of Alsace-Lorraine, moreover, and still more so those of Poland and Russia, were at least partly emptied in the same manner, without any attempt made to keep an exact number of these killings. Ternon and Helman speak of 12,000 patients actively murdered (shot, poisoned, etc.) in various Polish asylums, as well as 13,000 poisoned in the Obrzyce hospital (Obrawalde-Meseritz), which was converted into an extermination centre, and an unknown number of deaths from hunger in different asylums.[19] There was also an extermination programme for some tens of thousands of Polish tuberculosis patients, though it is unknown to what extent this was put into effect. It is believed that many of these were exterminated at the same time as the Polish Jews.[20]

For Russia, Ternon and Helman write that in the territories occupied by the Nazis, almost all establishments were 'liquidated' – a total of some 6,000 hospitals; 33,000 polyclinics, dispensaries and consultancy centres; 967 sanatoriums and 656 rest homes.[21]

It is hard to set a total figure. But if simple administrative indifference led to the death of 40,000 people in French psychiatric asylums, the systematic desire to exterminate not only mental patients but also the handicapped, malformed infants, and

18 Y. Ternon and Y. Helman, *Le Massacre des aliénés*, pp. 175–8.
19 Ibid., pp. 196–7.
20 F. Bayle, *Croix gammée contre caducée*, pp. 849–57.
21 Y. Ternon and Y. Helman, *Le Massacre des aliénés*, p. 197.

various 'deviants' (including the senile) certainly reaped far more victims in Germany, Austria, Poland and Russia than the figure of 70,000 that is often cited. In all probability, it was some hundreds of thousands – and we should not forget that, unlike Jews and Gypsies, these people had no possibility whatsoever of avoiding extermination by means of flight.

It should finally be noted that, given the public protests in Germany, news of which rapidly spread to other countries, it is hard today to claim that all this took place in secret, that the existence of gas chambers and Nazi extermination methods was unknown – even if it was not yet Jews who were affected. By way of proof, we need only cite an article published by *Reader's Digest* in June 1941 – a magazine with an extremely wide circulation.[22] This was the work of William L. Shirer, who reported from Germany until the end of 1940. His article explains not only that the Nazis were gassing mental patients, but also that three of the four gassing centres then active could be identified, those of Grafeneck, Hartheim, and Sonnenstein, and that in November 1940 the journalist's German informants maintained that 100,000 people had already been exterminated – a figure that Shirer deemed an exaggeration. The *Reader's Digest* article was a 'condensed extract' from Shirer's *Berlin Diary*, published in 1941 in New York. Here are a few key passages from the book:

> Berlin, 21 September [1940]. X came up to my room in the Adlon yesterday, and after we had disconnected my telephone and made sure that no one was listening through the crack of the door to the next room, he told me a weird story. He says the Gestapo is now systematically bumping off the mentally deficient people of the Reich. The Nazis call them 'mercy deaths'.
>
> Berlin, 25 November [1940]. I have at last got to the bottom of these 'mercy killings'. It's an evil tale.
>
> The Gestapo, with the knowledge and approval of the German government, is systematically putting to death the mentally deficient population of the Reich. How many have been executed probably only Himmler and a handful of Nazi chieftains know. A conservative and trustworthy German tells me he estimates the number at a hundred thousand. I think that figure is too high. But certain it is that the figure runs into thousands and is going up every day …

22 W. J. Shirer/Berlin Diary, ' "Mercy deaths" in Germany', *Reader's Digest*, June 1941, pp. 55–8.

> Of late some of my spies in the provinces have called my attention to some rather peculiar death notices in the provincial newspapers ... But these notices have a strange ring to them, and the place of death is always given as one of three spots: (1) Grafeneck, a lonely castle situated near Münzingen, sixty miles southeast of Stuttgart; (2) Hartheim, near Linz on the Danube; (3) the Sonnenstein Public Medical and Nursing Institute at Pirna, near Dresden. Now, these are the very three places named to me by Germans as the chief headquarters for the 'mercy killings' ...
>
> X, a German, told me yesterday that relatives are rushing to get their kin out of private asylums and out of the clutches of the authorities. He says the Gestapo is doing to death persons who are merely suffering temporary derangement or just plain nervous breakdowns.
>
> What is still unclear to me is the motive for these murders. Germans themselves advance three:
>
> 1. That they are done to save food.
> 2. That they are done for the purpose of experimenting with new poison gases and death rays.
> 3. That they are simply the result of the extreme Nazis deciding to carry out their eugenic and sociological ideas.
>
> The first motive is obviously absurd, since the death of 100,000 persons will not save much food for a nation of 80,000,000. Besides, there is no acute food shortage in Germany. The second motive is possible, though I doubt it. Poison gases may have been used in putting these unfortunates out of the way, but if so, the experimentation was only incidental. Many Germans I have talked to think that some new gas which disfigures the body has been used, and that is the reason why the remains of the victims have been cremated. But I can get no real evidence of this.
>
> The third motive seems most likely to me. For years a group of radical Nazi sociologists who were instrumental in putting through the Reich's sterilization laws have pressed for a national policy of eliminating the unfit. They say they have disciples among many sociologists in other lands, and perhaps they have.[23]

The *Reader's Digest* article added the following detail:

> In Berlin, when Dr Robert Ley, chief of the Nazi Labour Front, announced projects for Old People's Homes, more than one worker remarked: 'No, thank you. Better to starve on the streets than to be bumped off at Grafeneck.'

23 W. J. Shirer, *Berlin Diary* (New York: Knopf, 1941), pp. 569–71, 574.

This corresponds exactly to Theo Lang's statements about the extension of the extermination to old people who were brought into hospices. At all events, in June 1941, in other words even before the public protest of Mgr von Galen, and six months before the Wannsee conference at which the extermination of the Jews was 'officially' decided, the gassing of the mentally ill in Germany was well known in the United States, and reported in magazines with the widest circulation.

The question this raises is that of the relationship between this extermination and eugenics, i.e., the sterilization of mental patients, the handicapped, and so forth.

First of all, it is necessary to make clear that it is an abusive use of the term 'euthanasia' to use it to describe this extermination. The original Greek word meant a 'happy death'. It is applicable to putting an end to the suffering of patients whose life expectancy is very short, which was in no way the case here. It is also the term used by the RSPCA and its counterparts for the elimination of animals it cannot care for. To speak of euthanasia in the present instance, as is done by historians such as F. Bayle, W. Dressen, R. Hilberg, H. Friedlander, B. Massin, and others is almost scandalous. Gassing, the intracardiac injection of phenol, death by starvation, etc., has nothing 'happy' about it. It would clearly be shocking to apply this term to the extermination of the Jews, and even lead to the prosecution of a writer who did so. Here, however, it appears normal and shocks no one.

Once the question of the proper term is settled, we can move to the relation of this extermination to eugenics. First of all, it is true that not all eugenicists supported this extermination; all its supporters, however, were eugenicists. And the victims essentially belonged to groups that eugenic legislation targeted – with the exception of the elderly, who were evidently added to the category of the useless. One particular feature shows the length to which these theories were taken: in 1944 the Nazis considered extending extermination to common-law criminals who were ugly (whatever their crime), this ugliness being considered proof of their insufficient humanity.[24]

The 'biological' motivations for the extermination of mental patients, handicapped people, and others were a poor concealment for economic motivations. These 'useless' people were expensive

24 R. Hilberg, *The Destruction of the European Jews*, p. 1067.

to maintain, and economizing was necessary in wartime. Indeed, those responsible for their elimination calculated that, assuming an average life expectancy of ten years for their victims, gassing them had saved 885,439,800 marks.[25]

We should not believe that the biological motives for eugenics were ever really taken seriously by politicians (or even biologists), and that they did not also coincide with economic concerns. Thus the vigorous Major Leonard Darwin (Charles's son, and long-time chair of the Eugenic Education Society) could write in 1922:

> Political authorities should take into account the enormous burden that degenerates impose on the nation. The sums spent on legislation, criminal justice, and the police exceed £48 million per year. And that is not the whole charge ... If the community had to pay less for degenerates of all kinds, healthy men would have less to pay ... Every rise in taxation is a step towards the degeneration of the race.[26]

With both eugenics and 'euthanasia', the object was to create savings, just as much – if not more – as it was to prevent the degeneration of the race. The two concerns had the advantage of perfectly coinciding, with the biological logic masking the economic one. War simply exacerbated this process, and in a timely fashion sterilization was supplemented by extermination – which spilled over to certain categories not susceptible of procreation, such as the feeble-minded and the elderly, and subsequently to undesirable races. To this should be added, as Shirer already noted, the genetic fanaticism that was characteristic of the era, and which the Nazis turned into a political principle.

The 'euthanasia' of the mentally ill was not a specifically German invention, a refinement of Nazi cruelty or sadism. Hitler did not invent very much; in most cases he was content to take up ideas that were in the air and to pursue them to their logical conclusions. Euthanasia and profound meditation on 'lives not worth living' were commonplaces of the time – and not only in Germany, even if the Nazis made great use of them and conducted propaganda around this theme. More or less all countries, in Europe and the United States, saw organizations campaigning

25 W. Dressen, 'Euthanasia', in E. Kogon et al., *Nazi Mass Murder*.

26 L. Darwin, 'Practical Eugenics', retranslated from E. Apert et al., *Eugénique et sélection* (Paris: Alcan, 1922), pp. 196–8.

for the legalization of euthanasia. This was inspired by a kind of symmetry: eugenics (good birth) and euthanasia (good death) were the conditions for a good life – more important than social and economic reforms.

Ternon and Helman as well as Weindling cite a number of books, societies, and legislative projects along these lines.[27] The year 1895 saw the publication in Göttingen of a book titled *Das Recht auf den Tod, Soziale Studie* (The Right to Death, a Social Study), in which Adolf Jost appealed for the right of sick people to choose death, as much in their own interest as in that of society. In 1920 a book with the evocative title *Die Freigabe der Vernichtung lebensunwerten Lebens. Ihr Mass und Ihre Form* (Liberalization of the Extermination of Lives Not Worth Living. Its Extent and Form) was published in Leipzig – the work of a lawyer, Karl Binding, and a psychiatrist, Alfred Hoche. This invoked the case of patients unable to give consent (such as incurable idiots) and those for whom the decision should be referred to a specialized committee – to a certain extent the solution adopted by the Nazis. In 1922, Ernst Mann's book *Die Erlösung der Menschheit vom Elend* (Mankind's Deliverance from Misery) demanded the painless elimination of the mentally ill, the moribund, individuals tired of life or suffering from various diseases, and children who were weaklings or affected with incurable diseases. The Monist League, founded by Haeckel and the great chemist W. Ostwald, also championed euthanasia.

Germany was certainly not the only site of these ideas. In Switzerland, a doctor on a committee for the canton of Berne proposed the possibility of allowing mental patients to die. In Denmark, legislation to this effect was proposed in 1924. In Britain, a movement in support of euthanasia emerged in 1932 under the leadership of Dr Killick Millard, and the Voluntary Euthanasia Society was founded in 1935, chaired by Lord Moynihan (who also headed the Royal College of Surgeons). In 1936 a law was proposed for the voluntary euthanasia of the incurably ill, but was rejected. In the United States, the Euthanasia Society of America was established in 1938, a year after a law for voluntary euthanasia had been presented to the Nebraska state legislature.

---

27 P. Weindling, *L'Hygiène de la race*, vol. 1, pp. 220ff.; Y. Ternon and S. Helman, *Le Massacre des aliénés*, pp. 33ff., 42ff.

In 1942, the very year the wild euthanasia began in Germany, the *American Journal of Psychiatry* published a debate between two psychiatrists on the subject of euthanasia of the mentally ill.[28] Leo Kanner was opposed to it, but Foster Kennedy favoured it in some cases. Kennedy rejected euthanasia for individuals who had been in a good mental state before falling ill (in their case, cure was always possible); on the other hand, he supported it for mentally defective children who had reached the age of five without hope of improvement. This euthanasia would have to be requested by those responsible for the children in question, who would undergo three examinations. The debate was perfectly civilized, with perfectly defensible arguments. The only problem was that in Germany in 1939, the decision on the 'euthanasia' of the mentally ill had already been taken, and Kennedy's arguments closely repeated those which K. Brandt and V. Brack, the two individuals chiefly responsible for this policy, presented in their defence at the Nuremburg Tribunal.

We should note that Foster Kennedy had been president of the Euthanasia Society of America before resigning the post in 1939 because the society campaigned only for voluntary euthanasia, as against the compulsory euthanasia that he called for. In 1936, the Nazis had given him a doctorate *honoris causa* at Heidelberg University, along with H. H. Laughlin.[29] We should remember that at the time of this debate, in 1942, Americans were aware of the extermination of mental patients by the Nazis, if only through Shirer's article in *Reader's Digest*. Kanner also mentions Shirer in his argument against Kennedy's case for euthanasia.

A number of those politically responsible for 'euthanasia' in Germany were tried at Nuremberg in 1946–7. Dr Leonardo Conti, minister of public health, committed suicide; K. Brandt and V. Brack were hanged in 1948. Some of the doctors involved in this extermination were tried later, in some cases being condemned to death (certain of these had their sentences commuted to life imprisonment and were released after a few years). The great majority got off very lightly. Dr Hermann Pfannmueller, for example, director of the Eglfing-Haar asylum, who specialized

28 F. Kennedy, 'The problem of social control of the congenital defective. Education, sterilization, euthanasia', *American Journal of Psychiatry*, 1942, 99, pp. 13–16; L. Kanner, 'Exoneration of the feeble-minded', ibid., pp. 17–22. See also, in the same issue, the editorial comments on pp. 141–3.

29 S. Kühl, *The Nazi Connection*, pp. 86–7.

in the extermination of children by starvation (cited among those responsible for euthanasia in Theo Lang's report; see box pp. 198–9), was condemned to six years' imprisonment by a Munich court in 1949 for homicide with attenuating circumstances (he was ill), with the years he had already spent in prison deducted from this sentence.[30] Most of those implicated were not even prosecuted, as the greater part of the archives of asylums and extermination centres had been destroyed.

This question does not seem to have aroused great emotion, indeed hardly even much curiosity. The majority of victims here were German; they were sick, handicapped, abnormal, and so on; their value was seen as low; and there were more important things to be done. The word 'euthanasia' is a euphemism, and was clearly used to euphemize this extermination. Apart from the massacre of Jews, the general public were far more shocked by the Nazi medical experiments conducted on prisoners in the concentration camps.

It is not even certain whether this extermination of the sick, handicapped, and abnormal is included in the definition of crimes against humanity. At all events, eugenics is not. We can take the two cases, and try to compare them with the definitions of crimes against humanity and genocide given in the box below:

'Crimes against humanity: namely, murder, extermination, enslavement, deportation, and other inhumane acts committed against any civilian population, before or during the war; or persecutions on political, racial or religious grounds in execution of or in connection with any crime within the jurisdiction of the Tribunal, whether or not in violation of the domestic law of the country where perpetrated.'
(Article II.6(c) of the Charter of the International Military Tribunal, 'in pursuance of the Agreement signed on the 8th day of August 1945')

'In the present Convention, genocide means any of the following acts committed with intent to destroy, in whole or in part, a national, ethnical, racial or religious group, as such:
(a) Killing members of the group;
(b) Causing serious bodily or mental harm to members of the group;

30 Y. Ternon and S. Helman, *Le Massacre des aliénés*, p. 213; F. Bayle, *Croix gammée contre caducée*, pp. 840–1. See also p. 785 of Bayle's book, with the evidence of L. Lehner at Nuremberg on Pfannmueller's methods.

(c) Deliberately inflicting on the group conditions of life calculated to bring about its physical destruction in whole or in part;
(d) Imposing measures intended to prevent births within the group;
(e) Forcibly transferring children of the group to another group.'
[Article 2 of the UN Convention on the Prevention and Punishment of the Crime of Genocide, adopted 9 December 1948. The UN's earlier definition of genocide, given in UN resolution 96 (I) of 11 December 1946, was yet more general, and included 'political and other' groups as well as racial and religious ones.]

According to the internationally accepted definitions of 1945 and 1948 (but, importantly, not that of 1946), therefore, eugenics falls under the heading of a crime against humanity only if sterilization is considered an 'inhumane act'; there does not seem to have been any jurisprudence in this sense. And it falls into the category of genocide only if sterilization is conducted on a racial, religious or political basis. In other words, the sterilization of Gypsies (on the assumption that they form a race) or Jews (a religion) or Communists (political) could be considered genocide, but not that of mental patients, the handicapped, alcoholics or homosexuals – except in so far as members of the latter categories also belonged to one of the former, and were sterilized on these grounds.

'Euthanasia' does not fall under the definition of genocide, as it does not concern a racial, religious or political group. It could fall under the very vague definition of a crime against humanity. It should be noted, however, that apart from a few cases immediately after the war that were impossible to ignore, those responsible were not really pursued. This was undoubtedly because a large number of their victims were Germans (and therefore on the side of the conquered), and also because there were no organizations in a position to assert their rights, as was done in the case of the extermination of Jews. Very likely, again, it was because the definition was sufficiently vague and elastic to be either applied or not applied, according to the particular case. It indeed seems that, apart from Karl Brandt and the few in high positions who were judged at Nuremberg, those doctors subsequently prosecuted for 'euthanasia' were not judged for crimes against humanity.

In fact, the extermination of the sick, handicapped, abnormal, etc., was simply very quickly forgotten. And this is still the case today. When, by chance, it is recalled as more than a vague addendum to the extermination of Jews and Gypsies, it is generally to see it as a preparation for the latter. In other words, the extermination of sick people is still not ascribed importance in itself, but only for what it presages. This is another way of reducing the victims to negligible entities.[31]

31 Ternon and Helman cite L. Poliakov, for whom 'Germany's mental patients served as a "test bench" for the European Jews' (*Le Bréviaire de la haine* [Paris: Calmann-Lévy, 1951], p. 209), and Simon Wiesenthal's assertion that 'the euthanasia centres were regular schools of murder. I deal only with Hartheim (*The Murderers Among Us* [London: Heinemann, 1967], p. 272). They ask whether there may not have been a 'diabolical plan' by which sick people served to 'condition to murder' the torturers destined for the premeditated extermination of millions of Jews and Slavs; in this light, Operation T4 would have been stopped for this deliberate purpose (rather than under the pressure of public opinion), in order to 'transfer these pupils in killing to the East' (Le Massacre des aliénés, p. 146). It is hard to see in what way killing the sick was a better conditioning to murder than killing Jews or Gypsies. Besides, this explanation forgets that the 'genetically incorrect' had long been an explicitly designated target of all eugenicists, in Germany and elsewhere, even before the 'racially incorrect'. Although certain murderers of the mentally ill subsequently went on to work in the extermination camps, these camps drew a far larger personnel who had not been 'conditioned to murder' in that way. Finally, the official ending of Operation T4 in no way brought a halt to the extermination of the ill (not even to mention the case of children, who were not affected by T4). This kind of explanation is, above all, evidence of the difficulty of fitting the extermination of such a large number of sick people, handicapped, etc., into the a priori explanation of Nazism already decided on (see the following chapter).

12

The Resurgence of Eugenics

In 1927 the US Supreme Court ruled that sterilization was not a 'cruel and unusual punishment', and that eugenic legislation was not unconstitutional; not until 1942 did this stance weaken. But eugenics, as we said, did not immediately disappear after the Second World War. In 1948, Japan adopted such legislation, and from 1949 there were even attempts to restore the Nazi eugenic legislation of 1933 in West Germany.[1]

It was not the horrors of Nazism that led to the disappearance of eugenics, but rather the development of molecular genetics, which, if its beginning can be dated to 1944, only really got under way in the 1950s. Exactly the same thing then happened to eugenics as had earlier happened to biological sociologies – which is not surprising, since the theoretical (or at least, ideological) basis is the same in each case.

The development of molecular genetics relegated phenomenalist and mathematical methods to the background, along with the anthropological fable that depended on them – that of the degeneration of the human species. Genetics and evolutionism were now far better equipped from the scientific point of view, to the extent that they could dispense with the ideological buttress of social Darwinist doctrines. Eugenics gradually disappeared from scientific discourse, and thus from a political discourse that appealed to it. Only a few dinosaurs of pre-war genetics continued to hang on to it until their death – Muller, for example. Eugenic legislation came to be applied less often. And a silence fell on the question.

Eugenic activism found an outlet in this situation, reorienting itself to a new form of Malthusianism: control of world population, thanks in particular to Frederick Osborn (a former colleague

1 S. Kühl, *The Nazi Connection*, p. 104.

of Laughlin's) and again with the support of the Rockefeller Foundation.

To illustrate this reorientation, we can cite William Vogt's 1948 book *Road to Survival*, which enjoyed a certain success in the United States and was even translated into French. This is also a genuine curiosity, and it is hard to understand the relative oblivion into which it has fallen. The degeneration of humanity was now replaced by an apocalyptic description of the economic situation (all the themes of modern ecology are already present – it is no longer humanity that is degenerating, but the earth), coupled with a no less apocalyptic description of overpopulation. By protecting the weak and keeping them alive, society and medicine are multiplying populations beyond the level that the earth, already ecologically damaged, can support. Starting from this basic postulate, some of the most common eugenic themes are recycled, including resort to the Kallikaks and the Jukes, famous families of 'degenerates' that American eugenicists invoked in their quest for legislation (see pp. 226–7 below); these 'genetically incorrect' now become 'ecologically incorrect'. Here is a sample of Vogt's prose:

> The modern medical profession, still framing its ethics on the dubious statements of an ignorant man who lived more than two thousand years ago – ignorant, that is, in terms of the modern world – continues to believe it has a duty to keep alive as many people as possible. In many parts of the world doctors apply their intelligence to one aspect of man's welfare – survival – and deny their moral right to apply it to the problem as a whole. Through medical care and improved sanitation they are responsible for more millions living more years in increasing misery ...
>
> Then the French pharmacist, Louis Pasteur, gave to the world an understanding of microbes and their part in disease. In Europe life expectancies had been climbing, what with better diets, improved sewage disposal and water supply, more abundant food, a rising material standard of living. The control of a long series of diseases came within man's grasp, and the most effective remaining check on populations began to disappear ...
>
> Large-scale bacterial warfare would be an effective, if drastic, means of bringing back the earth's forests and grasslands ...
>
> With world populations as well as our own climbing, there is no likelihood of a decreased need for food for many decades to come; food – which means our land – has become a major political weapon in a world struggle ...
>
> The Jukeses and the Kallikaks – at least those who are obtrusively

> incompetent – we support as public charges. We do the same with the senile, the incurables, the insane, the paupers, and those who might be called ecological incompetents, such as the subsidized stockmen and sheepherders. These last, in so far as they deteriorate and destroy the grasses, expedite erosion, and contribute to flood peaks, are worse than paupers. They exist by destroying the means of national survival; were we really intelligent about our future, we should recognize such people as ecological Typhoid Marys – the source of environmental sickness with which they are infecting us all ...
>
> Every grain of wheat and rye, every sugar beet, every egg and piece of veal, every spoonful of olive oil and glass of wine depends on an irreducible minimum of earth to produce it. The earth is not made of rubber; it cannot be stretched; the human race, every nation, is limited in the number of acres it possesses. And as the number of human beings increases, the relative amount of productive earth decreases by that amount ...
>
> We, as the nation with greatest total wealth, are, of course, the number-one victim ...
>
> We are in a position to bargain. Any aid we give should be made contingent on national programs leading towards population stabilization through voluntary action of the people ...
>
> The greatest tragedy that China could suffer, at the present time, would be a reduction in her death rate.[2]

It was this neo-Malthusian movement that laid the basis for campaigns of sterilization in the Third World, a sterilization that served as an outlet for the surgical frenzy of the eugenicists. And this was also the context in which the research that led to modern means of contraception was conducted.

It seems that there has been an active desire to forget this page of history. Orthodox Communists, for example, kept silent, though they might have made use of this as an argument during the Cold War, being among the few to officially oppose eugenics. Certainly, the sterilization and extermination of the mentally ill were the subject of two articles in the French Communist periodical *La Raison* – appealing to Stalin and to Lysenko's biology. But all this was soon buried. In the same period, a pamphlet on 'bourgeois science and proletarian science', written by intellectual spokesmen of the French Communist Party and essentially devoted to Lysenko's ideas, dispatched eugenics in a single paragraph and

2 W. Vogt, *The Road to Survival* (London: Gollancz, 1949), pp. 61, 87, 140, 145, 193–4, 224.

had not a word to say on the role it had played in Lysenko's own campaign to take control of genetics in the USSR.[3] Apart from asserting a distinction between two different sciences, as the title indicates, the pamphlet confines itself to such bland statements as: 'Soviet biology seeks to develop agricultural productivity without limit', or: 'The cows of Karavajevo, with their monumental udders, were rejected because they had not issued from "pure breeds", even though the whole world knows that pure breeds exist only in the mind.'

Some fifteen years later, *The Brown Book: Nazi and War Criminals in the Federal Republic and West Berlin*, a GDR publication denouncing former Nazis still in service in the West, mentioned very few doctors: only Verschuer, and two or three others of minor importance, while eugenics and euthanasia were not even mentioned.[4] The Catholic Church remained similarly silent, despite its long opposition to eugenics. The former champions of eugenics, whether in the democratic countries or in Nazi Germany, clearly had every interest in avoiding any reminders of this aspect of their lives. And as eminent geneticists of that era were still active, it was impossible to count on denunciations by scientists or historians of science.

At all events, and whatever the reasons, what eugenics had been in the first half of the century was very quickly and almost totally forgotten – an oblivion from which we have only recently begun to emerge.

Here is a recent flagrant example of this forgetfulness, from a supposedly well-informed individual. In 1998, in an article on the UNESCO declaration on the human genome, Noëlle Lenoir (who chaired the UNESCO bioethics committee) wrote that it was not surprising that the Human Rights Declaration of 1948 had said nothing about genetics, and that the biological aspect appears there only in terms of the question of Nazi human experiments or the theories underlying the extermination of Jews and Gypsies.[5] Mme Lenoir seems to have forgotten here that even though genetic engineering was indeed fairly undeveloped at that time (limited to the transformation of streptococcus), geneticists had developed human applications of their discipline long before and had, since

3 F. Cohen, J. Desanti, R. Guyot and G. Vassails, *Science bourgeoise et science prolétarienne* (Paris: Éditions de la Nouvelle Critique, 1950).

4 *The Brown Book: Nazi and War Criminals in the Federal Republic and West Berlin* (Dresden: Verlag Zeit im Bild, 1965).

5 *Le Monde*, 9 December 1998.

1907, introduced these into the legislation of many countries in the form of eugenic laws leading to the forced sterilization of hundreds of thousands of people across the world. That in 1937 Germany extended the field of this legislation to racial questions, and the mixed-race children of the Ruhr and Rhineland were sterilized in the name of eugenics. And that finally, that in the summer of 1939, malformed infants were eliminated, followed by mental patients, handicapped people, alcoholics, uneducable children, and others – a goodly number of them having already been sterilized, and who were in any case all deemed to be 'genetically incorrect'.

Today, all this is well known to historians of biology. And at the end of the war, when it was all fresh, it could not be ignored by those with political responsibility. There were thus very good reasons why the Universal Declaration of Human Rights, adopted in 1948, should have included a few allusions to genetics and its eugenic applications. But there were also very good reasons why it remained largely silent on this question.

First of all, eugenic legislation remained in force in several countries; it was public knowledge that the Nazi law of 1933 had been based on American legislation, and it was not politic to link eugenics too closely with 'euthanasia'. Besides, almost all competent scientists had supported this legislation and continued to do so: sterilization continued until the 1960s. Eugenics could scarcely have been questioned in the 1948 declaration, when Julian Huxley had just been appointed head of UNESCO and the Nobel Prize awarded to Hermann Muller. In 1948, had eugenics been criminalized, there would have been a fine range of people occupying the bench of the accused – almost all the luminaries of the genetics of the day. Perhaps it would also have been necessary to recall the actions of the Carnegie Institute and the Rockefeller Foundation in this field.

The argument was very simple: it was not eugenics that was in question, but simply its perversion by the Nazis. (This is what Goldschmidt said, as cited on p. 190 above, and it is still said today by those seeking to revive eugenic discourse and even eugenic practice.) Here is what Jean Sutter wrote in 1950:

> Hitler's political action was largely inspired by eugenics, as is clear from the eleventh chapter of Mein Kampf, titled 'People and Race'. He had read in prison the treatise on human heredity by Baur, Fischer and Lenz. He knew of the work of Ploetz and Lenz's Deutsche Gesellschaft für Rassenhygiene,

> and had understood its significance. He even attended eugenic meetings before 1933. Intellectual circles preoccupied with these questions had enthusiastically welcomed him, as is attested by the dithyrambs that appeared in specialist periodicals on the occasion of his fiftieth birthday in 1939. Herr Fischer, for example, expressed himself in the following terms: 'What state would we be in today if this immense action of racial policy, demographic policy and social hygiene had not been victorious?', and continued: 'Everything must be undertaken to preserve our people in good physical and moral state: the right of individual freedom must yield to the duty to maintain the race.' Speaking of Hitler, Fischer said: 'Humanity must take part in his work.' Hitlerite eugenics was completely divorced from Galton's direction in that it was put uniquely at the service of a race that, as it happens, was no more than an artificially created concept, based on the erroneous views of romantic philosophers such as Gobineau and Vacher de Lapouge.[6]

It is uncertain whether attempts to shift responsibility, such as Sutter's attempt here, convinced a wide public, even if Gobineau and Vacher de Lapouge had long been favourite punching bags. It was all very awkward. It was preferable to look elsewhere and keep a discreet silence on the question of eugenics. Geneticists, even Nazis such as Verschuer, slipped through the net as long as they did not directly have blood on their hands (in Verschuer's case, this was only via Mengele), and the official history of genetics settled for a few blank shots. The victims were forgotten in the definition of crimes against humanity – they were marginalized, their numbers minimized, and their elimination charitably euphemized by calling it 'euthanasia'.

The overvaluation of genetics

When molecular genetics ran into theoretical difficulties in the 1970s, and still more so in the following decade, it sought to escape these by turning towards fields where theory was secondary, either because these were purely technical (the deciphering of genomes) or because only a minimal theoretical foundation was involved (genetic engineering is largely empirical).[7] In this way, the dominant theoretical framework in molecular genetics, that of

6 J. Sutter, *L'Eugénique, problèmes, méthodes, résultats*, p. 236.

7 On these theoretical difficulties, see A. Pichot, *Histoire de la notion de gène* (Paris: Champs-Flammarion, 1999).

the genetic programme, could be maintained despite all its defects. It was geneticists themselves who reoriented their discipline in this direction. Very understandably so, and even excusably, whatever we may think of this way of avoiding the need to deal with the theoretical problems. What is less understandable, and emphatically less excusable, is that in order to justify the new orientation, these geneticists deliberately overvalued their results, and still more so the possibilities of their application. The theoretical difficulties should rather have made them cautious and modest.

Famous biologists then began to make statements promoting the applications of genetics to social organization – surprising statements, to say the least. There certainly still existed in this field a tradition that had not completely disappeared; Frank MacFarlane Burnet, who received the Nobel Prize in Physiology or Medicine in 1960, is a caricature of this, rehearsing as he did the worst mistakes of Darwinian sociology from the first half of the century.[8] One might believe that this approach would have remained marginal, instead of which there has been a resurgence of such ideas.

Francis Crick, for example, joint winner of the Nobel Prize in 1962, along with James Watson, for their discovery of the structure of DNA, did not hesitate to declare many years later: 'No newborn infant should be declared human until it has passed certain tests regarding its genetic endowment and that if it fails these tests, it forfeits the right to live.'[9]

What Crick certainly did not know is that this proposal repeats almost word for word a project dreamed up in 1895 by Alfred Ploetz, founder of the German Society for Racial Hygiene, according to which each newborn infant would undergo a medical examination (the term 'genetic' was not yet used) in order to establish its right to live; he even added a further examination at the age of sixteen, to find out whether the individual in question was entitled to marry.[10]

The objection could indeed be made that Crick's proposal was simply a particular case, coming moreover from a person who had never been seen as especially cultivated. Yet it expresses

8 F. M. Burnet, *Credo and Comment* (Melbourne: Melbourne University Press, 1979).

9 Cited by P. Thuillier, 'La tentation de l'eugénisme', *La Recherche*, 155, May 1984, pp. 743–8 (quote from p. 744).

10 Cf. J. Roger, 'L'eugénisme, 1850–1950', pp. 424–5.

very well the general climate that was established (or rather that resurged) when geneticists, supported by the media, set out to overvalue their discipline at a time when it was lost in conceptual blind alleys. This is the same general pattern as at the end of the nineteenth century, when Darwinian biology made up for its scientific inadequacies by resorting to social pseudo-applications that served as an ideological buttress.

Since the present overvaluation of genetics is in no way underpinned by a theoretical advance in the discipline, but on the contrary rests on difficulties in this field, it inevitably ends up with eugenics, given that the latter, as a negative measure, has no need for an elaborate theoretical support. This eugenics is no longer a population eugenics, like that before the Second World War, but rather an individual eugenics. Population eugenics corresponded to the phenomenalist and mathematical approaches of the genetics of the time. Individual eugenics relates to the present molecular approaches, today's genetics being concerned with the 'physiology' of heredity, rather than the statistics of phenotypes. At this individual level, eugenics is founded on possibilities of prenatal detection of hereditary diseases, possibly to be followed by abortion.

The aim is not to improve humanity (the population concern) but to prevent the birth of individuals whose life is not worth living. This was precisely one of the themes developed by Nazi biology, but which it sought to resolve by euthanasia; the phrase sometimes used today, in the case of prenatal detection followed by abortion, is 'euthanasia of the foetus'.[11] This change of level has made it possible to say that this is no longer eugenics in the proper sense of the term. But in reality these new measures correspond completely to the definition of the eugenic project: to ensure the production of 'well-born' individuals. The only difference lies in the genetic foundations: today molecular, previously phenomenalist and mathematical.

The geneticist Daniel Cohen, moreover, creator of the charity Téléthon, recognized this perfectly well when he set out to find a certain charm in Galton's old eugenics, which he reproached only for a racist connotation that was 'in the end very marginal' – the hundreds of thousands sterilized, then exterminated in a good many cases, are for him a mere detail – and to welcome the fact that China intended to supplement its programme of quantitative

11 L. Milliez, *L'Euthanasie du foetus* (Paris: Odile Jacob, 1999).

control of births by a qualitative control.[12] He supplements this in his book with the following refrain, already heard a thousand times, and simply updated to today's taste:

> Down with natural selection, long live the human control of life! What good is it to conceal the fact? It is clear that man, in a more or less near future, will have the power to modify his genetic heritage. And the apprehension that evoking such a situation arouses scarcely seems justified ... I am persuaded that the future man, who has perfectly mastered the laws of genetics, can be the artisan of his own biological future, and not of his degeneration.[13]

While awaiting this future, Daniel Cohen has nothing better to propose than a eugenics that has the advantage of not requiring that one has 'mastered the laws of genetics'. In other words, the freedom of artificial selection designed to replace the dictatorship of natural selection.

At the same time, the fable of degeneration has been revived, not in its old form, too hollowed out to be of use, but in a modern one. Geneticists have announced with trumpet calls in the media that humanity is threatened by several thousand genetic diseases.[14] All the same, as the new eugenics no longer bears on a population but on the individual, this threat is not accompanied by a discourse about degeneration, but rather on the way in which future parents can avoid, thanks to prenatal detection, having a child affected by one or other of these diseases – while awaiting the possibility of curing them in an indefinite future.[15]

In crude terms, it is no longer a question of waving the worn-out

---

12 D. Cohen, *Les Gènes de l'espoir* (Paris: Lafont, 1993), p. 229.

13 Ibid., pp. 262–3.

14 For example, 'Le marché des gènes', *Sciences et Avenir*, 565, March 1994.

15 Officially, work on hereditary diseases aims at perfecting gene therapies. However, these have met with considerable theoretical and technical obstacles, and at the present time no way out of these can be seen. Added to which are the problems posed by the small number of individuals affected by each of these diseases. Is it possible, given the technical difficulty and the procedures to be followed, to develop viable therapies for such small numbers? Especially as hereditary diseases vary greatly in kind, and it is not certain that the methods used for one would be fully applicable to another. This difficulty is only increased by the fact that, thanks to detection and therapeutic abortion, the number of those affected is likely to decline further. In such conditions, can anything more be done than a few experimental procedures for the simplest cases? It seems evident that knowledge of the genetic basis of a hereditary disease allows a better understanding of it, which may permit envisaging a non-genetic therapy, by way of medication, that would often seem more promising.

spectre of degeneration (a catastrophism that is now ineffective), but rather provoking among future parents, and especially pregnant women, who are particularly sensitive to this subject, the fear of a malformed baby – proposing (for sale?) detection kits as the solution to their anxiety. The threat is to the individual, not to the population; so, likewise, is the remedy. As we have seen, this is simply a rhetorical procedure, since the essential tenor of the discourse is the same as in the first half of the century (simply adding a mercantile aspect, the interest of biotechnological commerce, where formerly it was the social cost of 'genetically damaged' individuals that was invoked).

We have to recognize, however, that whereas population genetics was completely ineffective as far as the genetic composition of the population was concerned, this individual eugenics does produce a result at the individual level, to the extent that it effectively avoids the birth of sick individuals, although this individual effect still has no impact at the level of population.

In parallel with this, various intellectuals have offered their versions of the issue. In the great majority of cases, these have been critical; they clearly remember the 1930s better than do the geneticists. Two strands, however, stand out as exceptions.

One of these can be illustrated, in France, by articles published in the periodical *Esprit* by P.-A. Taguieff (an authority cited by D. Cohen,[16] and whom one would not expect to find in this role). Taguieff clearly has no awareness of the true reasons why genetics has reoriented towards these techniques, and confines himself to supplying them with intellectual support. Confident in progress and science, he adopts a positive and progressive attitude. Added to which is a refusal to divinize life and declare the genetic heritage untouchable, which he considers is a religious and thus retrograde position. Criticism of the technical drift of genetics, according to him, amounts to a new kind of irrational, or even irrationalist, religion, to be placed on the same level as the Catholicism, which is clearly as hostile to this new eugenics as it was to the old.[17]

This is the discourse of the eugenicists of the first half of the twentieth century, updated to the contemporary idiom, with the aforementioned transition from the population level to that of the individual, along

16 D. Cohen, Les Gènes de l'espoir, pp. 227, 229.

17 P.-A. Taguiefff, 'L'eugénisme, objet de phobie idéologique', *Esprit* 156, November 1989, pp. 99–115; 'Retour sur l'eugénisme. Question de définition (réponse à J. Testart)', *Esprit* 200, March–April 1994, pp. 198–214. See also the controversy with Testart in *Esprit* 199, February 1994, and 205, October 1994.

with a confirmation that there is no question of state planning of reproduction – as if social and economic pressure, combined with that of the media, were not as constraining as legislation, and as harsh as the dictatorship of natural selection. The whole thing is accompanied by the reminder that eugenics is not necessarily racist or Nazi (repeating the arguments of Goldschmidt and Sutter, see pp. 190, 217–8, 256 and 288 above).

I admit I have difficulty understanding the purpose for which this type of discourse is designed, if not to maintain the most questionable assertions of certain geneticists – questionable from both a scientific and a moral point of view. Can a geneticist at the same time wave the absurd threat of thousands of hereditary diseases, vaunt 'new look' eugenics, and have an interest in a biotechnology company able to produce kits for genetic diagnosis?[18]

As for the second strand of ideologists to show a positive reaction towards the resurgence of eugenic doctrines, these are associated rather with the far right, even with neo-Nazism. This is the case with Jean-Marie Le Pen's praise of Carrel, for example (the starting-point of this pseudo-controversy). Or again, in 1987, at the time the Téléthon was launched, it was the case with the republication of an apologia for Nazi eugenics, written in the 1930s by the racial theorist Hans Günther – apparently a perfectly legal thing to do, and one which failed to arouse any protest.[19]

The pattern of the early twentieth century has thus been reconstituted, without any awareness of what is happening. On the one hand, the retrograde Catholic opponents; on the other, geneticists and their progressive champions (in general, anti-papist), to whom can be added discreetly the theorists of Nazism, also champions of eugenics. The whole edifice silences those who try to make audible the voice of reason.

Let us now try to return to this reason, and assess, independent of any moral aspect, the validity of the new project of individual eugenics.

Since it appeared at the point when the molecular genetics that underlies it was experiencing theoretical difficulties, this assessment must necessarily take account of these difficulties – something that

18 For the business links of geneticists, see, among others, 'Le marché des gènes', *Sciences et Avenir*, 565, March 1994; B. Andrieu, 'Le capital génétique', La Pensée, 306, 1994, pp. 75–89; J. Rifkin, *The Biotech Century* (New York: Putnam, 1998).

19 H. F. K. Günther, *Platon, eugéniste et vitaliste* (Puiseaux: Pardès, 1987).

neither the geneticists themselves have done, having no interest in acknowledging difficulties, nor have the various ideologues, lacking as they do any understanding of the question. I shall not go into these theoretical difficulties of genetics here; I simply refer the reader to my work on the history of the notion of the gene (see note 7 on p. 218 above), and shall thus confine myself to the practical aspect of things.

Let us take a look first of all at these 'thousands' of genetic diseases that threaten us – five thousand, according to one authority. As I have noted on a number of occasions,[20] if each of these diseases killed a hundred people per year in France (which is very few in terms of the total population), they would alone be responsible for almost the entire annual number of deaths, which is just over half a million. In fact, however, some 175,000 of these deaths are due to cardio-vascular diseases, and 150,000 to cancers. It is not that geneticists have lied on this subject; they have simply taken up the figures given in such standard sources as *Mendelian Inheritance in Man*.[21] From a medical point of view, this simply makes no sense.

We can then turn to the effectiveness of individual eugenic measures, starting from the numbers associated with an example of a genuine hereditary disease, Friedreich's ataxia (monogenetic and recessive, and quite rapidly fatal), the 'gene' for which was discovered in 1996. (To be correct, it is a 'pathogenic mutation' that was discovered, rather than a 'gene'.) This discovery was quite naturally announced in the press, along with a commentary specifying that one person in 120 was a bearer of this 'gene'.[22] This means that there are some half a million bearers in France, who are not necessarily affected by the disease, since the mutation is recessive, requiring therefore two 'genes'. For a naive reader, this seems a very high figure, considerably enhancing the importance of the discovery reported. It also has the benefit of scaring future parents; might they not be among these half-million unheeding bearers?

20 A. Pichot, *L'Eugénisme, ou les Généticiens saisis par la philanthropie* (Paris: Hatier, 1995), p. 58; Pichot, 'Racisme et biologie', *Le Monde*, 4 October 1996.

21 V. A. McKusick, *Mendelian Inheritance in Man: Catalogs of Autosomal Dominant, Autosomal Recessive, and X-Linked Phenotypes* (Baltimore: Johns Hopkins University Press, 1992), cited in J.-C. Kaplan and M. Delpech, *Biologie moléculaire et médecine* (Paris: Flammarion, 1993), p. 266.

22 For example, *Le Monde*, 9 March 1996.

A brief calculation, however, shows how in these conditions, out of slightly more than 700,000 annual births in France, there would be some twelve or thirteen infants affected by Friedreich's ataxia (only an approximate figure, but at this level of rarity any calculation can only be approximate). If the population is in balance, twelve or thirteen people can be expected to die of the disease each year.

Approximate as they are, these figures – half a million healthy carriers, thirteen deaths, and thirteen new cases per year – indicate the relative interest in therapy (thirteen cases per year) and detection (500,000 carriers) for the pharmaceutical industry. Let us leave aside here the question of therapies (see note 15 on p. 221 above) and instead focus on eugenics based on prenatal detection, possibly followed by abortion.

Assuming that the half a million individuals carrying the anomaly could be successfully detected (which would require tens of millions undergoing a genetic test, at a cost of several hundred euros per test), the result would merely be the avoidance, via prenatal diagnosis and abortion, of the birth of a dozen sick children each year.[23] It is clear that no one would embark on a systematic detection programme on this basis, and that only families known to have already presented cases of this disease would be enrolled for such tests. This might succeed in avoiding the birth of a few sick children each year, and reassure a few other families when the test result was negative. Over the years, and assuming a database and a follow-up of families at risk, this figure might be increased to a dozen or so – if such a database were authorized, given the weak results that can be envisaged.

A genetic diagnosis of this kind, despite its evident interest for the families affected, would thus have practically no noticeable effect on public health. Besides, we cannot really say that it would significantly increase the number of abortions, which in France is about 220,000 per year. These therapeutic abortions, where a choice is made of which foetuses will survive and which will not, should be differentiated from regular abortions that are no more than belated contraception, and where a choice of this kind is not involved; but this is a moral question, not a technological one.

---

23 There would still remain those cases that appear as a result of new mutations in families previously unaffected, and that could not therefore be detected.

The same argument can be made for the majority of serious hereditary diseases, i.e., those capable of being detected in this way. These diseases all affect very small numbers of people. The most common are cystic fibrosis (some 350 cases each year in France, of varying severity), Duchenne muscular dystrophy (about a hundred cases), and haemophilia (about seventy cases). On the scale of a country such as France, most such diseases manifest no more than a few dozen cases each year, and often fewer. Their possible prenatal diagnosis, followed by abortion, would have no effect whatsoever on public health, and still less so on the genetic heritage of the human race, give that these diagnoses and the follow-up involve only a small fraction of the world population with the financial and technological means to implement them. Nor would it have a large effect on the number of abortions. There is thus absolutely no common measure between the social and medical import of these genetic procedures and the media barrage that surrounds them.

All that remains is the moral problem they raise, but it is not for me to decide whether a woman should or should not keep an embryo that is known to be affected with a serious illness. The elements on which one bases the decision as to whether a life is worth living or not are elusive, to say the least, and it is by no means certain that genetic criteria are always the most relevant.

The objection could certainly be made that this statistical approach is somewhat inhumane, and neglects the individual dramas involved when certain hereditary diseases strike particular families. To which I would respond that, given the very small numbers, this field covers nothing other than particular cases, and these particular cases, however spectacular, can never be viewed as examples from which general conclusions and laws can be drawn. Sadly, this has always been the case here, and the exploitation of such particular cases has always been, and remains, par for the course.

At the origins of American eugenic legislation we find particular cases that were carefully presented as examples – so particular, indeed, that we still know the names of the individuals in question. Above all, there is the particular case of the celebrated Martin Kallikak, who, during the American Civil War, had a child with a feeble-minded girl. In 1912, out of this child's 480 descendants, only 46 were deemed to be normal – though the reliability of these figures is certainly suspect. There was also in the United States the particular case of the Searing family, of whom, out of 600 individuals studied, 66 died before the age of 3, and 45 between 4 and 25; there were 24 cases of mental illness, 17 indigent, 16 alcoholics, 31 tubercular, 10 abnormal, 10 epileptic, and so on. Not to mention the particular case of the Juke family, of whom 300 died at an early age, 300 were beggars (who spent a total of 3,600 years in shelters), 440 were affected with various diseases (venereal, in particular), 130 delinquent (including 7 murderers), and of whom half the women were prostitutes.[24] These cases seemed sufficient to prove the hereditary character of anything one liked, and justify any legislation for eugenic sterilization.

At the origin of the extermination of malformed infants and the mentally ill by the Nazis, we also find the particular human drama of the Knauer family, who had an incurable child for whom they requested and were granted *Gnadentod* (mercy killing) – heralding the murder of hundreds of thousands who had not been asked. Nazi propaganda films such as *Erbkrank*, which Laughlin wanted to distribute in the United States, exhibited sufficiently terrible particular cases to justify eugenics, or even the 'euthanasia' of these sick people.

Charitable spectacles such as Téléthon likewise display particular cases, showing sick children (the most presentable, sufficiently damaged to arouse pity but not so damaged as to arouse disgust – not wise to shock prospective donors) rather than venturing to provide actual figures, and also discussing genetic therapy, while they know very well that once the 'gene' is discovered, what will follow are detection tests and therapeutic abortions.

The pinnacle of the particular is reached with diseases that are so rare, and affect so few individuals, that they are known as 'orphan diseases' – a label that sounds straight from an advertising agency. Such cases are so particular that they have scarcely been studied, not just because such diseases would not be profitable for the pharmaceutical industry but also because it is inherently difficult to study a disease of this kind, let alone envisage a therapy, when there are so few known cases. Medicine has never performed miracles, even with sufficient financial resources.

It is undoubtedly because of the speed with which it became clear that the applications of genetics to genuinely hereditary diseases are extremely limited (in terms of both therapy and eugenics) that the attempt has been made to extend them to a broader field, one more susceptible of valorizing the discipline and

24 J. Sutter, *L'Eugénique, problèmes, méthodes, résultats*, pp. 121–3.

supporting it in its theoretical difficulties. Herein lies the origin of the 'pan-geneticism' that is fashionable today.

First of all, there is the supposed genetic therapy for cancer, which would certainly concern a far larger number of people, but which seems no easier from a technical point of view, than genetic therapies designed for hereditary diseases. Cancer has been over-invoked to argue for many things, in genetics as elsewhere. The founding work of modern genetics, *The Mechanism of Mendelian Heredity*, which presented the work done by Morgan and his team on *Drosophila*, was published in 1915. Three years later, there appeared an article on a hereditary tumour in *Drosophila*.[25] This was the first of a long series, which today is beyond counting. Yet the advance in therapies for cancer has been due to clinical research rather than to basic research of this kind.

Then there are the supposed genes for intelligence, schizophrenia, longevity, homosexuality, alcoholism, conjugal fidelity, and so on. We have seen how Davenport extended heredity to behaviour with his study of nomadism; all these genes are no more than the modern equivalent of Davenport's extension. The approach is so worn-out that it does not even merit discussion.

More delicate are questions of genetic predisposition and predictive medicine. Not a week goes by without the announcement of a newly discovered gene involved in one disease or another, whether diabetes, obesity, cardio-vascular disease, etc. Not that the gene in question is said to determine these disorders single-handedly (as would be the case with a genuinely hereditary disease), but that it is deemed to be part of the determination process, predisposing the individual who carries it to the disease in question. Such discoveries are inevitably accompanied by a commentary stating – in case we have forgotten – that they open the way to far-reaching perspectives for the development of treatments.

It is in this domain of 'genetic predispositions' rather than in that of genuinely hereditary diseases that there could be a real danger of eugenics – on the one hand, because these predispositions are far more vague than established hereditary conditions; on the other hand, because they involve far more people. Take the predisposition to obesity: with more than 20 per cent of American children considered obese, plus countless women trying to become

25 M. B. Stark, 'An hereditary tumour in the fruit fly Drosophila', *Journal of Cancer Research*, 1918, III, pp. 279–300.

slim, genetic control of body weight is far more interesting to the pharmaceutical industry than all the hereditary diseases put together. We can understand from this what are the real results expected from basic research in genetics.

For predispositions as for genuinely hereditary diseases, given that the possibilities of treatment necessarily lag behind those of diagnosis, it cannot be ruled out (if the option proves commercially attractive) that certain people will envisage a eugenic solution, this time without the limits previously described, and on a far broader scale. It will certainly be said that this scenario is unlikely, but in this field unlikelihood does not count for much; we need only cast a glance at the recent past to be convinced of that.

These genetic predispositions are considered to fall within the framework of predictive medicine. To take the definitions given by Jacques Ruffié, the issue essentially arises in cases where a genetic constitution renders the individual in question particularly susceptible to develop a certain disease under certain environmental conditions:

> Predictive medicine ... consists in predicting, at birth or even before, situations of risk that subjects may encounter in the course of existence (that is, during the full unfolding of their genetic programme), according to two sets of factors:
> (a) The constitution of their genetic inheritance, which may be more or less capable of responding to certain environmental conditions;
> (b) The types of demands or aggressions they will experience from this environment, and the possibility they will have of dealing with this according to their innate aptitudes ...
>
> Predictive medicine always contains an element of chance, being conditioned by two sets of factors, genetic and environmental, which may or may not combine. It assesses a risk and indicates the conditions under which the disease may appear. It creates the possibility of avoiding pathogenic situations. It does not establish a disease but defines a possibility or a probability of its appearance.[26]

This is worryingly similar to the arguments of Verschuer and his ilk – arguments that were by no means specifically Nazi but sought to establish a hereditary predisposition for all kinds of disease, infectious or otherwise, for example tuberculosis (which

26 J. Ruffié, *Naissance de la médecine predictive* (Paris: Odile Jacob, 1993), p. 60.

also results from the combination of a pathogenic element and a terrain on which it is susceptible of developing). Ruffié's is simply a modernized version, adapted to molecular genetics. In itself, this project – like Verschuer's own – is not to be condemned *a priori*. The question is what is done with it, for not only does it confront a number of difficulties, it also arouses strong distrust on the part of anyone with even a slight acquaintance with the history of these doctrines.

Medicine has been familiar with the importance of terrain ever since Hippocrates, but it is obviously far easier for it to act on the pathogenic element, and whenever possible, that is what it does. Knowledge of the terrain will at best lead to the adoption of a more specific treatment, adapted to the particular case. But is genetic analysis really necessary for this? Apart from a few very simple particular cases, the association between a certain genetic factor and a certain predisposition will most likely be no more than empirical. In this case, a refined diagnosis could certainly be based on other criteria, equally empirical but far simpler, which would have the advantage of taking into consideration not only the hereditary factors but also the characteristics inherent to the specific history of the individual, which equally constitute a 'terrain'. Conditions of life, and still more so their combination in an individual history, do not just come into play as 'triggers' of a disease for which the individual has a genetic predisposition; they are capable of leading to predispositions as much as genes may do. By their exclusive attachment to the dualism of hereditary versus environmental factors, and by neglecting individual history, predictive medicine reproduces a stereotype of genetics (which confuses memory and history) and raises a strong risk of getting bogged down in the same difficulties as genetics did. It is hard, in fact, to see how preventive medicine can lead to concrete applications that are genuinely consistent. Besides, if we refer to the book by Ruffié cited above, it is notable that the sections devoted to prediction are practically all written in the future tense, and occupy only a few paragraphs out of 475 pages (only the case of psychiatry is developed a bit, extending to five pages).

When action on the pathogenic element is not possible (in instances when no treatment exists), predictive medicine will at best, and in a very theoretical way, lead only to a possible improvement in prediction. Certainly this is not negligible, yet apart from a few very simple particular cases, knowledge of a predisposition will remain essentially empirical, and the predisposition known in this way will be simply statistical. There is a risk, then, of falling not only into the dualistic stereotype of hereditary/acquired, but also into a paralogism of Weismann's that sees any acquired characteristic as acquired only because there exists a hereditary predisposition to acquire it:[27] any disease appears only because there is an adequate hereditary terrain for it. On this basis, anything is possible, including and above all a eugenics which has the advantage of not requiring well-developed knowledge.

However well founded its basic principles might appear, predictive medicine is thus fraught with uneasy implications. The medical ideal is first of all to heal. Prevention certainly has its place, but medicine must take patients as it finds them, not modify them or sort them.

As far as the 'social' counterpart of this predictive medicine is concerned – that is, the application of its supposed predictions to insurance, employment, and so on – it is as difficult to draw a firm conclusion for that as it is for the properly medical aspect. *A priori*, it is best to be distrustful, as this kind of assertion has a past. Here, for example, is what Verschuer declared in 1938, at an international medical congress on occupational accidents and diseases, held in Frankfurt:

> Only in very rare cases can we envisage the cause as purely external (an occupational accident or lesion) . . . In other cases, this occupational accident or lesion are only triggers for a hereditary pathology . . . In Germany, the results of genetic research have already been integrated into measures of public health. Prophylaxis of hereditary diseases by sterilization, prohibition of marriage and marital counselling is regulated by specific legislation.[28]

There is nothing new under the sun, in this field at least. History repeats itself, and if you know the original version, the copy is rather less interesting, being generally a farce instead of a tragedy. We know what the database of the Eugenics Record Office was used for, likewise the anthropological studies of the German population made by Fischer, and those of Ritter on the Gypsies. In their own time, these were seen as the latest thing in science and public policy. And now, in our time, there is a project under way to draw up a similar database on a hundred thousand Icelanders

27 A. Weismann, *Essays on Heredity and Kindred Biological Problems*, Chapter 7.

28 Cited in E. Klee, *La Médecine nazie et ses victims*, p. 328.

(apparently without even a backward glance at the million files that the Cold Spring Harbor Laboratory compiled on American families).

In a general sense, the medical applications of genetics are undoubtedly far less important, and far more difficult, than is claimed – with eugenics being an easy option in this field. The media hype around such work is above all an operation of public relations, designed on the one hand to regild the escutcheon of a discipline beset by theoretical difficulties, and on the other to facilitate the financing of work that, it is hoped, will first of all conceal these difficulties and subsequently lead it out of its conceptual cul-de-sacs (a quite vain hope, in my opinion). In fact, the main result of this hype is ideological: the creation of a state of mind that reduces everything to heredity, with all the risks inherent in that reductionism.

III. TAXONOMY, EVOLUTION AND RACISM

13

The Classification of Races

After eugenics comes racism. Not that the latter makes its appearance at a later time, but rather that the particular racism at issue here, i.e., modern racism, is fashioned after the model of eugenics. Eugenics presents itself as the purification of society of all those individuals who are 'undesirable' and 'inferior' from a biological or psychological point of view. Among these undesirables and inferiors necessarily figure those of particular races, with particular hereditary characteristics that are not deemed 'eugenic'. Besides, one of the leitmotifs of the late nineteenth and early twentieth centuries was that racial characteristics were hereditary. Though this idea seemed to go without saying, the idea was stressed by authors of the time, most likely because it implied that racial questions fell less within the anthropological or social domain than that of genetics, and thus eugenics. In Germany, racial hygiene combined both eugenics and racism in variable proportions.

The main focus of our interest here will be on the relationship between this racism and taxonomy, rather than on the notion of race set forth in the various anthropological theories that have succeeded one another over the last hundred and fifty years. These theories, various and highly diverse, are often quite removed from what is customarily seen as science.

We shall start by correcting two conceptions of biological racism that are more or less mythological. The first of these relates to Gobineau, the habitual target who is supposed to have founded modern racism. The second relates to the legend according to which Darwinism, by replacing an essentialist conception of species (and hence of race) by a kinship between living beings, supposedly relativized taxonomic categories and in this way opposed racism.

Gobineau's racism

Though it is unwarranted to see Gobineau as a forerunner of eugenics, there is no doubt he was racist. This is quite unchallengeable: his conception of history is almost entirely based on a notion of race. On the other hand, it is highly questionable whether this racism bears any relation at all to biology, let alone to a biological conception of society.

What Gobineau's *Essai sur l'inégalité des races humaines* has in common with Darwin's *Origin of Species* is that both books are more often appealed to than actually read. Admittedly, both are very long, deeply boring and somewhat incoherent. The result is that all kinds of marvellous ideas are ascribed to Darwin that he never actually had, while Gobineau is blamed for abominable theories that have nothing to do with what he wrote. One is the object of hagiography, the other serves as a foil, blamed for all the crimes of the century.

In actual fact, Gobineau's racism is far from the worst kind, and has little in common with a biological notion of race. Rather than having a connection with taxonomy, it relates to a social order, even an order of the world, understood along the lines of the Indian caste system. This racism, indeed, fits into the context of the Aryan myth, which in turn refers to India. The inequality of races constructed on this caste model displays a double inequality: an inequality in the sense of difference and an inequality in the sense of hierarchy.

On the one hand, each race is supposed to have received from nature the particular gifts that characterize it, hence the biological differences between them. On the other hand, this inequality in the sense of difference is the substructure for a hierarchy within civilization. The white race, and the Aryan in particular, is evidently assumed to be the superior race, with the highest civilization; it is even the only civilized race. The black race, no less evidently, is assumed to be the one least capable of civilization. The yellow race stands between these two. But there are no precise and measurable criteria for civilization (Gobineau actually sees the black race as originating artistic creation), nor even a linear progression.

This hierarchy of races carries neither a logic of exploitation (as is the case with colonial racism), nor a logic of extermination (as is the case with Nazi racism). Gobineau inclines rather to a kind of paternalism on the part of the superior race towards the inferior ones, somewhat akin to the pattern of the Jesuits in South America. It cannot really be said that he shows contempt for 'inferior' races. His *Nouvelles asiatiques*, one of the gems of French literature, actually attests to a certain sympathy for the civilizations and customs described in it. Only the first of these stories – the weakest of them – contains an allusion to his racial theories, and what it particularly condemns is interbreeding, rather than a particular race as such.

With Gobineau, the racial hierarchy is certainly a biological one, to the extent that it relates to characteristics that are supposedly hereditary, but it has nothing in common with an evolutionary hierarchy running from black (the least evolved, close to the ape) to Aryan (the summit of evolution), with a whole series of gradations in between – Gobineau was an opponent of Darwinism. And it is in no way accompanied by a progress in the degree of humanity, from ape to superman.

Gobineau rejected any comparison of the 'lower' races to apes. For him there was a fundamental difference between man and animal, and no hierarchy in the degree of humanity.[1] This enabled him to reconcile his racism with his Catholicism (the doctrine of the church saw all men made in the image of God, whatever their race). According to Gobineau, men of every race had a soul, and it was this soul alone that the church took into account, not their degree of civilization or their intellectual level. It was in this way that the church could claim to be catholic (in other words, universal), and Christianity could claim to address all peoples and races without distinction.[2]

Moreover, for Gobineau the human species was a single species – a question much debated in his time. He puts forward as arguments the possibility of fertile crossings between different races (evidence of belonging to one and the same species), and, more surprisingly, the fact that the Bible speaks of only a single creation of man – Adam and Eve – not of the creation of different races.[3] Gobineau thus subscribed to monogenism, a single origin for the human species. But clearly he was not very firm in this conviction, yielding to the authority of the Bible rather than being convinced by scientific argument.

At all events, racial difference for him was not original, as it is in contemporary polygenetic theories, even evolutionary ones. For

1 J. A. de Gobineau, *Essai sur l'inégalité de races humaines* [1853–5] (Paris: Belfond, 1967), p. 95.
2 Ibid., pp. 87ff.
3 Ibid., p. 133.

Gobineau, the differentiation of races was due to the dispersion of primitive men into different regions with different climates, at a time when conditions of life were more extreme, and so could strongly mark the individuals subjected to them. (This was indeed one of the explanations of the origin of species that was current at the time.) On the basis of this, these differentiated races, now permanently distinct and characterized indelibly in their forms and features, remained separate, keeping only their common origin and the possibility of fertile crossings.[4] Since, according to Gobineau, Adam was white, this formation of races could already be considered, at least for the coloured races, as a kind of degeneration.

If the difference between races is thus only vaguely explained (Gobineau clearly finds it an awkward question), their hierarchy is explained only too clearly. Gobineau takes it for granted, on the simple basis of the difference between civilizations, that these races each developed in its own geographical space – the whites being the most civilized, the yellows rather less so, and the blacks very little.

This hierarchy of races is based on the pattern of an aristocratic hierarchy, or the hierarchy of Indian castes. It appeals to 'fixist' conceptions of biology and society. The white race, and the Aryan in particular, is not the summit of evolution but rather the aristocracy of humanity. The yellow race is its bourgeoisie, and the black race its plebs – but also its artist. Each has its place in the order of the world, and each should remain there.

Here, by way of illustration, is Gobineau's Indian ideal – certainly racist, but above all aristocratic and conservative:

> The long persistence of India is simply the benefit of a natural law which has only rarely been able to operate freely. With the dominant race being eternally the same, this country has always possessed the same eternal principles; whereas everywhere else, groups mingling without restraint or choice have rapidly followed one another, and not managed to give their institutions permanence, because they have themselves rapidly disappeared in the face of successors equipped with new instincts.[5]

The disturbance par excellence of this human and social order is interbreeding; this is what lies at the origin of humanity's

4 Ibid., pp. 149ff.
5 Ibid., p. 387.

degeneration. Not simply because interbreeding, by mixing a higher with a lower race, lowers the former (despite raising the latter), but rather because it alters both, and thus alters the order in which both are inscribed. The mestizo is rather equivalent to the *déclassé*, with no proper place anywhere. Interbreeding is for the order of humanity what *mésalliance* is for the aristocratic social order or the Indian caste system. For Gobineau, moreover, generalized interbreeding is concomitant with the generalization of democracy. It is the formulation in racial terms, and at the level of humanity, of the disturbance of the social order and the mingling of classes brought about by the Enlightenment and the Revolution.

The model is perfectly clear, and illuminated by a detail in Gobineau's own biography. His mother, Anne-Louise-Magdeleine de Gercy, had the reputation of being a loose woman, and Gobineau was by no means sure of his paternity. She herself, moreover, was supposedly the daughter of an illegitimate son of Louis XV. A familiar drama of aristocratic pretensions, in which Gobineau could claim a royal ancestor, despite belonging to a family whose nobility was uncertain and financially ruined – degenerate and doubly illegitimate.

Gobineau's arguments were not highly original. He borrowed much from all kinds of historians and anthropologists of his time, so many of whom shared more or less the same ideas. According to M. Lange, his main source was a book by the doctor and physiologist (and painter) Carl Gustav Carus: *Über ungleiche Befähigung der verschiedenen Menschheitsstämme für höhere geistige Entwicklung* (On the Unequal Capacity of Different Human Races for Higher Mental Development), published in Leipzig in 1849.[6] Vacher de Lapouge, for his part, cited as Gobineau's forerunner the *Allgemeine Kulturgeschichte der Menschheit* (General Cultural History of Mankind) by the anthropologist Gustav Friedrich Klemm (Leipzig, 1843–52).[7] These kinds of ideas were sufficiently widespread at the time; Gobineau did not invent them and was not even the first to systematize them. We should also note that slavery was not abolished in French

6 M. Lange, *Le comte Arthur de Gobineau, étude biographique et critique* (Strasbourg: Publications de la Faculté des Lettres de l'Université de Strasbourg, 1924), p. 118.

7 G. Vacher de Lapouge, *Race et milieu social* (Paris: Rivière, 1909), p. 289.

territories until 1848, only five years before the first volume of Gobineau's opus.

Starting from the basis of the inequality of races and interbreeding/degeneration, Gobineau goes on to explain the history of humanity by constructing anthropological fables, perfectly comparable with those that we have already cited apropos Darwinian sociology, except that their motive force is interbreeding rather than struggle and competition, and their result is degeneration rather than progress. In these fables, races come into contact in one way or another, they interbreed, and they degenerate. It may be war that brings them into contact, but then, as against what happens in the Darwinian fables, it is not the strongest that wins out, since the defeated or weaker side ends up overcoming its conquerors by interbreeding with them and thus leading them to degenerate.

Just as with most of the Darwinian fables, the result is always the same, but instead of this being the triumph of the stronger, and progress,[8] it is inevitably the degeneration of humanity – a very literary degeneration, as Gobineau's anthropological fable keeps a certain style: 'Nations, or rather human flocks, suffering under a dismal somnolence, will then live submerged in stupidity, like the buffalo ruminating in the ponds of the Pontine marshes.'[9]

Gobineau's book is not a work of racial or racist anthropology, but rather an epic poem, the sad epic of humanity in full degeneration. It is a description of the end of the world or, more precisely, the end of a particular world, that in which Gobineau had his place. Here is his opening sentence: 'The fall of civilizations is the most striking and at the same time the most obscure phenomenon in the whole of history.'[10] And here are his final paragraphs (compare them with the quotation on pp. 14–15 from Vacher de Lapouge, who imitated Gobineau even in his style):

> We cannot pretend to calculate with any accuracy the number of centuries that divide us from this certain conclusion. Yet it is not impossible to attempt a first approximation . . . We would be tempted then to assign the domination of man on the earth a total duration of between twelve

8 Darwinian fables that are exceptions to this rule, such as those of Vacher de Lapouge (see quotation on p. 45 above), are generally marked by Gobineau's influence.

9 A. de Gobineau, *Essai sur l'inégalité de races humaines*, p. 871.

10 Ibid., p. 39.

> and fourteen thousand years, divided into two periods: the one that has already passed was that of the youth, vigour, and intellectual grandeur of the species; the other, which has begun, will see its halting procession into decrepitude.
>
> Even if we stop at the age that necessarily will shortly precede the last sigh of our species, and restrain our gaze from those ages possessed by death, when the globe, become silent, will continue – but without us – to describe in space its inflexible orbits, I do not know whether it is correct to describe as the end of the world that less distant epoch that will already see the complete abasement of our species. No more will I maintain that it would be easy to interest ourselves with a residual affection in the destiny of a few handfuls of individuals stripped of strength, beauty and intelligence, if we did not recall that they will still retain at least their religious faith, the last link, sole memory, and precious heritage of better days.
>
> But even religion has not promised us eternity; and science, showing us that we have had a beginning, seems also to assure us that we shall have an end. We should therefore be neither surprised nor moved to find one further confirmation of a fact that could not be seen as doubtful. The saddening prediction is not death, but rather the certainty of only reaching this in a degraded state; and perhaps even this shame in store for our descendants would leave us insensitive, if we did not perceive, in secret horror, that the rapacious hands of destiny are already raised over us.[11]

Darwin described the origin of species, Gobineau the end of humanity. For Darwin, we descend from the ape; for Gobineau, we degenerate as we descend. Their two major works were almost contemporary, 1853–55 and 1859 respectively, but they are completely incompatible. Nothing would be more foreign to Gobineau than social Darwinism, and vice versa, no matter that each was racist in his own way. One saw decadence, the other progress; one was aristocratic and Catholic, the other bourgeois and secular (or at least Protestant); one was a humanist and idealist, the other a scientist and positivist.

There is indeed no doubt that Gobineau was a racist, especially in his conception of history, and that he was to eventually achieve celebrity as a theorist of racism, though less in France than in Germany – particularly by way of the Gobineau Vereinigung, a society founded by Ludwig Schemann in 1894, twelve years after Gobineau's death. It is also clear that the Nazis acclaimed him. But this was only by way of a distortion of his ideas by H. S.

11 Ibid., pp. 872–3.

Chamberlain, Vacher de Lapouge and other thinkers of the same style. Indeed, those who have not read Gobineau commonly ascribe Chamberlain's theories to him, as we shall see.

Even if Gobineau professes Aryan superiority clearly enough, his work was not anti-Semitic. Jews were not an inferior race, but rather a great race which, like all others (though not more than others), had degenerated by interbreeding:

> If the sons of Abraham had been able to maintain, after their descent from the Chaldean hills, the relative purity of race that they brought with them, there can be no doubt that they would have preserved and extended the preponderance that they were seen to exercise, at the time of the father of their patriarchs, over the more civilized populations of Canaan – richer, yet less energetic, because they were darker. Unfortunately, against their fundamental prescriptions, and despite the successive defences of their law, despite even the terrible examples of reprobation recalled by the names of the Ishmaelites and Edomites, illegitimate and rejected descendants of Abraham's line, the Hebrews did not refrain from allying outside their own kin . . . The resulting mixture penetrated into the members of Israel through all their pores. It is true that the principle remained; that later on Zerubbabel pronounced severe penalties against men who married gentile women. But the integrity of the blood of Abraham had nonetheless disappeared, and the Jews were as sullied by the Melanian alliance as the Chamites and Semites among whom they lived.[12]

Gobineau did not proclaim the advent of the superman but rather the end of civilization, and despite his Germanophilia, the English seem less degenerate in his book than do the Germans – a point that Hitler must have noted.

It is quite absurd to maintain that Gobineau's racism has anything in common with biology and science. His races are inspired by Indian castes, and have as much relationship to taxonomy as the blue blood of aristocrats has with haematology. If any proof of this were needed (evidence as well of his pervasive theme of interbreeding), here is what he writes on consanguinity:

> There was a time, not long ago, when prejudice against consanguine marriage was such that there was a question of enshrining this in law. To marry a first cousin was tantamount to affecting all your children right away with deafness and other hereditary afflictions. No one seemed to

12 Ibid., pp. 269–70.

> reflect that the generations preceding our own, much given to consanguine marriage, experienced none of the morbid consequences that are claimed to arise from this; that the Seleucids, Ptolemies and Incas, who all took their sisters as wives, enjoyed excellent health and acceptable intelligence, not to speak of a beauty that was generally out of the common.[13]

Gobineau was displaying here the opinion of an aristocracy among whom endogamy was quite usual, with everyone being interrelated by half a dozen branches of the family tree. He was totally unconcerned with genetics, or of such antecedents of it as existed in his time.

We shall return to the subject of interbreeding after examining the question of races in evolutionary taxonomy.

Classification to order

After Gobineau's racism, let us explore the ideas of biology on the question of races and species – both in general, and in the particular case of man.

In this field, as we have said, the common opinion is that of the prevalence of an essentialist and fixist notion of species (and thus of race), which Darwin swept away with his conception of evolution, establishing a kinship among different living forms. At one blow, with the essentialist definition of species and races overthrown, all men became brothers or cousins, bringing an end to racism. (The love of animals, as our lesser brothers, was established at the same time – see the quotation from Haeckel on p. 62).

Historical reality shows a rather different picture. Here, moreover, is Vacher de Lapouge's ironic comment on this matter: 'All men are brothers, all animals are brothers, and men's brothers, and fraternity extends to all living beings; but to be brothers doesn't stop some eating others. Fraternity, indeed, but woe to the vanquished! Life is only maintained by death. To live you have to eat, to kill in order to eat.'[14]

It is true that the idea of the constancy of living forms prevailed for a long while, vaguely inspired as it was by Aristotle. At the same time, however, the notion of species remained very poorly defined, and belief in its constancy was balanced by belief in the possibility

---

13 Ibid., foreword to 2nd ed., pp. 34–5.

14 G. Vacher de Lapouge, *L'Aryen, son rôle social* (Paris: Fontemoing, 1899), p. 512.

of heterodox crossings that generated both monsters of all kinds and intermediate forms. It was only in the mid-eighteenth century that Linnaeus established principles of classification that were rather more scientific and standardized, initially for plants (on the basis of the form of their reproductive apparatus) and then, with greater difficulty, for animals (on the basis of vaguer criteria). Only at this time were taxonomic categories defined – the five levels of class, order, genus, species, and variety, of which genus and species were the most important. For Linnaeus, the species was fixed and created by God, the genus still more so; indeed, his initial view was that the genus was fixed, while there was the possibility of a certain variation of species within the genus. The two categories above genus (class and order) were more arbitrary, as was that below species (variety or race – the two words are equivalent). Linnaeus even saw varieties (races) as artificial products, totally different from the species created by God.[15]

Linnaeus thus had a realist or 'essentialist' conception of species: species are real and exist in themselves, created by God in the form they have maintained; they correspond to essences, and it is possible to classify them in a natural and non-temporal order. It is good form to mock Linnaeus's fixism, but to define species and establish classifications it was necessary, at least initially, for these species to be stable natural entities. Only after living forms had been given a bit of order was it possible to establish their relationships and variations. Linnaeus's fixism was in no way a reactionary or religious state of mind, simply an indispensable stage in the work of taxonomy. As for races (or varieties), he nowhere said that these corresponded to essences; in Linnaeus's taxonomy, only genera and species were 'real' entities of this kind.

Even this, moreover, did not mean that with Linnaeus the genus and species were well defined by strict criteria, especially in the case of animals, or that living beings could be perfectly classified. The very idea of such classification was not always well received (in particular by Buffon), as it involved many difficulties and required a certain arbitrariness.

Lamarck criticized the fixity of species at the beginning of the nineteenth century, questioning at the same time their definition and their principles of classification. According to him, living beings

---

15 For a résumé and references to the different Linnaean conceptions on this question, see D. A. Godron, *De l'espèce et des races dans les êtres organisés de la période géologique actuelle* (Nancy: Grimblot-Raybois, 1848).

formed a continuum that the observer partitioned for convenience into distinct portions, giving these different names. Far removed from Linnaeus's realism, Lamarck thus had a nominalist notion of species; this was no more than a name given by the observer to a portion of the continuum that living beings formed. It was not Darwin who abandoned the essentialist conception of species; this had already been done by Lamarck several decades before.

Moreover, this continuum of classification, from the simplest to the most complex, was also a temporal continuum, translating a process of formation, since Lamarck saw the simplest species as generating more complex ones, and so on.

It was Cuvier who rebelled against Lamarck's transformism and sought to (re)-establish fixism. He drew his arguments from his experience in comparative anatomy and taxonomy. For him, living beings were divided into various branches, corresponding to types of biological organization which could not be derived from one another. His principle of the correlation of forms, moreover, saw each part of a living being as a close function of the other parts, so that the form of the being was a whole, whose parts had scant possibility of variation without jeopardizing the functional unity of the whole. Consequently, for Cuvier, the species was necessarily something fixed, and his conception of it was a realist and not a nominalist one.

Even so, Cuvier did not define the species by an essence. His fixism was structuralist rather than essentialist and creationist. The form of the living being was a functional whole, and this meant that it had to obey constraints that prohibited, or at least very strongly limited, the possibilities of variation. Cuvier's biology was based on the structure of the living being, but it was unable to explain either the origin of this structure or its transformation, whether in ontogenesis or phylogenesis.

Unable to define the species structurally (which is almost impossible) or to conceive of an essence for it, Cuvier classically resorted to the resemblance among individuals and their capacity for procreation with one another. Sexual reproduction was in some sense the test of the structural identity of two progenitors; fertilization required that the two parents have comparable structures. They therefore had to belong to the same species, but they could be of different races. For Cuvier, differences between races were superficial, affecting only parts of the body that had little importance for the animal economy – parts therefore more apt to vary than those that were closely integrated into the functional totality of the body.

The debate continued after the deaths of Lamarck (1829) and Cuvier (1832). Transformism and the nominalist conception of the species steadily gained ground, though without a decisive victory. Cuvier's arguments were scientifically respectable, and there were as yet no known mechanisms that could explain the transformation of species. The debate rather ran into the sand, or at least slipped into the background, the descriptive and structural disciplines of anatomy and taxonomy becoming rather secondary as cellular theory, physiology and biochemistry advanced to the forefront of biologists' considerations. The early nineteenth century saw the triumph of taxonomy and comparative anatomy, in the work of Lamarck, Cuvier, and Étienne Geoffroy Saint-Hilaire. But as the century advanced and the broad lines of classification were established, scientists became less concerned with the form and classification of living beings, and increasingly concerned with their functioning.

In 1859, Darwin's theory swung biology towards transformism in a decisive way, by offering a likely explanation – though one that was still very shaky, for lack of a theory of inheritance and variation. Until the theory of mutations was developed in 1901–3, this Darwinian evolution was a gradual and continuous evolution, requiring a continuum of living beings partitioned in a more or less arbitrary way into distinct species. In other words, it took over Lamarck's nominalism lock, stock and barrel. Darwin and Darwinism did not invent anything in this respect.[16]

In this continuistic conception, the distinction between race and species – which Linnaeus had insisted on, seeing one as an artificial creation and the other as a natural entity – tended to disappear. Races in this view were nascent species, species in the process of differentiation, and it became hard to say at what point a race became a species distinct from the one from which it had issued.

In all this, apart from the ancestral Linnaeus, there was no trace of an essentialist conception of the species. And no essentialist conception of race ever existed in taxonomy, even in Linnaeus. Cuvier's fixism was not really essentialist but rather structuralist, pertaining to species and not to race. Darwin's struggle against

---

16 After 1901, the De Vries mutations, which are sudden variations in form, were to establish a certain discontinuity in evolution, along with a quasi-essentialist conception of the species; the transition from one species to another was by way of a sudden leap, which in De Vries's view was always of major amplitude. But this conception was rapidly attenuated and continuity re-established, mutation à la De Vries being replaced by modern mutation of an amplitude far too low to bring about by itself a change in species.

an essentialist conception of species and race is a mere legend, invented along with the complementary legend of his struggle against creationism. Darwin's opposition to a racism based on an essentialist conception of race and species is equally imaginary.

Let us look now at how man appears in these classifications. It was Linnaeus who integrated man into the classification of animals, ranking humanity together with apes and monkeys in the order of primates – and thus provoking a certain reaction from the church, as well as from some of his colleagues, including Buffon. No subsequent effort was made to shift man from this place, apart from some abortive attempts to separate other primates and place them in a distinct order – Johann Friedrich Blumenbach (1752–1840) tried to create a 'two-handed' order for man and a 'four-handed' order for other primates, a principle also adopted by Cuvier – or else to separate man completely from the rest of nature. Thus Isidore Geoffroy Saint-Hilaire (1805–61) and Jean Louis Armand de Quatrefages de Bréau (1810–92) proposed to establish a fourth kingdom specially for man, on top of the three traditional kingdoms – mineral, vegetable and animal – according to the principle that the mineral is inert; the vegetable lives; the animal lives as well as feels; and man lives, feels and thinks.

For Linnaeus there was only one human genus with a single species, itself divided into six races. Four of these corresponded to four continents, whose names they bore: *Homo sapiens europaeus*, *asiaticus*, *americanus* and *afer*; the two others, *ferus* and *monstruosus*, respectively denoted wild men (moving on four paws, dumb and hairy) and forms considered 'monstrous' in that their proportions were 'altered': the Hottentots (in Linnaeus's example) and doubtless also the Pygmies.[17] Blumenbach recognized only five races (white, black, yellow, red, and Malay).[18] Cuvier adopted the principle of a single human species in a single genus, itself the sole genus in the order *Bimane* (two-handed), but he recognized only three quite distinct races (white or Caucasian, yellow or Mongolian, and black or Ethiopian).[19]

17 C. Linné, *Systema naturae per regna tria naturae secundum classes, ordines, genera, species cum characteristibus, differentiis, synonymis, locis* (10th rev. ed.) (Holm: Laurentii Salvii, 1758–9), vol. 1, pp. 20–4.

18 J. F. Blumenbach, *De l'unité du genre humain et de ses variétés* (Paris: Alut, An XIII [1804]), pp. 281–2.

19 G. Cuvier, *Le règne animal distribué d'après son organisation, pour servir de base à l'histoire naturelle des animaux et d'introduction à l'anatomie comparée* (Paris: Déterville, 1817), 4 vols, vol. 1, pp. 94–100.

In the course of the nineteenth century, certain classifications (that of Quatrefages, for example, a follower of Cuvier) reduced the genus *Homo* to a single species, whilst others classified it as containing several species (Haeckel, who counted twelve human species divided into thirty-six races, as discussed below).

In evolutionary theories, different human races (or species) had either a single common origin (monogenesis) or several (polygenesis). It was possible to be a monogenist and maintain either the existence of a single human species or of many (all derived from the same origin). On the other hand, polygenists generally maintained the existence of several human species, rarely a single one.

Proposals in this field were not always clear or well justified. Darwin, for instance, in *The Descent of Man*, successively explained the arguments both of those who championed the existence of a single species and of those who championed several, but did not take a position of his own.[20] He described races as 'semi-species' and seemed to see the most distinct among them as different species. This did indeed tally with his theory that races were nascent species, in the course of differentiation, so that there were no very clear limits between them and species. He did, however, maintain that men, whatever their species or race, all had a common ancestor, thus clearly opting for monogenesis:

> Through the means just specified, aided perhaps by others as yet undiscovered, man has been raised to his present state. But since he attained to the rank of manhood, he has diverged into distinct races, or as they may be more fitly called, sub-species. Some of these, such as the Negro and European, are so distinct that, if specimens had been brought to a naturalist without any further information, they would undoubtedly have been considered by him as good and true species. Nevertheless all the races agree in so many unimportant details of structure and in so many mental peculiarities, that these can be accounted for only by inheritance from a common progenitor; and a progenitor thus characterized would probably deserve to rank as a man.[21]

Carl Vogt (1817–95), on the other hand, was a polygenist. He justified his position on the basis of the present situation,

20 C. Darwin, *The Descent of Man, and Selection in Relation to Sex* (Harmondsworth: Penguin, 2004), Chapter 7.

21 Ibid., p. 678.

in which he saw three distinct lineages of ape, each tending towards a more perfected form – in other words, three lineages capable of generating three different human species. He concluded that the different human races could only have been formed in this way:

> As regards the notion of the species, we shall stick to the proposition that the human genus is made up of different species, these being as distinct among themselves, if not more so, as most of the primate species, so that, if the principles of systematic zoology are of any value, they must be applied to human beings as well as to other primate genera . . .
>
> These facts, which cannot be denied, precisely prove what we maintain, in other words that different parallel series of apes have at their summit forms having a higher development, superior types gravitating towards the human. If we extend in thought the development of these three anthropomorphic types through to the human type that they do not attain and will never attain, we obtain, arising from these three parallel series of apes, three original human races: two dolichocephalic races issuing from the chimpanzee and gorilla, and one brachycephalic one issuing from the orang-utan. That emanating from the gorilla would be more remarkable in terms of the development of the teeth and the thoracic cavity, that issuing from the orang-utan by the length of the arm and the reddish-blond colour of its hair; and finally that arising from the chimpanzee by its dark colour, its weaker bones and its less massive jaw.
>
> If we take into account the development of the apes in parallel series, each crowned by one higher form, nothing justifies the admission of a single intermediate form between man and the apes, since we can recognize three different paths by which transitional forms have been able to lead to the present state of creation.[22]

Contrary to the prevailing legend, it was not Darwin but Lamarck, back in 1809, who first asserted man's descent from the ape. Indeed, Lamarck did no more than logically extend the procedure of Linnaeus, who had brought them into proximity in his classification, and the comparative anatomy of the time, particularly that of Petrus Camper. Darwin did not mention this in *The Origin of Species*, but waited until 1871 to put forward his position in *The Descent of Man*, by which time the whole

22 K. Vogt, *Leçons sur l'homme, sa place dans la création et dans l'histoire de la terre* (Paris: Reinwald, 1878), pp. 298, 629–30.

world (more or less) had extrapolated his theory to the origins of humanity.

The only problem is that, if already before Darwin it was generally agreed that man descended from the ape, there were many conflicting views as to how this happened, and that was still the case after Darwin had tackled the question. Each school of thought searched the palaeontological record (some human fossils had already been discovered), comparative anatomy (which showed the kinship between man and certain animal forms), and even embryology (according to the formula that saw ontogenesis as repeating phylogenesis, so that traces of ancestral animal forms could be found in the development of the human embryo). Some of these views were rather curious: Haeckel, for example, rejected Blumenbach's distinction between 'bimanes' and 'quadrumanes', arguing that in certain primitive tribes the big toe was opposable to the other four, and the foot prehensile, as with the apes.[23] In any event, a number of different genealogies were established, all of them relating man to the ape in one way or another, and all placing the different human races in a hierarchy, from the least evolved (and closest to the ape) to the most evolved (generally the white European).

The diagrams on p. 251 presents the most celebrated of these genealogical trees, that presented by Haeckel. In his terminology, the *Ulotrichi* and *Lissotrichi* are respectively the human races with crinkly hair and those with smooth hair; *Engeco* is the chimpanzee, *Satyrus* the orang-utan, and *Hylobates* the gibbon.

Haeckel also proposed a genealogical classification of different human species and races. His criteria were extremely vague, involving hair type, skin colour, skull shape and various biological characteristics, but added to these were all kinds of other considerations of an intellectual, linguistic and social nature. Haeckel's classification is a taxonomic fable; it is enough to know that he distinguishes thirty-six human races divided into twelve different species. The diagrams on pp. 252 and 253 present his table of human taxonomy and his human genealogical tree.

Among his twelve species, two stand out in the firmament of evolution, the Indo-Germanic and the Semite. The diagrams on pp. 254 and 255 give their specific family trees.

23 E. Haeckel, *The History of Creation* (London: H. S. King, 1876), pp. 397–8.

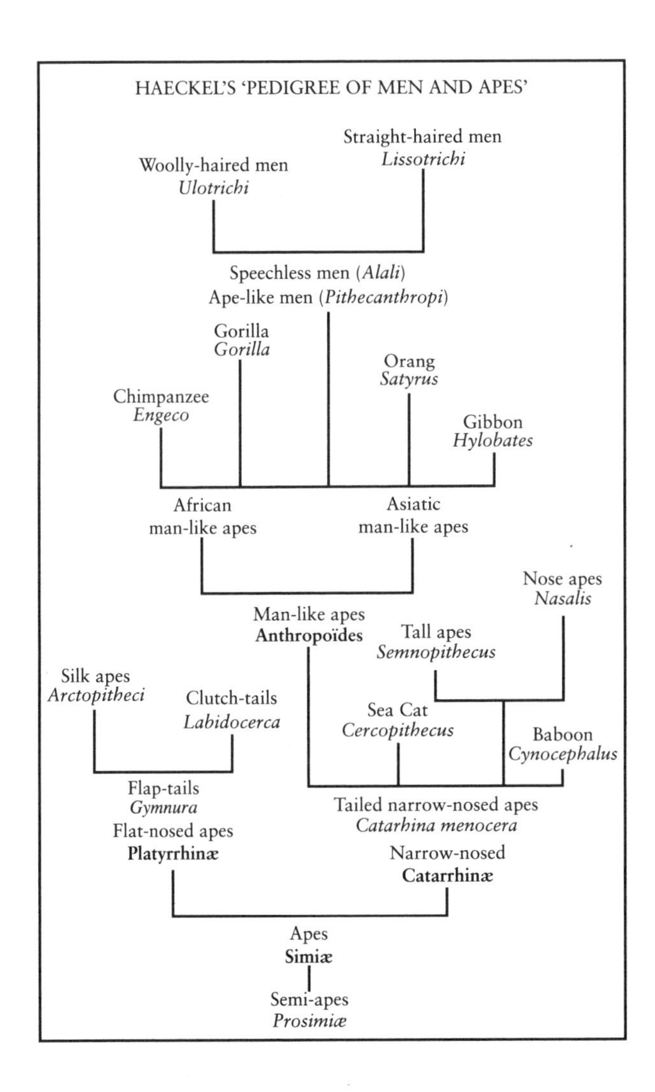
HAECKEL'S 'PEDIGREE OF MEN AND APES'
Straight-haired men
Lissotrichi
Woolly-haired men
Ulotrichi
Speechless men (*Alali*)
Ape-like men (*Pithecanthropi*)
Gorilla
Gorilla
Orang
Satyrus
Chimpanzee
Engeco
Gibbon
Hylobates
African
man-like apes
Asiatic
man-like apes
Nose apes
Nasalis
Man-like apes
Anthropoïdes
Tall apes
Semnopithecus
Silk apes
Arctopitheci
Clutch-tails
Labidocerca
Sea Cat
Cercopithecus
Baboon
Cynocephalus
Flap-tails
Gymnura
Tailed narrow-nosed apes
Catarhina menocera
Flat-nosed apes
Platyrrhinæ
Narrow-nosed
Catarrhinæ
Apes
Simiæ
Semi-apes
Prosimiæ

HAECKEL'S 'SYSTEMATIC SURVEY OF THE TWELVE SPECIES OF MEN AND THEIR THIRTY-SIX RACES'			
	Species	*Races*	*Home*
1, 2. Lophocomi	1. Papuan *Homo papua*	1. Negritos	Malacca, Philippine Islands
		2. New Guinea men	New Guinea
		3. Melanesians	Melanesia
		4. Tasmanians	Van Diemen's Land
	2. Hottentot *Homo hottentotus*	5. Hottentots	The Cape
		6. Bushmen	The Cape
3, 4. Eriocomi	3. Kaffre *Homo cafer*	7. Zulu Kaffres	Eastern South Africa
		8. Bechuanas	Central South Africa
		9. Congo Kaffres	Western South Africa
	4. Negro *Homo niger*	10. Tibu negroes	Tibu district
		11. Sudan negroes	Sudan
		12. Senegambians	Senegambia
		13. Nigritians	Nigritia
5-9 Euthycomi	5. Malay *Homo malayus*	14. Sundanesians	Sunda Archipelago
		15. Polynesians	Pacific Archipelago
		16. Natives of Madagascar	Madagascar
	6. Mongolian *Homo mongolus*	17. Indo-Chinese	Tibet, China
		18. Coreo-Japanese	Corea, Japan
		19. Altaians	Central Asia, North Asia
		20. Uralians	North-western Asia, Northern Europe, Hungary
	7. Arctic Men *Homo arcticus*	21. Hyperboreans	Extreme N.E. of Asia
		22. Eskimos	The extreme north of America
	8. American *Homo americanus*	23. North Americans	North America
		24. Central Americans	Central America
		25. South Americans	South America
		26. Patagonians	The extreme south of South America
	9. Australian *Homo australis*	27. North Australians	North Australia
		28. South Australians	South Australia
10-12 Euplocomi	10. Dravidas *Homo dravida*	29. Tamils	Further India
		30. Todas	Nilgherri
		31. Dongolese	Nubia
	11. Nubian *Homo nuba*	32. Fulatians	Fulu-land (Central Africa)
		33. Caucasians	Caucasus
	12. Mediterranese *Homo mediterraneus*	34. Basque	Extreme north of Spain
		35. Hamo-Semites	Arabia, North Africa, etc.
		36. Indo-Germanic tribes	South-western Asia, Europe, etc

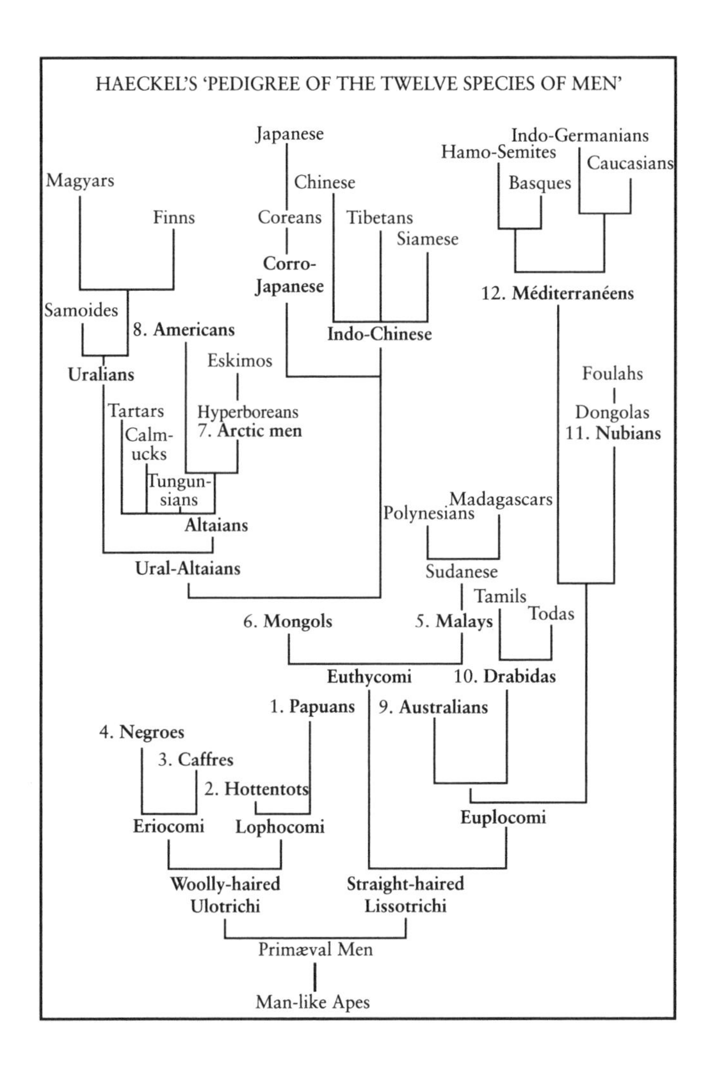

HAECKEL'S 'PEDIGREE OF THE TWELVE SPECIES OF MEN'

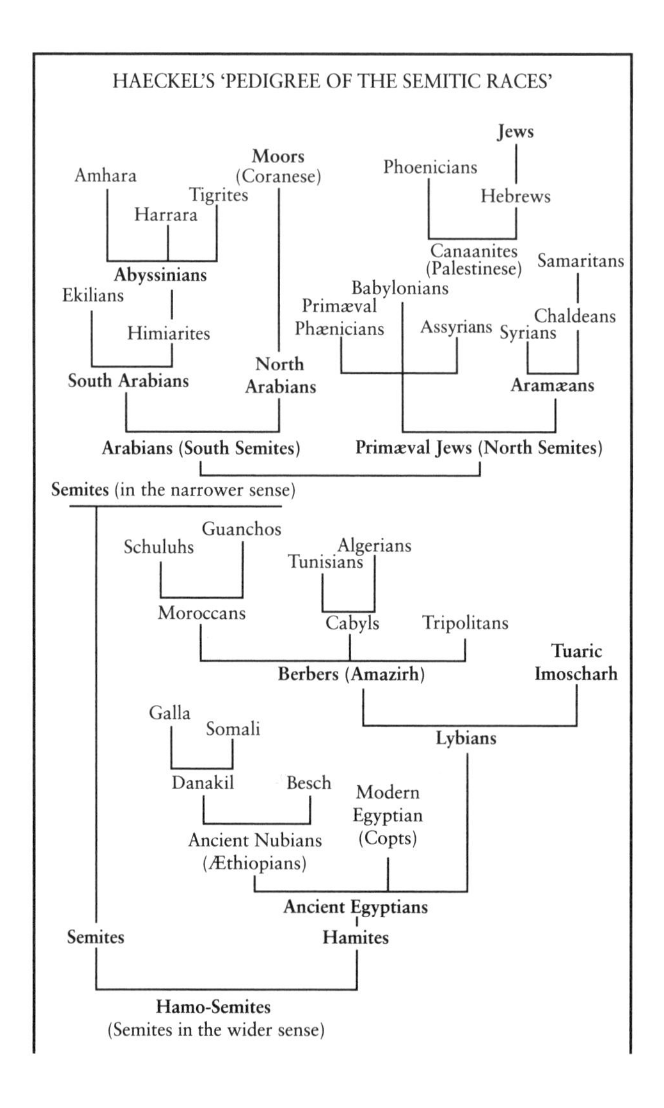
HAECKEL'S 'PEDIGREE OF THE SEMITIC RACES'
Jews
Moors
(Coranese)
Phoenicians
Amhara
Tigrites
Hebrews
Harrara
Canaanites
(Palestinese)
Samaritans
Abyssinians
Babylonians
Ekilians
Primæval
Phænicians
Chaldeans
Himiarites
Assyrians
Syrians
North
Arabians
South Arabians
Aramæans
Arabians (South Semites)
Primæval Jews (North Semites)
Semites (in the narrower sense)
Guanchos
Schuluhs
Algerians
Tunisians
Moroccans
Cabyls
Tripolitans
Tuaric
Imoscharh
Berbers (Amazirh)
Galla
Somali
Lybians
Danakil
Besch
Modern
Egyptian
(Copts)
Ancient Nubians
(Æthiopians)
Ancient Egyptians
Semites
Hamites
Hamo-Semites
(Semites in the wider sense)

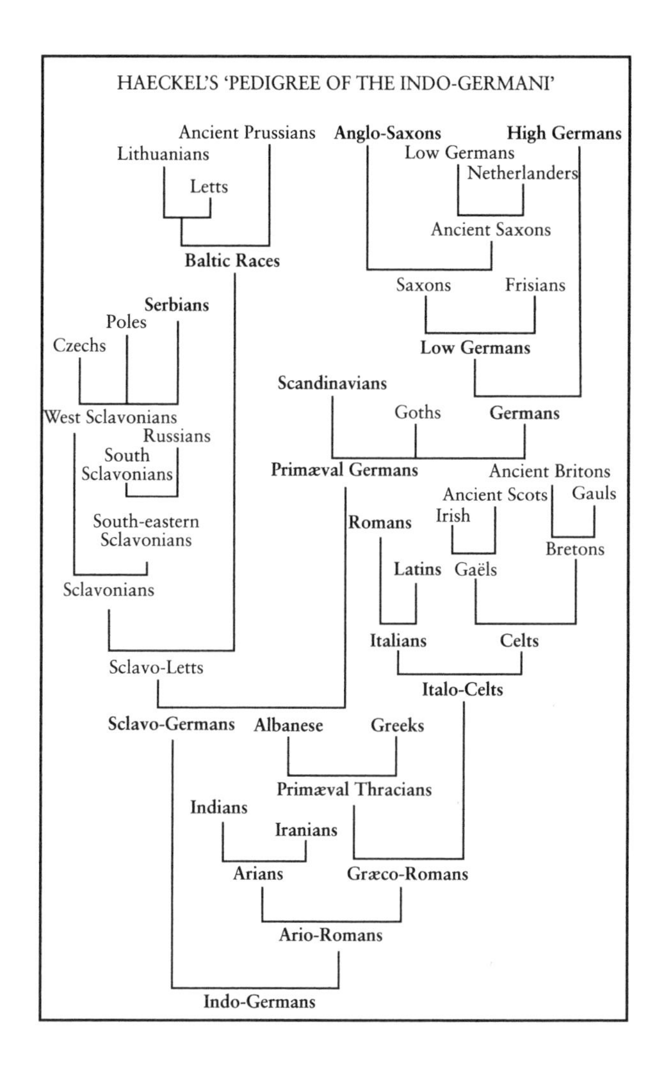
HAECKEL'S 'PEDIGREE OF THE INDO-GERMANI'
Ancient Prussians
Anglo-Saxons
High Germans
Lithuanians
Low Germans
Netherlanders
Letts
Ancient Saxons
Baltic Races
Saxons
Frisians
Serbians
Poles
Czechs
Low Germans
Scandinavians
West Sclavonians
Goths
Germans
Russians
South Sclavonians
Primæval Germans
Ancient Britons
Ancient Scots
Gauls
South-eastern Sclavonians
Romans
Irish
Bretons
Latins
Gaëls
Sclavonians
Italians
Celts
Sclavo-Letts
Italo-Celts
Sclavo-Germans
Albanese
Greeks
Primæval Thracians
Indians
Iranians
Arians
Græco-Romans
Ario-Romans
Indo-Germans

Darwinism brings us very clearly into a hierarchical ranking of races according to their degree of evolution. This hierarchy has no need of an essentialist definition of species and race; it fits very well into a continuity of living forms, a gradation running from the least to the most evolved. It has nothing in common with the fantasies of a littérateur like Gobineau, but was established by recognized scientists, consecrated by their professional institutions, and widely distributed to the general public. (Haeckel's *History of Creation* was a successful work translated into several languages.)

This hierarchy does not appeal to degrees of civilization, inevitably subjective, but rather to the scientific fact of biological evolution. Haeckel and the other authors in this style would evidently find it very hard to explain the exact nature of the degree of biological evolution according to which they classified races, but this question did not even enter their minds. In reality, like Gobineau, they ranked civilizations in a hierarchy, and they transposed this hierarchy into the domain of biology by giving these civilizations a hereditary biological foundation. For Haeckel, the superiority of Germans and Anglo-Saxons was manifestly inspired by the greater advance of the Industrial Revolution in the countries in question. And yet, no matter how fictitious this degree of evolution as criterion for a hierarchy of races, it fitted perfectly into the social Darwinist ideology of the time, lending support to the racism inherent in it by imparting a semblance of scientific foundation – something that the anti-Darwinian Gobineau totally lacked, inasmuch as his hierarchy of races was modelled on the fixed and conservative social hierarchy of the aristocracy.

With the evolutionary hierarchy of races, the rejection of interbreeding found the biological justification that Gobineau lacked. Interbreeding had a counter-evolutionary effect, since by mixing races that were unequally evolved, it was retrograde in terms of biological evolution. (For a long while, biological evolution remained an end in itself for progressivism; in 1944, Julian Huxley still sought to make evolution the basis of ethics – whatever opened the prospect of evolution was ethical.)

Finally, the evolutionary aspect gave racism a characteristic that was totally absent in Gobineau: the greater or lesser humanity of different races. The degree of evolution as a criterion of racial hierarchy patches over any difference in nature between man and animal – a difference that Gobineau recognized, criticizing those who, even before Darwin, brought man and the apes together.

'Scientific' racial theories could thus introduce a gradation in human character, equating the more human with the more evolved.[24] It was precisely on this point that they came up against religious notions. The book by Carl Vogt (a professor at Geneva) cited above ends with a criticism of Swiss pastors who challenged his interpretations, concluding that 'it is better to be a perfected monkey than a degenerated Adam' – a theme also rehearsed by Clémence Royer, the French translator of *The Origin of Species*, whose preface opposed 'Christianity as religion of the fall' to 'Darwinism as religion of progress'.

The relationship between man and the apes is understood variously by different authors; and there are two kinds of argument, either biological or psychological. Here, for example, is what Haeckel writes on this subject: there are greater mental differences between the 'higher' human races and the 'lower' ones than between the latter and the anthropoid apes:

> An impartial critical examination likewise confirms Huxley's law: the psychological differences between man and the anthropoids are less than those between the latter and the lower primates. This physiological fact corresponds precisely to the anatomical observations that show us the differences of structure of the cerebral cortex, that 'organ of the soul' whose importance cannot be denied. The great significance of this finding becomes still more palpable in the light of the extraordinary differences of mental life in the human species. At the summit we see a Goethe and a Shakespeare, a Darwin and a Lamarck, a Spinoza and an Aristotle – and at the bottom of the scale, we find the Weddas and the Akkas, the Australians and Dravidians, the Bushmen and the Patagonians! Mental life presents incomparably greater differences when we move from those magnificent minds to these degraded representatives of humanity than it does between the latter and the anthropoids.[25]

The majority of authors, however, refer less to psychology (always seen as a branch of biology) than to biology in the narrower sense. There are those who accept a quasi-continuity between the anthropoid apes and the 'lower' human races. An example of this is Ludwig Büchner (1824–99), who in 1870 also used arguments

24 By and large: the white European is Man, the other races less so. The Aryan, of course, tends towards the superman. See also note 51 on p. 173 above, where Vacher de Lapouge imagines a crossing between man and anthropoid apes to produce 'sub-men' to be used as slaves.

25 E. Haeckel, *Origine de l'homme* (Paris: Schleicher, [c. 1900]), pp. 24–5.

of comparative anatomy as well as those of embryology (the Negro brain resembling both that of the ape and that of the European neonate):

> Man is thus essentially distinguished at first glance from his cousins the anthropoid apes by his prominent forehead, broad and strongly developed. In this respect, however, the Negro serves as a transition between man and animal, his forehead being narrow and receding, which coincides with a weak development of the anterior cerebral lobes; also in the Negro, the general conformation of the brain and the whole structure of the body offer numerous simian analogies. In the predominance of its longitudinal diameter, in the imperfection of its convolutions, in the flattened and narrow character of the anterior hemispheric extremity, in the rounded form of the cerebellum, in the large size of the cerebellar vermis, in the relative thickness of the pineal gland, the Negro brain is, according to Huschke, inferior in type, imperfectly developed; it is reminiscent on the one hand of the brain of the European neonate, and on the other of that of those animals most closely related to man. In general, cerebral differences between the lower and the higher races are identical to those observed between the human and simian brains . . . The observations of the same researcher [R. Wagner] have established an important fact, namely that in the human embryo of five to six months, the brain has a conformation completely analogous to that of the lowest primates. This is supporting evidence of the old transformist proposition, according to which the human embryo, in the successive phases of its development, temporarily reproduces the lower animal types.[26]

Other authors did not accept this continuity between the apes and the 'lower' forms of man – Darwin, for example, when he tackled the question of man in 1871. In his view, the gap between ape and man was explained by the fact that the intermediate forms were now extinct. This gap, he believed, had a high chance of reappearing, however, as a consequence of the probable extermination of the 'lower' human races as well as the apes:

> The great break in the organic chain between man and his nearest allies, which cannot be bridged over by any extinct or living species, has often been advanced as a grave objection to the belief that man is descended from some lower form; but this objection will not appear of much weight to those

26 L. Büchner, *L'Homme selon la science* (Paris: Reinwald, 1870), pp. 257–8.

> who, from general reasons, believe in the general principle of evolution. Breaks often occur in all parts of the series, some being wide, sharp and defined, others less so in various degrees . . . But these breaks depend merely on the number of related forms which have become extinct. At some future period, not very distant as measured by centuries, the civilized races of man will almost certainly exterminate, and replace, the savage races throughout the world. At the same time the anthropomorphous apes, as Professor Schaaffhausen has remarked, will no doubt be exterminated. The break between man and his nearest allies will then be wider, for it will intervene between man in a more civilized state, as we may hope, even than the Caucasian, and some ape as low as a baboon, instead of as now between the negro or Australian and the gorilla.[27]

Vogt, a few years previously, found a substitute for this missing link between man and ape: the idiot, whose brain development was supposed to have halted at a stage in phylogenesis corresponding to one of those disappeared intermediate forms that Darwin spoke of:

> I am not afraid to say, despite Bischoff and Wagner, even J. Müller, that the microcephalic, the idiots by birth, constitute a set of transition states between man and ape, as complete a series as could be desired . . . Innate idiotism is evidently an arrested development of the brain, bearing essentially on its anterior part. The skull is shaped to the form of the incomplete brain . . . The impression that these individuals produce is decidedly simian, to the point that the public authorities have themselves used this expression. The arms appear disproportionately long, the legs short and weak. The head is completely that of an ape, the skull covered with tufted woolly hair; the forehead is lacking, the hollow eyes shine from behind prominent bony rings; the nose is broad and open; the lower part of the face projects forward like a muzzle; the teeth are obliquely placed ... It is enough to compare the skulls of the chimpanzee, the Negro and the idiot, as we have done here, to see that the skull of the idiot occupies in every respect an intermediate place between the two others.[28]

Vogt is in total agreement here with Büchner in terms of the application to man of the principle that ontogeny recapitulates

---

27 C. Darwin, *The Descent of Man, and Selection in Relation to Sex*, pp. 183–4.

28 K. Vogt, *Leçons sur l'homme, sa place dans la création et dans l'histoire de la terre* (Paris: Reinwald, 1878), pp. 258–63.

phylogeny, and that the Negro is thus halted at a stage of development that the white race has left behind:

> Young orang-utans and chimpanzees are intelligent and pleasant animals; they learn easily, understand quickly, and can be civilized to a high degree. After their transformation [i.e., puberty] they become nasty, irritable, wild and inaccessible to any domestication or improvement.
>
> The same is true of the Negro; as a child he is no way inferior to the white in respect of his intellectual faculties; all observers agree in recognizing that the young Negro plays with as much animation as white children, understands as easily and rapidly as they do, and is also just as docile. In places concerned with their education – not where they are deliberately raised like cattle, in the slave states of America for example, so that it can later be said that they are incapable of anything – it is soon recognized in schools that black children are in no way inferior to white, and even surpass them in the rapidity with which they learn and in their docility, so that they are often employed as monitors. But as soon as young Negroes reach the fatal period of puberty, with the disappearance of the cranial sutures and the projection of the face, we see the same phenomenon as among the apes. The intellectual faculties now remain stagnant, and the individual, as likewise the whole race, becomes incapable of any progress.
>
> From the intellectual point of view, the adult Negro simultaneously resembles the child, the woman and the old man in the white races.[29]

These are not claims made by political extremists or militant racists, but rather by eminent scientists. Vogt was a German liberal, opposed to Bismarck. Büchner was a materialist, a man of the left, very far from a reactionary in the style of Gobineau – though the latter was incomparably more moderate in his racism. This kind of theory was not the preserve of any particular political current, but could be found on all sides. It is paradoxical, in fact, that it was generally the more reactionary thinkers who displayed more moderate views on this question. The idealist Quatrefages, for example, a disciple of Cuvier and an anti-Darwinist, was incomparably less racist, and maintained the unity of the human species. He acknowledged a *de facto* inferiority of certain civilizations, but did not rule out – and even tended to believe – that these 'lower' races, placed in different and more favourable circumstances, could develop perfectly well. And he concludes: 'Finally, and thinking of our own past, we should guard

29 Ibid., p. 253.

against refusing the other races aptitudes that remained hidden for centuries among our own ancestors before they developed, aptitudes that are still in a latent state in too large a number of our compatriots and contemporaries.'[30]

Some of Quatrefages's arguments against the theories whose broad lines we have just presented are fairly close to those that could still be made today. It is permissible, then, to believe that these racist theories were not inevitable, and were not necessarily taken for granted in their time. Yet this does not prevent the poor reputation that Quatrefages has today as a retrograde spirit, whereas Darwin, Büchner, Vogt and the like are seen as the vanguard of biology in the late nineteenth century.

It is clear in any case how misguided is the legend that biological racism originated in a supposed essentialist and fixist taxonomy, which Darwinism brought to an end by demonstrating the relativity of the taxonomic categories and the kinship of all living beings. All these authors (Darwin, Büchner, Vogt, etc.) were racist, but in no way did they depend on the fixism of Cuvier or an essentialist notion of species and race. They were all evolutionists, and they all had a nominalist conception of species and race; continuity between living forms and their genealogical relatedness fitted very well with an evolutionist hierarchical ranking, which became a hierarchy of absolute value once evolution was seen as the general reference point for everything.

In sum, the origin of racism does not lie in the recognition of different races, but in their hierarchical ranking. And this hierarchy, when it is evolutionary, has no need for an essentialist realism of species and race; it is quite compatible with nominalism, and an affinity and continuity of living forms. Moreover, far from Darwinism having had the anti-racist effect that legend attributes to it, it actually gave biological support to racism (which it certainly did not invent), as well as a biological foundation for the hierarchical ranking of races, which had previously been based uniquely on the 'degree of civilization' of the peoples concerned.

Interbreeding

For Gobineau, interbreeding was to be condemned for itself, as a kind of perturbation of the natural and social order, but this

30 J. L. A. de Quatrefages, *L'Espèce humaine*, 6th ed. (Paris: Germer-Baillière, 1880), p. 336.

condemnation was not given any biological justification. In the evolutionary classification of races, interbreeding is condemned because it supposedly has a counter-evolutionary effect. At least this is the logical conclusion of such arguments. On all sides, biological theories condemned interbreeding as such, considering 'half-breeds' as inferior to each of the two races from which they sprang – for example, crossings between black and white would be inferior to both white and black.

This kind of theory seems to have been particularly characteristic of physical anthropology in the late nineteenth and early twentieth centuries. Anthropology classified human races as a function of a certain typology – bearing on skull shape in particular, but also on a wide range of other characteristics. Interbreeding was thus conceived as a disturbance of the type, a kind of unbalance.

These typologies of physical anthropology were challenged early on (see pp. 290–1 below for a statement by Gumplowicz in 1883), and the classification of races was thus thrust back onto genetics. This was undoubtedly one of the reasons why there was so much insistence at this time on the hereditary character of race, as can be seen on p. 235 above. The practice of criticizing interbreeding on the basis of physical anthropology's criteria, however, was more or less maintained. To a certain extent, such criticism survived the typologies from which it arose, and served to justify racial segregation.[31]

In parallel with these anthropological theories, moreover, there also existed biological theories that sought to condemn interbreeding as such (not merely as being counter-evolutionary) or even to justify the benefits of consanguinity (returning here to Gobineau). This was the case, for example, with Vacher de Lapouge, who was a lawyer and a dilettante rather than a recognized biologist, but one of those who attempted to combine Gobineau and Darwin. He extended to the interbreeding of races the characteristics of the interbreeding of species, i.e., either the resulting infertility of the process or the malformation of hybrids. All that he needed for this was to view races as quasi-species, benefiting from the fact that evolutionism had abolished the strict divide between race and species.

According to Vacher de Lapouge, the 'half-breed' was 'disharmonic' by virtue of his mixed parentage, hence inferior.

31 See, for Germany, 'Anthropologie raciale et national-socialisme: heurs et malheurs du paradigme de la "race"', in J. Olff-Nathan (ed.), *La Science sous le Troisième Reich* (Paris: Le Seuil, 1993), pp. 197–262.

The black was inferior to the white but was nonetheless 'harmonic', whereas the 'black-white half-breed', because of this 'disharmony', was inferior not only to the white but also to the black.[32] Vacher de Lapouge extended this theory to a number of cases. For instance, he claimed that mixed-race women had an asymmetrical uterus, which rendered them sterile.[33] Or again, if, as against what happens with females of other species, women do not become pregnant every time they have intercourse, this was because the human species had become very interbred, which had weakened its fertility – for the same reason of asymmetry.[34] Here is Vacher de Lapouge's general principle (O. Ammon shared this idea of the inferiority and disharmony of 'half-breeds', but without developing the line of argument):[35]

> The theory of infertility for lack of reciprocal accommodation, such as I have developed most recently in my study on Les Lois de l'hérédité, follows necessarily from all that we know of the mechanism of the formation and development of the blastoderm. When the two elements of fertilization do not proceed from subjects of the same race, and still more so not from the same species, numerous failures are to be expected. Each brings different tendencies in embryogenetic development, whether in terms of rapidity or the point at which phenomena are expressed. If in the segmentation of the fertilized ovum certain cells or groups of cells divide more rapidly than others, this will soon lead to disorder. This can be directly observed when the attempt is made to artificially fertilize eggs of certain ascidians or echinoderms with the sperm of other species . . . The same happens if the parts of the embryo organized according to hereditary tendencies that impose somewhat different morphologies on them are not in agreement . . .
>
> Crossings therefore do not yield good physical results. The mental results, however, are still worse. The incoherence of the half-breed is not only somatic; it exists also in terms of intelligence and character. It establishes a remarkable contrast between the man of the pure race, having throughout his life the same ideals and tendencies, his mentality a single block, and the half-breed, torn between different tendencies, fatally condemned to an unravelled existence and changing behaviour by the multiplicity of mental heredities that dominate him by turns. The superiority of the Yankee, the

32 G. Vacher de Lapouge, *Les Sélections sociales* (Paris: Fontemoing, 1896), p. 181.

33 Ibid., p. 168.

34 Ibid., pp. 174ff.

35 O. Ammon, *L'Ordre social et ses bases naturelles, esquisse d'une anthroposociologie* (Paris: Fontemoing, 1900), pp. 188–9.

> Englishman, the Hollander, or the Scandinavian over the Frenchman, Italian, Spaniard or South American is not just the result of a superiority of race, but also of a purity of blood. The former are both of European race and practically pure, whilst the latter have little European blood, and their degree of inferiority corresponds to their degree of interbreeding. The French proceed from two main races, the Italians and Spaniards from three, the South Americans from four or five.[36]

In 1896, this view of Vacher de Lapouge went against what was by then the current trend in biology. He is well aware of this himself, and criticizes the trend that condemned consanguinity, emphasizing that 'consanguinity, like other unions, gives good results when the authors are good, and bad ones when they are bad', likewise that 'consanguinity between perfect individuals is on the contrary the means to fix perfection'.[37] This is not completely false, but in no way justifies his opinion on interbreeding.

Brazil, a country of interbreeding between several races, was for Vacher de Lapouge and thinkers of his ilk the very model of degeneration, as seen in the above quotation. But this was not a unanimous view. Quatrefages, for instance, used the Brazilian example to criticize Gobineau and his habit of linking interbreeding with degeneration:

> Even though modern crossings go back no more than three centuries, they have already produced results that show beyond a doubt how races can emerge from interbreeding that are remarkable from every point of view. The province of São Paulo in Brazil gives a striking example. It was peopled by Portuguese and Azoreans from the Old World, who interbred with the Gayanazes, a peaceful tribe of hunters, and the Carijos, a warlike agricultural race. From these regularly contracted unions emerged a race whose men have always been distinguished by their fine proportions, their physical strength, their indomitable courage and their resistance to the worst fatigues. As for the women, their beauty gave rise to a Brazilian proverb that attests to their superiority. This population has given proof of initiative in every field . . . 'Today,' F. Denis tells us, 'the most fortunate moral development, along with the most remarkable intellectual movement, seems to belong to São Paulo.'[38]

---

36 G. Vacher de Lapouge, *Les Sélections sociales*, pp. 171–2, 183–4.
37 Ibid., pp. 157, 194.
38 J. L. A. de Quatrefages, *L'Espèce humaine*, 6th ed., pp. 209–10.

Quatrefages was certainly not completely free from racial prejudice, but he saw only one human species, with the fertility of crossings the criterion of this. Crossing between races was then perfectly possible, without any biological ill effect. For him, the situation of inferiority that people of mixed race often suffered was entirely due to social questions (birth outside marriage, abandonment of the black mother by the white father, etc.). He even conceived interbreeding as a means of raising 'lower' races to the level of 'higher' ones ('the mulatto can equal the white in all respects; his intelligence is equal to ours').[39] Racist theories in this field were thus in no way an unavoidable phenomenon of the era and its prejudices.

Apart from the theses of Vacher de Lapouge, there were also works by recognized biologists that sought to justify the rejection of interbreeding by biological arguments. This was particularly the case in the United States, the only country at the time to have both a fairly large mixed-race population and a scientific infrastructure sufficient for this kind of research. (Brazil was more mixed but lacked scientists, and, as we have just seen, it aroused diverse reactions.) Thus in 1911, in *Heredity in Relation to Eugenics*,[40] and in 1917, in 'The effects of race intermingling',[41] Davenport claimed that crossings between races resulted in disharmonic products (the disproportion of various organs, functional difficulties between these organs, psychological problems, etc.). Similar ideas can still be seen in a later work, *Race Crossing in Jamaica*, which refers to a supposed disharmony in people of mixed race and their inferiority to the races that constitute them.[42]

These works of Davenport's lacked any theoretical foundation, and were based solely on phenomenalist and mathematical studies, of the kind we have already discussed in relation to nomadism (see p. 140ff. above). Davenport did, however, make an attempt at rationalization in this case: since races had been formed by the selection of those characteristics best adapted both to the external environment and to one another, an individual from a given race formed a coherent whole (this was an adaptation of Cuvier's principle to Darwinism), so that interbreeding risked producing a

39 Ibid., p. 211.

40 New York: Holt, 1911.

41 Proceedings of the American Philosophical Society, 1917, 56, pp. 364–8.

42 C. B. Davenport and Morris Steggerda, *Race Crossing in Jamaica* (Washington, DC: Carnegie Institution, 1929), pp. 470–1.

child who combined in a disharmonic fashion elements that might not be mutually compatible, coming as they did from two totalities that were each coherent, but different from one another (the same explanation given by Vacher de Lapouge). As we have already seen in the case of eugenics, Darwinian genetic theories were an inexhaustible sophistry, on the basis of which anything and everything could be justified. In a 1917 article, Davenport extended his conclusions to human society; in his view, precocious death, madness and criminality in the United States could be explained in terms of the disharmonies arising from interbreeding, and he explained in the same fashion the decline of the great civilizations of the past – it was an explanation typical of Gobineau, even if it is uncertain whether at that time Davenport had ever even heard of Gobineau.

Various contemporary writers challenged the validity of these works of Davenport's, but the same kind of theory persisted until the Second World War, at least among racist geneticists.[43]

Gobineau and Darwin

Racism was certainly not invented by Darwinian biologists, nor by Gobineau. Lukács sees social Darwinism as the middle ground between Gobineau and Nazism, as far as issues of race are concerned.[44] This assumes, however, that Darwinism needed Gobineau to conceive its racist anthropological fables, something that is by no means certain. To be sure, evolutionary racism can be considered the bourgeois version of Gobineau's aristocratic racism – one being progressive and activist, the other decadent and fatalist. But Darwinian racial classifications, such as those of Haeckel, had their own logic and could readily dispense with Gobineau.

Gobineau and Darwin were close contemporaries, and there is no reason to grant Gobineau any priority in matters of racism, unless it is for the sake of imagining a pure and immaculate original Darwinism that Gobineau allegedly perverted. This would be all the more dubious in that the success of Darwin's book was immediate, whereas it was much longer before Gobineau enjoyed his much lesser success.

---

43 For the history of these theories in the United States, see W. B. Provine, 'Geneticists and the biology of race crossing', *Science*, 1973, 182, pp. 790–6.

44 G. Lukács, *The Destruction of Reason* (London: Merlin Press, 1980).

Gobineau did not even appear in the fourth (1870) edition of G. Vapereau's *Dictionnaire universel des contemporains*,[45] a publication that, as its name implies, covered all prominent figures of the time. His supposed precursors (see p. 239 above), such as Carus and Klemm, were each given an extensive biological note; they were luminaries of the era, even if they are almost forgotten today. Darwin, Büchner and Vogt were likewise well covered, as famous scientists, and Quatrefages as well. Haeckel does not appear, but in 1870 he was only thirty-six years old, and his ideas became known in France only in 1873, with the book by Dumont cited above (pp. 7–8). In any event, fifteen years after the publication of his book on the inequality of human races, Gobineau was still a low-ranking diplomat who wrote novels and essays, too insignificant to figure among the 'notable persons of France and foreign countries'.

In 1880, Gobineau did merit a short notice in the fifth edition of this *Dictionnaire*; Haeckel now appeared as well, with an entry twice as long. He was not, however, presented as a theorist of racism, but rather as a 'French diplomat and littérateur' who 'has made a name for himself by works of history, criticism, philosophy and epigraphy, relating to the countries where he has lived'. Gobineau did not reappear in the sixth edition of 1893, eleven years after his death.

According to Michel Lémonon, the first edition of Gobineau's *Essai sur l'inégalité des races humaines* had a print run of five hundred copies at the author's expense (its four volumes appeared between 1853 and 1855). The second French edition dates from 1884. The German translation was finished in 1893, but did not find a publisher until 1897 – thanks to the action of the Gobineau Vereinigung founded by Schemann in 1894. About a thousand copies of each volume were now printed, the fourth and final volume in 1901. Lémonon estimates that Gobineau's *Essai* must have sold some four thousand copies in German between 1897 and 1934.[46] (This would in all likelihood mean a smaller number of readers; given the length and dullness of the work, many must have let it drop from their hands.) Compare this with Haeckel, whose *Riddle of the Universe* sold four hundred thousand copies in Germany and was translated into almost every language.

45 G. Vapereau, *Dictionnaire universel des contemporains, contenant toutes les personnes notables de la France et des pays étrangers* (Paris: Hachette, 1870).

46 M. Lémonon, 'Le Rayonnement du gobinisme en Allemagne,' typescript thesis, University of Strasbourg II, 1971 (2 vols), vol. 2, pp. 247, 249, 253, 257, 403, 418.

Gobineau eventually acquired a certain celebrity, a literary one in particular. It is probably this that helped make him into an emblem, among racists and anti-racists alike (not that either camp bothered to read him). Even so, his ideas did not have as great a resonance as was subsequently believed. Thus in 1900, in the preface to his translation of Otto Ammon's *L'Ordre social et ses bases naturelles*, H. Muffang wrote that Gobineau was 'much forgotten today, and greatly misunderstood'.[47] It was only in the early twentieth century that he really began to gain a certain reputation in France – among such figures as Charles Péguy, Marcel Proust, Daniel Halévy and the historian Robert Dreyfus, these names showing how Gobineau's Aryanism was not at that time viewed as anti-Semitic.[48] The First World War, with the theories of Germanic racial superiority that it brought to the fore (see the quote from Boutroux on pp. 32–3), undoubtedly contributed to Gobineau's subsequent reputation. In the 1930s his aristocratic, Catholic and reactionary ideas made it possible to confuse him with the would-be aristocrats of Action Française (when he wrote at the time of Louis Philippe and Napoleon III), and make him a target for both sincere anti-racists and for those who wanted the role of biologists in racism to be forgotten. All the more so, in that Nazism had in the meantime recuperated him – if considerably modifying his ideas.

In any case, this 'success' of Gobinism was subsequent to that of evolutionary racism, and undoubtedly secondary. His famous *Essai* seems to have served, above all, to integrate Aryan theses into evolutionary racism, and it is likely that without this it would have remained no more than a literary curiosity (see below).

There were no affinities at all between Gobineau and Haeckel, even if their careers overlapped slightly and they were aware of each other by reputation. One, however, was French and the other German, thus on opposite sides in the war of 1870–1. Gobineau, as a professional diplomat – despite his Germanophilia – was no friend of Bismarck's Prussia and preferred Austria. The Bismarckian Haeckel was a progressive, committed to the Protestant and scientistic *Kulturkampf*, whereas the Catholic Gobineau was a *décadentiste* and proud of having had an ancestor who was particularly active in the Saint Bartholomew's Day

47 O. Ammon, *L'Ordre social et ses bases naturelles*, preface, p. viii.

48 A. Combris, *La Philosophie des races du comte de Gobineau et sa portée actuelle* (Paris: Alcan, 1937), p. 182.

massacre. One was anti-Darwinian and the other the universal popularizer of Darwinism. They were opposed in everything, and the racism of each was quite different. Nevertheless, this did not prevent hybridization.

These two forms of racism had different principles of hierarchical ranking, and different rejections of 'miscegenation'; one was fixist and the other evolutionist. Yet there were points on which they agreed, in both their forms and their foundations, with the result that a number of authors – Vacher de Lapouge, Ammon, Chamberlain – could happily mix them.

These mixtures preferentially referred to Haeckel and Darwin rather than to Gobineau, given that science now had a far greater weight than literature. Thus, against all evidence, it was possible for Chamberlain to write:

> My own views on race – as every reader has been able to confirm – are completely contained within the circle of ideas that forms the field and the atmosphere of the natural science: that is their element. There are only very rare passages where I am close to Gobineau, and touch on his world. What I know of the question, what I think of it when my thought embarks on a theory of the facts, all this is simply the scientific inheritance of a century of assiduous work – the century stretching from Blumenbach to Ujfalvy – and the master I appeal to above all, as we have seen, is Charles Darwin . . . And far from taking over Gobineau's thesis, I attack it everywhere that I can.[49]

It must be said that Chamberlain was the apostle of the Germanic race (for Gobineau, the English had been the last Aryans not to have been too interbred), and campaigned for the establishment of a new Aryan aristocracy, whereas Gobineau had been pessimistic, fatalistic, and a prophet of decline.

Vacher de Lapouge did not share this attitude of more or less complete rejection of Gobineau, being almost as pessimistic as his mentor, and likewise seeing the Anglo-Saxons as the last true Aryans. He viewed him as a precursor of his own anthropological sociology, but a very vague one, since 'Gobineau was in no way a scientist, more of a learned traveller',[50] whereas he himself had scientific pretensions, and appealed to biology in the same way as did Chamberlain.

---

49 H. S. Chamberlain, *La Genèse du XIXe siècle* (Paris: Payot, 1913), vol. 2, p. 1387. This additional text, titled 'Wagner, Gobineau, Chamberlain', does not appear in the English editions.

50 G. Vacher de Lapouge, *Race et milieu social*, p. 172.

Let us now take a look at what the cross between the littérateur Gobineau and the scientist Haeckel brought about. We shall take Vacher de Lapouge as our example here rather than Ammon or Chamberlain, since Vacher de Lapouge is always cited along with Gobineau as the originator of modern racism – most likely because he wrote better and was far more clear and direct than either Ammon or Chamberlain, who were just as racist as he.

Vacher de Lapouge, as we have said, was an ideologist, viewed nowadays as an extremist. He was soon expelled from his only teaching post, at the University of Montpellier, and then worked for many years as a university librarian, first at Montpellier, then at Rennes and finally at Poitiers. He certainly attracted a readership, but only drawn from a public who already shared his ideology. His success, moreover, should not be overestimated; thus *L'Aryen, son rôle social* reproduced the 'free lectures' that Vacher de Lapouge had given at Montpellier in 1889–90, but was published only in 1899 by an imprint that clearly specialized in this kind of literature. As for the lectures on *Le Sémite, son rôle social*, these were not even published.

Did Vacher de Lapouge have an essentialist view of the human species and its races, which lay at the origin of his racism? As a start, here is what he wrote about racial purity. He did not believe in it, and sought at most to minimize the effect of interbreeding and the 'dilution of foreign blood':

> Whatever population we study, we can be sure that it is in no way pure, and that each individual carries in their veins, in very varying proportions, the blood of quite different races. I have shown elsewhere that each one of us owes their origin, in the twentieth generation, to more than a million ancestors, and inherits from each of these less than a millionth part . . . If we go back to the time of Jesus Christ, the number becomes quite fantastic: 18,014,583,333,333,333 . . . These absurd figures prove two important points: 1) the theoretical impossibility of absolute purity of race, as just one accidental crossing is needed in the most sequestered population to introduce a dilution, no matter how small, with the blood of all foreign races; 2) the prodigious number of consanguine crossings that must have taken place, given that the theoretical number of ancestors shows by its absurdity how often the same individuals must appear in the genealogical table. It must be noted, however, that isolated and accidental crossings do not have a serious impact: the dilution of foreign blood very quickly reduces the disturbing hereditary influence to an infinitesimal order of magnitude, such that this is only a theoretical consideration. On the other hand, the

> repetition of the same ancestors in different ancestral branches assures these individuals a cumulative hereditary influence, which subsequently has a preponderant value . . . Thus except from the point of view of avatism, the mixture in very unequal proportions, with an extreme predominance of one element, may be assimilated in practice to purity of blood. A sixteenth part of foreign blood, therefore, is already very little, and a hundredth part scarcely counts at all.[51]

And here is how he criticizes the position of someone like Quatrefages, who maintains the unity of the human species (along the lines of Cuvier), with a diversity of races within this single species. Vacher de Lapouge is completely Darwinian here. Races are nascent species, in a perpetual state of becoming; there is no reference at all to any kind of racial essence, nor even an essence of the species:

> Another question that has been much discussed is whether we should talk about a number of human species or human races. Earlier anthropologists, and Quatrefages in particular, were passionate in their discussion of this question of terminology. This was a residue of the great struggles between adherents of the Bible and rationalists, monogenists and polygenists. All this scholasticism is now in the past, and Darwinism has brought everyone into agreement. Whether the point of differentiation is placed more or less far in the past, whether these convergences are even ascribed certain present resemblances, a common ancestor must always be admitted. Even if they are different, and may indeed issue from different species, the human species nonetheless represent branches of a single stem; all are in a perpetual state of becoming, and races are species in the process of formation; genera are highly differentiated species, isolated as a group from their kin, whose evolution has taken another course.[52]

This did not stop Vacher de Lapouge from being completely racist. He was stuck on blonde dolichocephalics and less keen on dark-haired brachycephalics. It is interesting therefore to see how he defined these:

> The dolichocephalic has great needs, and works constantly to satisfy them. He is better at gaining wealth than preserving it, accumulating and losing it with equal ease. Adventurous in temperament, he risks everything, and

51 G. Vacher de Lapouge, *Les Sélections sociales*, pp. 3–4.
52 Ibid., p. 11.

his boldness assures him incomparable successes. He fights for the sake of fighting, but never without an afterthought for profit. Any land is his, and the whole globe his country. His intelligence varies greatly according to the individual, from dullness to genius. There is nothing that he dares not think or want; and wanting, for him, means going for it right away. He is logical when it suits him, and never minces words. Progress is his most intense need. In religion he is Protestant, in politics all he asks of the state is respect for his activity, and he seeks rather to raise himself up than to do others down. He sees his personal interests clearly, even at a great remove, and likewise those of his nation and his race, which he boldly prepares for the highest destinies.

The brachycephalic is frugal, laborious, and economical. He is remarkably cautious, and leaves nothing to chance. While not lacking courage, he does not have a taste for war. He loves the land and his native soil. Though rarely stupid, he still more rarely shows great talent. The ambit of his aims is very limited, and he works patiently to realize these. He is very distrustful, but easy to woo with words, as his exact logic takes no trouble to see what lies beneath them; he is a man of tradition, and of what he calls good sense. Progress does not seem necessary to him, he is mistrustful of it, he wants to remain like everyone else. He loves uniformity. In religion he is happy to be Catholic; in politics he has only one hope, the protection of the state, and only one tendency, to bring down to his level everyone higher than himself, without feeling the need to raise himself up. He sees his personal interest very clearly, at least in a limited time; he also sees and promotes the interests of his family and those around him, but the borders of his country are often too wide for his vision. Among those of mixed race, the spirit of egoism is reinforced by the energetic individualism of the dolichocephalic, the feeling of family and race are neutralized and reduced; combined with a stronger cupidity, this leads to all the vices for which our bourgeois are reproached, and finally to the elimination of what the English call 'self-restraint'.[53]

Today this sounds like a horoscope comparing the respective characters of people born under Leo and Gemini. In any event, there is not the least trace here of an essentialist definition of race. It is not even biological in the strict sense – apart from the incantatory evocation of the cephalic index. These are simply political criteria disguised as racial ones, with the help of a few considerations about skull shape and skin colour. If this needs to be rubbed in any further, here is another quote:

53 Ibid., pp. 13–14, 17–18.

> In the brachycephalic countries above all, France, Germany or Russia, public service swallows a large number of families that, if not eugenic, are at least intelligent. The blonde dolichocephalics, with their more bold and independent nature, are less attracted to public service, which is why this is less developed in England and the United States. In countries where the brown-haired dolichocephalics are numerous, these do not behave the same as the blonde ones; Italy, Spain and Latin America swarm with functionaries.[54]

This shows how far some authors took the relationship between Darwinian concerns and Anglo-Saxon liberalism. With Vacher de Lapouge, the political discourse is quite transparent beneath the biological language. Like Gobineau, he viewed England and the United States as the great reservoir of blonde dolichocephalics – as well as the cradle of the political views that he championed. (At that point in time, Aryan theories leaned more towards the Anglo-Saxons than the Germanic barbarians. If certain qualities of the latter were recognized, the belief was that, especially in southern Germany, these had notably degenerated since the period when they ravaged the Roman empire; the blonde beast had not yet been born.) Here again is Lapouge:

> From earliest times through to our own day, the history of the British Isles reflects the spirit of the dominant races. Originally, on this remnant of the Anglo-Scandinavian massif that was the cradle of the European race, this either dominated or lived alone ... The great fortune of the Isles is the fruit of the singular chance that the brachycephalics never set foot there. The future will tell whether, alone in Europe and perhaps the whole world, this privileged land will remain in the power of the Aryans or will follow the common fate. If this should happen, it would be wrong to speak of a degeneration of the English people; we should rather say that they had lived their life.[55]
>
> Aryan colonization has covered North America with its swarms, though less so Mexico. The race Europaeus is on home ground there, more robust and exuberant than anywhere else. At the present time its centre is no longer the North Sea, but the Atlantic. The United States and Canada face England, northern Germany and Scandinavia. And the young Gallo-Saxons

54 Ibid., p. 353. Ammon, on the other hand, is very fond of German officials (*L'Ordre social et ses bases naturelles*, pp. 292ff.).

55 G. Vacher de Lapouge, *Les Sélections sociales*, pp. 95–6.

> of America, legitimate descendants of both Gauls and Germanics, have the edge over the purely Germanic Anglo-Saxons by their boldness and vigour. The same race is being formed in Australia, a mixture of Scots, Welsh, Irish – all Gallic in origin and Celtic in language – and Anglo-Saxons.[56]

We shall return later to how Vacher de Lapouge saw the confrontation between Aryans and Jews for 'world domination', and the extent to which this confrontation was based on racial criteria.

To conclude, it is always possible to maintain, as many historians do, that Vacher de Lapouge was, along with Gobineau, the source of biological racism. But Darwin, Haeckel, Büchner and Vogt, to cite only the eminent names already discussed, were not far behind, even if they used different terms. In Vacher de Lapouge's case, the political foundations are very visible; the racial discourse is transparent, and crudely stuck on to the political argument, which is what has discredited it. The Darwinian biologists, on the other hand, appealed to science and packaged their ideological presuppositions rather better.

56 G. Vacher de Lapouge, *L'Aryen, son rôle social*, p. 344.

14

Gumplowicz and Racial Struggle

Besides hierarchical racial classifications of the kind made by Haeckel, there is also the dimension of taxonomy, i.e., the dynamic of the appearance and disappearance of human races, the evolution that created these races and continues to modify them. Modern 'scientific' racism did not rest content with the hierarchical ranking of races; it sought to promote a certain number of social measures in accordance with this hierarchy. And it justified these measures by 'naturalizing' them, in other words by integrating them into the natural movement of the appearance and disappearance of races. In this case, the model is Darwinian in the pure state, and we find the biological sociologies that were already discussed in Part One. We are back once again with a 'scientization' of sociology through the naturalization of society, in this case a naturalization that proceeds by way of race.

The question of the appearance of races was little developed, as at this time Darwinism did not have a theory of variation that could explain the appearance of a particular characteristic that was subsequently subject to selection. The explanation was thus extremely vague, even on the part of Darwin himself. The disappearance of races, on the other hand, could be perfectly well explained by the new theory; these came into competition and mutual struggle, and the strongest one triumphed – this being the very yardstick of evolution and progress. Here is what Darwin wrote on the subject – it is not unfavourable physical conditions such as climate that bring an end to certain races, but rather competition and struggle:

> Man can long resist conditions which appear extremely unfavourable for his existence. He has long lived in the extreme regions of the North, with no wood for his canoes or implements, and with only blubber as fuel, and melted snow as drink ...

> Extinction follows chiefly from the competition of tribe with tribe, and race with race. Various checks are always in action, serving to keep down the numbers of each savage tribe – such as periodical famines, nomadic habits and the consequent deaths of infants, prolonged suckling, wars, accidents, sickness, licentiousness, the stealing of women, infanticide, and especially lessened fertility. If any one of these checks increases in power, even slightly, the tribe thus affected tends to decrease; and when of two adjoining tribes one becomes less numerous and less powerful than the other, the contest is soon settled by war, slaughter, cannibalism, slavery, and absorption.[1]

Colonization is used here to provide examples that illustrate the war of the races. Here again is how Darwin explained the disappearance of 'lower races' colonized by the Europeans; it is probable indeed that this disappearance served as a model for the theoretical explanation:

> When civilized nations come into contact with barbarians the struggle is short, except where a deadly climate gives its aid to the native race. Of the causes which lead to the victory of civilized nations, some are plain and simple, others complex and obscure. We can see that the cultivation of the land will be fatal in many ways to savages, for they cannot, or will not, change their habits. New diseases and vices have in some cases proved highly destructive; and it appears that a new disease often causes much death, until those who are most susceptible to its destructive influence are gradually weeded out . . . It further appears, mysterious as is the fact that the first meeting of distinct and separate people generates disease. Mr Sproat, who in Vancouver Island closely attended to the subject of extinction, believed that changed habits of life, consequent on the advent of the Europeans, induces much ill health. He lays, also, great stress on the apparently trifling cause that the natives become 'bewildered and dull by the new life around them; they lose the motives for exertion, and get no new ones in their place'.[2]

At this time of writing, there was no concealment of the extermination of indigenous peoples in the colonies. Colonizing enthusiasm and belief in its civilizing virtues did not blind the

1 C. Darwin, *The Descent of Man, and Selection in Relation to Sex*, pp. 211–12.

2 Ibid., p. 212. Taken from G.M. Sproat, *Scenes and Studies of Savage Life* (London: Smith, Elder, 1868), p. 284.

colonists to the consequences of their actions. This extermination was recognized and deliberate, but considered natural. More precisely, it was naturalized by Darwinism. Wallace used a comparison with plants to proceed to a similar naturalization (see p. 56 above). Darwin, for his part, took as his example the extinction of animal species, and, as evidence of the validity of this comparison, pointed out that the indigenous people of New Zealand themselves compared their fate to that of the marsupial rat:

> Finally, although the gradual decrease and ultimate extinction of the races of man is a highly complex problem, depending on many causes which differ in different places and at different times; it is the same problem as that presented by the extinction of one of the higher animals – of the fossil horse, for instance, which disappeared from South America, soon afterwards to be replaced, within the same districts, by countless troops of the Spanish horse. The New Zealander seems conscious of this parallelism, for he compares his future fate with that of the native rat now almost exterminated by the European rat.[3]

Not only was it perfectly well known that colonization could bring about the extinction of the colonized populations, but, as we saw on pp. 127–8 above in the case of Vacher de Lapouge, this was even the basis for conceiving a eugenic method based on the disappearance of these populations through alcohol, disease, and so on.

Let us turn now to what became of this struggle of races in 'biological sociologies' in the full sense. Here again we find the familiar anthropological fables. Developed by biologists such as Darwin, Wallace, and Haeckel, these fables were taken over and amplified by the sociologists. For the latter, races were the basic elements of society, and the struggle between them the motor of social processes, even of history itself.

The first major sociologist to theorize this racial conception of society and man – or at least the most famous to do so – was Ludwig Gumplowicz (1838–1909), whom we can take as a model for ideas of this kind, and whom we already introduced in Part One. His fundamental work, published in 1883, bears a title that could not be more explicit: *Der Rassenkampf* (The Struggle of Races).[4]

3 Ibid., pp. 221–2.
4 L. Gumplowicz, *La Lutte des races* (Paris: Guillamin, 1893).

For Gumplowicz, the racialization of social processes was a corollary of the naturalization of society and the scientization of sociology (see Chapter One, pp. 16ff.). Social processes had to be understood and studied as natural processes. In his view, in the same way as astronomy has elements (stars, planets, etc.) and interactions between them (physical forces), so sociology has races as its elements, and struggle as the interaction between them. The social dynamic was thus understood along the lines of the model of biological evolution, in other words competition between variant forms of the species (in his sense, between races). We find again the omnipresence of struggle that we already noted in Part One, but here this is a struggle between races:

> In any natural process, two essential factors are observable . . . : on the one hand, heterogeneous elements, on the other hand, a reciprocal action that these elements exert on one another and that we ascribe to certain natural forces.
>
> These factors can be observed in the natural process of astronomy: the various celestial bodies exert on one another actions that we ascribe either to a force of attraction or a force of weight . . .
>
> Let us turn then to the sociological process. We consider this as being of a special kind, and believe that it differs from the four kinds of natural process mentioned above [astronomic, chemical, vegetable, animal], but remains nonetheless a natural process. It would not deserve this title, however, if one could not detect in it the two essential factors of the natural process, those that constitute its generic definition. We should thus be able to distinguish these here, and indeed we can. The heterogeneous elements here are the ethnic elements, the countless human bands that can be seen as the most remote beginning of the existence of humanity . . .
>
> It only remains to seek the second constitutive factor of any natural process: the determinate actions that these elements exert on one another, and in particular those actions that have the character of a natural and necessary regularity, a character of eternal duration . . .
>
> Well, then, we are fortunately in a position to establish a formula for the reciprocal action that the heterogeneous ethnic elements, or – if you prefer – social elements of different origin, exert on one another. This is a formula that cannot be denied a complete and almost mathematical certainty and generality, as it presents itself to us always and everywhere, in the most irrefutable manner, both in the field of history and in present-day life.
>
> This formula is very simple: every ethnic or social element seeks to make use for its aims of any weak element that exists in or enters its orbit of power. This proposition on the relationships that heterogeneous ethnic and

> social elements present among themselves, with the consequences deriving from it without a single exception, completes the solution of the riddle of the natural process of human history.[5]

The Darwinian model has been somewhat shifted here. Gumplowicz superimposes on it a very ancient pattern of thought, which is found throughout the history of science and goes back at least to Empedocles, with his physics made up of four elements (earth, water, air, fire) combined and divided by two forces (love and hate). If hate and struggle govern relationships between races, something has to be imagined to govern the relationships between individuals of the same race. This is what Gumplowicz calls 'syngenism' (from the Greek *syngeneia*, kinship), a kind of mutual recognition among the members of a single race, a sentiment of equality between them, which makes them unite – especially to struggle against outsiders.

This 'syngenism' is reminiscent of E. O. Wilson's sociobiological altruism, or Chamberlain's intuition of race (see pp. 170 and 84 above). It could indeed be the ancestor of these notions. But Gumplowicz was a more sophisticated author. His syngenism is not simply biological, but comprises all kinds of factors, such as a common language, religion, material interest, and so on:

> The most primitive sentiment that syngenism produces, a sentiment that undoubtedly exists prior to any social development, is the sentiment of connection among all members of the band ... This primitive syngenetic sentiment persists up to our time with the same nature and strength. It unites the members of consanguine circles and those groups of men aware of a common origin, or who believe in a common origin. This sentiment, in its nature, is a sentiment of natural equality, of identity of essence. Above all, and by virtue of a natural sympathy, it surpasses in intensity all social and humane sentiments. The physical substratum of this sentiment is the perceived fact of physical resemblance, along with that of intellectual resemblance; the idea of equality separates out from this ...
>
> Let us now examine this in rather more detail: first of all the different motivations, or, if you like, the material and moral substrata of these syngenetic sentiments; and secondly, the way in which social communities are formed according to the motivations at work. These motives, as they appear to us in life and history, include a great variety of factors, among which we can indicate the following as being the most important:

5 Ibid., pp. 156–9.

> 1) community of blood, produced by unobstructed interbreeding; 2) a common language; 3) a common religion, with common morals and customs attached to it; 4) common civilization and education; 5) common material interests. Each of these factors already possesses by itself the necessary strength to form a social group, by way of a syngenetic sentiment. It is clear then that, according to the number of factors, a social group may be maintained by more or less powerful syngenetic sentiments, since there are communities and groups that are linked either by one or another of these factors, or by several of them.[6]

Vacher de Lapouge takes up this idea of a solidarity among members of the group, but differentiates it according to whether the group is dolichocephalic (Aryan) or brachycephalic; the former have the solidarity of the wolf pack, the latter that of a flock of sheep:

> Refractory to the least attempt at authority, bridling at the least assault on his personal freedom, the Aryan willingly becomes the model soldier and submits to all civilian discipline when this becomes necessary. No other man is so keen on having his own house, separate from any other, yet no other affiliates as he does to a plethora of societies of all kinds. This intense solidarity gives the Anglo-American peoples a threatening power ... In man, the struggle for existence changes its character by the intervention of solidarity. The struggle of one against all, and all against one, continues, but it yields to the struggles of groups, in which individuals stand in solidarity against the common enemy. Man, in other words, associates for the purpose of struggle, but this intervening solidarity is only a means for success, as it is not solidarity with everyone, but only with those who share his interests ... By his aggressive way of conceiving solidarity, the Aryan possesses a crushing superiority over other races, and over the brachycephalic race in particular. The Aryan likes to put himself forward, the brachycephalic remains behind. The solidarity of the one is that of the pack of hounds chasing the boar, each pressing the other to charge first and counting on its companions to come to its aid if it meets with too great a resistance. The solidarity of the second is that of a flock of sheep, each seeking to hide behind its neighbour, and counting on this to pass unnoticed in the moment of danger ... These qualities of the Aryan are expressed in practice by an intense development of public liberties.
>
> The English and Americans are undoubtedly raptores orbis [the world's predators], the former in fact, the latter potentially, but at home and among

6 Ibid., pp. 242–3, 246.

themselves they are free. And it is precisely because the Aryan is born with the soul of a free man that he raises himself above those who have the souls of slaves.[7]

It would be wrong to believe that Gumplowicz's recognition of factors in syngenism that are not strictly biological has any moderating effect on his theory of racial struggle. His view of humanity is one of total darkness and violence, entirely based on struggle, murder, domination and exploitation:

> Nothing is more deeply rooted in the nature of the masses than the reciprocal desire for murder; and the most absurd pretext is always held sufficiently good and reasonable when it responds to the satisfaction of this need. Now, nothing develops the desire for murder with greater violence, nothing satisfies the conscience of the masses so much, as the idea of a difference of race, in the commonly accepted and false sense of this expression which is taken to mean a diversity of origin – in particular when this supposed fact is supported and maintained by a social or national difference. This idea therefore always supplies the masses with an excellent pretext to mutually exterminate one another, and to do so in all tranquillity of conscience.
>
> History and the present offer us the image of almost uninterrupted wars between tribes, peoples, states and nations. The aim of all wars is always the same, whatever the different forms under which this aim is envisaged and attained: it is to use the enemy as a way of satisfying one's own needs.[8]

We have seen that for Wilson and the sociobiologists, altruistic sentiment (for one's kin) is a ruse of the genes, designed to maximize their presence in the population. For Gumplowicz, syngenism could be a more sophisticated form of this, less purely biological and more social. But Gumplowicz goes further than Wilson, as he adds the symmetrical sentiment of hatred towards the enemy. And that is not all. What underlies the struggle of races is not just competition, hatred, murder and extermination; it is also the will to power, the desire to dominate. For Gumplowicz, domination and exploitation are added to hatred for the outsider, forming the historical axis around which rotate all those processes of which the motor is the struggle of races:

---

7 G. Vacher de Lapouge, *L'Aryen, son rôle social*, pp. 373–5, 399.

8 L. Gumplowicz, *La Lutte des races* (Paris: Giard, 1898), pp. 301–2 (n. 2), 175.

> The struggle of races for domination and power, this struggle in all its forms, whether avowed and violent or latent and peaceful, is thus the real propulsive principle, the motive force of history; but domination itself is the pivot on which all phases of the historical process turn, the axis around which these move, since social amalgamations, civilization, nationality and all the highest phenomena of history are only revealed in the wake of organizations of power and by means of such organizations.[9]

It is no longer a mere matter of survival, as in the Darwinian struggle. The issue is to dominate and exploit. With Gumplowicz, the struggle of races combines Darwin's biological struggle and Marx's class struggle. It is almost as if domination and exploitation are justified (even approved), and social classes, like occupational castes, are always deemed to have an ethnic origin. This is quite simply because the division into social classes has exploitation as its principle and aim, and because, by virtue of syngenism, the exploitation of one group by another is always conducted on a racial basis, even if syngenism is not just something biological – note here the way in which social groups can be formed out of several races:

> The coincidence in the population of a state between classes and occupational castes on the one hand, and ethnic or racial differences on the other, arises from the fact that it is only with the object of a political-economic division of labour that it has been necessary to forcibly organize domination. In order that agriculture will yield as much as possible and enable the proprietors to live freely, enjoying leisure or the exercise of preferred occupations, a large number of individuals must be used, or, as the socialists say, 'exploited', by a small number. Now as we have seen, and as we shall explain in more detail further on, it lies in the nature of man that all exploitation of other men, wherever it is forced to take place, always seeks its victims outside its own syngenetic circle.[10]

This medley of Darwinism and Marxism is an explosive one, as Gumplowicz is well aware. The box below contains a lengthy passage in which he explains how man differentiates himself from the animals by his desire to dominate and exploit, and how this domination and exploitation lie at the origin of civilization. The struggle of races here plays more or less the role that the

9 Ibid., p. 217.
10 Ibid., p. 210.

prohibition of incest does for Lévi-Strauss in the transition from nature to culture.

Certain passages in the text that follow irresistibly call to mind Nazi theories (society based on race, enslavement of inferior races, etc.). Gumplowicz taught public law at the University of Graz, which gives an idea of the formative education that Austrian politicians received in the early twentieth century.

Struggle, Domination, Exploitation and Civilization According to Gumplowicz

The most considerable difference between the majority of animals and man is that the former are not in a position to get services rendered to them by other individuals of their own kind or another; in other words, animals are not capable of domination.

The primitive human band does not utilize its own members, individuals of its own lineage: this band is made up of individuals who are perfectly equal and equally free, each of them seeing to the satisfaction of his own needs, whether in isolation or in common. As long as this band remains isolated, no civilization is possible, since even the most modest civilization, the first phases of development of even the most primitive civilization, are subject to a division of labour by virtue of which one person is charged with lower-grade and harder tasks, another with easier and higher-grade tasks (which include that of command).

The essence of this division of labour is that some people work for others; only this division enables those who are worked for to turn their minds to higher subjects, to reflect on higher things and incline to an existence 'worthy of man'.

Men would remain eternally in a state similar to that of the animals if they were all obliged to see to their own subsistence and could not turn themselves to any other occupation. So that they can raise themselves above the animal state, it is necessary for some of them to be freed by the work of others from the most absorbing cares, the most crushing tasks.

Now we know that no one submits voluntarily to the yoke of another; that no one freely undertakes inferior and heavy work to provide others with the satisfaction or even the possibility of living idly. This first step on the path of progress and civilization would never have been taken if it depended on the devotion of some people for others, on Comte's 'altruism'. We cannot expect from man in general, and particularly from men in the raw state of nature, this kind of devotion for unknown higher ends, still less a prophetic penetration, the foresight of a common future prosperity. Each person, concerned only for his immediate advantage, the immediate satisfaction of his needs, and his immediate convenience, would always choose the role of master, and never that of worker and slave. If events had depended on men's insight and goodwill, we should still today be at the stage in which we find the inhabitants of Tierra del Fuego, in the remotest point of South America.

Fortunately, the natural process of history does not depend on the good grace of individuals; nature appears to have taken precautions in this as in other matters. It has placed in men's hearts irresistibly powerful propensities that prevent this process from coming to a halt and that incessantly encourage development, in the same way as various physical forms perpetuate and accelerate the astronomical, chemical, vegetable and animal processes.

When humanity populated the globe in its countless syngenetic bands, the penchant for personal preservation and the egoism of various foreign bands, coupled with deep repulsion and inexorable hatred towards these foreign bands, shook up the great natural process of history. Who was to work for whom? Who had to render services to others? Who was to form the lowest rank so that others could rise to a higher degree of development of civilization? It was necessary that these questions should no longer depend on free will, on understanding or choice. They were soon resolved by way of natural necessity. In the 'struggle of races' for domination, the strongest band settled things to its own advantage.

This phenomenon is based on a natural law: we can show this by the entire course of known history and by events of the present day, just as the chemist can conclude from everyday observation that in remote ages water already evaporated under the influence of the sun.

There is another phenomenon that similarly presents a natural regularity that knows no exception: the penchant for personal preservation and egoism press one social group to make the weaker group serve its ends by force, to subject and dominate it, dictating and settling by violence a division of labour.

Domination, at the end of the day, is nothing more than a division of labour settled by greater strength, a division in which inferior and burdensome tasks fall to the dominated, while the higher and easier occupations (often confined to command and administration) are the appurtenance of the dominant. Just as it is impossible to imagine any civilization exempt from the division of labour, in the same way no fruitful division of labour is possible without domination, since, as we have seen, no one will submit voluntarily to the performance of inferior and burdensome tasks.

We are now able, envisaging nature from a teleological point of view, to admire the great 'wisdom' that it has shown in the way it resorts to those means that are most practical and appropriate to its goals ...

If men had had 'humane' feelings, if each had seen in the other his 'brother', many great works of civilization would not even have been undertaken.

Nature has wisely and carefully avoided the hurdle that such 'humane' feelings would have set against any development of civilization. It is true that primitive man was also endowed with 'humane' feelings; but he only had such feelings towards members of his own band. This syngenetic feeling, or syngenism, is one of those social laws that exist always and everywhere, even if in forms adapted to the most varying degrees of civilization, as history and observation have shown us. But side by side with this syngenism there was profoundly rooted in human nature hatred of the foreigner, execration of foreign blood, completely insensibility towards the sufferings of any social group of a different origin. Only this hatred of the foreigner enabled the preparation of civilization by settling the division of labour by force, and imposing on foreigners, when intellectual progress had reached the point of no longer consuming them, all the burdensome work that is necessary to prepare a life of civilization, to realize the works of civilization.

It is in this way that nature has facilitated and made possible, by the original heterogeneity of ethnic elements and by the feelings of hostility that reign between them, the organization of the authority of some over others, conditions sine qua non for a fruitful division of labour. This latter, in its turn, established the relationship of cause and effect between the organization of domination and the development of human civilization.

Let us now examine, in somewhat greater detail, what the division of labour imposed by violence consists of, appearing as it indeed does as an 'exploitation' of some people by others: the exploitation of those who work and are dominated by those who command and dominate, but not without services rendered in the way of reciprocity by the latter to the former. These services consist in maintaining the order of the state whose development ends up procuring certain advantages even to those who appear exploited, by enabling them to participate, in many respects, in the goods and benefits of civilization acquired by way of this order and its development.[11]

11 Ibid., pp. 232–7.

Such are the general principles of the sociology of racial struggle according to Gumplowicz. It is not hard to recognize in them the themes of altruism and egoism repeated by the entire Darwinian school (see Part One); the point here is simply that Gumplowicz adds to these their natural complements, carefully avoided by the Darwinian writers: hatred of the foreigner and war of the races.

As far as Gumplowicz's ideas on the relationships between sociology and politics are concerned (politics being the application of sociology), it is enough to refer to the quotation given on pp. 17–18. This will give the idea of what kind of politics his sociological theories support.

Starting from these sociological principles, it is clearly possible to read history in a very slanted fashion, or to write all sorts of anthropological fables. Such is the property of this kind of theory, in which explanation follows from reconstructions done according to certain *a priori* principles. It should, however, be recognized that Gumplowicz's analyses and fables are in general quite well conceived. They are far more subtle than the crude Darwinian fables we have explored above. Their greater power of seduction makes them all the more dangerous.

As it is impossible to go into all this in the framework of the present book (and I am not a historian of sociology), I shall just cite a passage by Lester F. Ward, which appeared a few years after the publication of Gumplowicz's book. Here he summarizes the origins of society, the formation of the state and the main social structures as arising from the struggle of races and the enslavement of the defeated race by the victorious race within an ethnically heterogeneous group (divided into castes that are at once racial, occupational and social):

> Gumplowicz and Ratzenhofer have abundantly and admirably proved that the genesis of society as we see it and know it has been through the struggle of races. I do not hope to add anything to their masterly presentation of this truth, which is without any question the most important contribution thus far made to the science of sociology. We at last have a true key to the solution of the question of the origin of society. It is not all, but it is the foundation of the whole ... It is the only scientific explanation that has been offered of the facts and phenomena of human history ...
>
> This process has been fully described and illustrated by both Gumplowicz and Ratzenhofer, and they not only agree as to what the successive steps are but also as to the order in which they uniformly take place. I therefore

> need only enumerate these steps and refer the reader to the works of these authors, especially to Gumplowicz's Rassenkampf [1883], and Ratzenhofer's Sociologische Erkenntnis [1898]. The following are these steps arranged in their natural order: 1. Subjugation of one race by another. 2. Origin of caste. 3. Gradual mitigation of this condition, leaving a state of great individual, social, and political inequality. 4. Substitution for purely military subjection of a form of law, and origin of the idea of legal right. 5. Origin of the state, under which all classes have both rights and duties. 6. Cementing of the mass of heterogeneous elements into a more or less homogeneous people. 7. Rise and development of a sentiment of patriotism and formation of a nation.[12]

Here we see a sketch of how, on the basis of heterogeneous races in mutual struggle, a more or less homogeneous state and nation are constructed, and finally a new patriotism (this being simply a form of syngenism).

As for the importance of this naturalistic explanation of state formation, along with the naturalization of society and the scientization of sociology that follows from it, here is what Ward says (in the midst of the Darwinian revolution in the sciences criticized by Novicow – cf. Part One above – and in characteristic spirit):

> The discovery of the true origin and nature of the state ... has brushed away a greater amount of error than almost any other established truth in science. All the old ideas of the origin of the state are placed by it in the same list as the geocentric and Ptolemaic theories of astronomy, the doctrine of phlogiston in chemistry, and those of special creation and the immutability of species in biology. There is no longer a social compact, no divine right of kings or of 'the powers that be', no abstract right. By a perfectly natural, evolutionary process society everywhere and always has worked out a regulative system which, while not an organism, may still be compared with the regulative system of the Metazoan body, and has precisely the same sanction as a positive fact. The state is a natural product, as much as an animal or a plant, or as man himself.[13]

To end our discussion of the explanatory power of these kinds of conceptions, here is something rather curious – the explanation of

12 L. F. Ward, *Pure Sociology* (London: Macmillan 1903), pp. 203–5.
13 Ibid., p. 549.

the ban on incest imagined by Vacher de Lapouge on the basis of principles that are clearly inspired by those of Gumplowicz:

> Among the ancient peoples of Europe, the prohibition of unions between close relations goes back a long way; it seems to have originated in the practice of exogamy, and to go back to the era of savage existence. In order to have full authority over a woman, the surest procedure was to take as wife a slave seized from a neighbouring tribe. This is the point of departure of marital authority, and its point of contact with political authority. The discredit attaching to other forms of union less advantageous to the husband led to the general custom of taking women who were purchased or seized. This seems to have led by survival to the prohibition of taking a wife from within the tribe, then within the family when the tribe became a people. The horror of incest would thus have a common historical origin with the conception of marriage by seizure or purchase, which, in a well-known evolution, produced our present legislation.[14]

There is one particular point that needs to be made here, as it distinguishes Gumplowicz from many racist theorists of his era: in his conception there is no hierarchy of races, at least not one given once and for all, and no more is there an associated discourse of racial purity. For him, the struggle of races does not lead to the decisive domination of one of these, which then reigns over others by remaining 'pure', if indeed it does not eliminate them. The struggle is, on the contrary, an unending process, and does not rule out the mixing of groups. To understand this, we have to examine the development of races in this process.

First of all, Gumplowicz is a vigorous polygenist (indeed, he criticizes Gobineau's monogenism).[15] Polygenism for him is a 'policy of nature' (the title of Chapter 11 of his book), a nature that displays in this way its 'wise foresight'. He absolutely does not believe that humanity had its unique origin in a primordial pair. In this way he is opposed both to Darwin's monogenism and to Haeckel's quasi-monogenism (as he himself indicates). In Books I and II of his work (pp. 41–154), he develops a lengthy critique of monogenism and his own arguments in favour of polygenism – arguments that are biological, linguistic (diversity of primitive languages) and religious (diversity of religions). As against an original biological unity of human races, he prefers

14 G. Vacher de Lapouge, *Les Sélections sociales*, pp. 325–6.
15 L. Gumplowicz, *La Lutte des races*, p. 37, n. 1.

an 'intellectual affinity' between them, an 'essential similitude of intellectual aptitudes' that he locates in the faculty of language;[16] this is then rather a convergence, a characteristic that different groups of different origin separately acquired.

It is these originally diverse human races (often referred to as 'multiple syngenetic bands') that engage in struggle when they come into contact. The result of these struggles is not necessarily physical elimination, as with Gumplowicz the logic of domination and exploitation often triumphs over that of extermination and murder. Simple murder may be followed by an early form of exploitation – cannibalism – and this in turn by slavery. The defeated and enslaved races gradually mix with the victorious race in a social group that is not homogeneous from the ethnic point of view; here we have the formation of social classes and occupational castes of racial origin, with the victorious race forming the dominant caste/class, and the defeated race the inferior ones. With the passage of time, a certain biological homogenization is achieved by interbreeding, so much so that this social group ends up forming a new race. This in turn can struggle against another race, which may have been formed in the same fashion, or may have arisen from the original polygenism. The overall pattern is that races change, their number tending steadily to decline from the original polygenism, so that, ideally, humanity will end up unified (a remote ideal).

It follows from this that present-day races, as against the original races (polygenism), are not strictly biological. They include, by virtue of the way in which they were formed, a certain number of historical and social factors. The unity of blood of a present-day race may even be no more than a consequence of these historical and social factors, if this unity results from a homogenization over history of two or more races that were initially different. Once achieved, however, this unity of blood becomes the cement of the group, the basis for a new 'patriotism'. We find here the principle of Gumplowicz's syngenism, which, as we have seen above, closely combines ties of blood and hatred for the foreigner with social, linguistic and religious factors. Race and nation(ality) then merge.

By way of comparison, this is what Vacher de Lapouge wrote on the nation and its formation; the influence of Gumplowicz is clear:

16 Ibid., p. 53.

> A nation is a set of individuals issuing from different races, but united by complex ties of family, and whose ancestors historically reacted on one another, subject to common processes of selection. It includes the living, the greater number of dead, and the posterity down the ages, since by its nature the nation necessarily stakes a claim to eternity and universality, in other words to remain alone and cover the entire globe with its descendants. The nation in process of formation includes various races, in different proportions, and distributed in a certain fashion in the social hierarchy. These individuals gradually give rise to a more compact group. Lineages are combined over the generations, and ramify and combine without end. The community of plasma is established in the entire mass, and there is no longer any individual who is not related in some degree to all others.[17]

This process of the formation of races is, according to Gumplowicz, the warp and weft of universal history:

> The formation of races, in other words the formation of ethnic units understood in the sense that we have indicated above, is the most important component of human history. This formation of races, with all its concomitant phenomena, is the essential thing in what is called universal history.[18]

Here we have a form of racism that not only does not appeal to an essentialist conception of races, but more or less abandons the notion of race in the strictly biological sense of the term; this racism is rather a kind of exacerbated xenophobia. Gumplowicz absolutely does not believe in attempts to characterize existing races in purely biological terms, the terms then current in physical anthropology:

> Let us turn now to the anthropologists. These claim to divide humanity into tribes and races according to physiological and anatomical criteria – or this is at least the procedure of the pure anthropologists among their ranks. Well, they are no more successful in these efforts than are linguists and historians. Anyone who has sought to draw on the results of anthropological research into the various human types will be aware of the sorry role played by the measurement of skulls, etc. There is an inextricable confusion here: the 'average' numbers and measures do not give tangible results. What one anthropologist describes as the German type is exactly

17 G. Vacher de Lapouge, *L'Aryen, son rôle social*, p. 366.
18 L. Gumplowicz, *La Lutte des races*, p. 255.

what another describes as typical of the Slavs. There are Mongolian types among the 'Aryans', and 'anthropological' criteria constantly lead Aryans to be confused with Semites and vice versa. From the physical point of view, the Gordian knot of humanity is highly complicated; the problem is insoluble in the physical domain, and we can only refer to social and national groups as they exist in fact. It is not physical circumstances that have had a decisive influence in the formation of these, but totally different ones.[19]

And then there are the conclusions that Gumplowicz draws for the notion of human races. He does not deny their existence, but mixes together biological and non-biological factors:

> The notion of race today can never and nowhere be simply a notion of natural science, in the strict sense of the term; its only existence now is as a historical notion. Race is not the product of a simple natural process, in terms of the significance that this word has always had until now; it is a product of the historical process, though this is also itself a natural one. Race is a unity that, in the course of history, has been produced in and through social development. Its initial factors, as we shall see, are intellectual ones: language, religion, customs, law, civilization, etc. It is only later that the physical factor appears: a unity of blood. This is far more powerful, it is the cement that maintains this unity.[20]

For Gumplowicz, therefore, race is not as biologically simple as for the majority of racist thinkers of his time. It is biological and socio-historical, but the socio-historical dimension is as natural as the biological one, by virtue of the naturalization of society and history. The criteria of racial differentiation are thus also themselves natural, in the same way as those that prevail in animals and plants. So much so, in fact, that race here, without being purely biological, is just as natural a classification as those of pure taxonomy.

This notion of race, which combines biological and socio-historical factors, is perhaps closer to that of Darwin himself than it is to that of the 'progressive' Darwinism that quickly gained the upper hand and was prevalent in Gumplowicz's time. All the more so, in that for Darwin, rather than racial mixing, it was the elimination of one race to the benefit of another in the competition

19 Ibid., pp. 190–1.
20 Ibid., p. 192.

for life that ensured evolution; for Darwin, races were 'nascent species' sorted by selection.

The characteristic 'progressivism' of the Darwinism of that time was likewise totally absent. Gumplowicz rejected any notion of progress. More exactly, for him progress was only local and temporary. History was not a linear movement, not guided by progress; it led nowhere, and was made up of interwoven cycles with no finality or end. The most it involved was a certain tendency towards uniformity, by a reduction in the number of ethnic groups starting from the original polygenism.

As a result, the race that was victorious under certain conditions at a certain time could very well be defeated at another time or under different conditions. There could be no hierarchy of races given once and for all, as history does not show the least trace of direction or teleology – not even a teleology of progress as in contemporary Darwinism. History and society were naturalized by their racialization; they became natural processes and therefore lacked any meaning, direction or significance. For Gumplowicz, nothing whatsoever has the least such value. (As in Horace's verse: 'Nature cannot distinguish what is just from what is unjust.')

This absence of racial hierarchy has nothing in common with any kind of humanism or egalitarianism. It results from the naturalization of society and history. Gumplowicz claims for his procedure a cold scientific objectivity (one might better say a coldness of spirit, as his approach is still far removed from science). An attitude of this kind, which derives from an absolute nihilism, is somewhat rare among the theorists of racism, who for the most part have certain values according to which they classify races, from lower to higher (their own) and generally in a progressivist Darwinian framework (progress as the elimination of the lower race by the superior one).

Ward, who appeals to Gumplowicz despite rejecting his polygenism, reintroduces an evolutionary and taxonomic aspect in his history of races. He sees this as a 'sociological tree' – in other words, an evolutionary tree similar to that of biological species, but indicating at the same time the genealogy of races and that of societies. The two are connected, given that race in these theories is the foundation of society. The result is that sociology becomes a kind of extension of evolutionary biology – something that E.O. Wilson's sociobiology sought to reinvent nearly a century later. There is truly nothing new under the sun in this field, where the same ideas are tirelessly revamped. Here again is Ward:

> Passing over, then, with these few hints, the field of zoology, let us rise at once to the plane of human history and see whether we cannot find a similar parallel here. We may look upon human races as so many trunks and branches of what may be called the sociological tree. The vast and bewildering multiplicity in the races of men is the result of ages of race development, and it has taken place in a manner very similar to that in which the races of plants and animals have developed. Its origin is lost in the obscurity of ages of unrecorded history, and we can only judge from existing savages and the meagre data of archaeology and human paleontology, how the process went on. But we know that it did go on, and when at last the light of tradition and written annals opens upon the human races we find them engaged in a great struggle, such as Gumplowicz has so graphically described. But we also find, as both he and Ratzenhofer have ably shown, that out of this struggle new races have sprung, and that these in turn have struggled with other races, and out of these struggle still other races have slowly emerged, until at last, down toward our own times and within the general line of the historic races, the great leading nationalities – French, English, German, etc. – have been evolved.[21]

No great deal of imagination is needed to understand that all these theories can be readily permutated, and that Ward's racial-sociological tree links up with Haeckel's genealogical tree of human races, with its evolutionary hierarchy, its irremediably lower races at the bottom of the classification and its higher races at the apex of evolution. The mixture between Haeckel's racial hierarchy and Gumplowicz's struggle of races, both equally 'scientific' and pitiless, was thus more or less inevitable. In point of fact, the majority of racist theorists propose both an evolutionary hierarchy and a struggle of races, in varying forms and with diverse intentions.

Gumplowicz was a Polish Jew, originally from Galicia, and as we have mentioned, he taught public law at the University of Graz in Austria. It may seem odd that a Jew should have invented this kind of theory, in an empire marked by anti-Semitism and shaken by quarrels among its heterogeneous nationalities – unless, by a rather simplistic social determinism, we see these phenomena rather than social Darwinism as the inspiration for Gumplowicz's theory. It is all the more curious that Gumplowicz should have had as a fellow-citizen and exact contemporary Georg von Schönerer (1842–1921), one of Hitler's models and head of the

21 L. F. Ward, *Pure Sociology*, pp. 76–7.

German national movement in Austria, which conducted violent campaigns that combined pan-Germanism (for the unification of the German peoples), anti-Catholicism (conversion of the Austrians to Protestantism, considered the proper German religion),[22] and inevitably anti-Semitism. This coexistence of the two theories may well seem quite strange, but as we have already seen in the case of eugenics, this history came to be seen in Manichaean terms only after the Second World War. Previously (and *a fortiori* in the late nineteenth century), roles were not distributed in such a clear-cut fashion, with the good on one side and the bad on the other. Theories intermingled, and it was not yet clear what was going to come out of this (even if some already suspected it would be nothing good – Gumplowicz and Vacher de Lapouge among them, as we shall see).

According to Daniel Gasman, Gumplowicz converted to Protestantism while remaining both philo-Semitic and anti-Zionist.[23] This suggests that he did not convert out of religious conviction, but rather to facilitate his university career – no rare thing in Austria at this time. Gasman ascribes to his Jewish (and Polish) origin the fact that Gumplowicz, though strongly influenced by Haeckel, did not accept Haeckel's assumption of the superiority of the Indo-Germanic race, preferring the non-

22 The inevitable Vacher de Lapouge has his own ideas on this subject: 'The adoption of Christianity by the peoples of Great Britain and the Germanic lands was the starting-point for a series of selections that I have examined in a chapter of the previous volume (Les Sélections sociales, chapter X). The orientation of this selection changed after several centuries with the adoption of Protestantism by almost all the dolicho-blonde populations, and this transition was a great benefit. The influence of Christianity had acted powerfully to reduce these peoples to inferiority, not by virtue of religious dogma but above all by its dangerous morality. The dolicho-blondes of the north had the double advantage of falling under the influence of the Catholic Church several centuries later, and of shaking this off several centuries ago. This advantage made itself clearly felt in the proportion of eugenics among the dolicho-blonde and the Catholic peoples of Europe. The religious selection, in any case, has long since been oriented in an altogether different direction in the Catholic and the Protestant countries, and the example of the Irish and Flemish, dolicho-blonde and Catholic, enables us to appreciate the practical import of this divergence by comparing these with the peoples of Scotland and Holland' (L'Aryen, *son rôle social*, p. 364). 'The Reformation can thus be viewed as an attempt to adapt Christianity to the hereditary tendencies of the Aryan race. Like freedom, it never managed to strike firm root among other races'. (Ibid., p. 387).

23 D. Gasman, *Haeckel's Monism and the Birth of Fascist Ideology: Social Darwinism in Ernst Haeckel and the German Monist League* (London: MacDonald, 1971), p. 48.

hierarchical and more fluctuating conception whose main lines we have presented above.[24] I am not certain this connection is a necessary one. I rather believe that Gumplowicz adopted this non-hierarchical conception of races because it was completely coherent and in conformity with the rest of his theory. As for his philo-Semitism, I am also unsure whether this was really so pronounced, at least at the theoretical level. Here, for example, is an extract of a study comparing Phoenicians and Jews; it is hard to maintain that this is a philo-Semitic text, especially when we know that for Gumplowicz, the merchant is always a foreigner who establishes himself in a social group in order to exploit it:[25]

> To the precise extent that the power of the Phoenicians grew, the trading policy of this people developed; this policy became a colonial one ...
>
> Nonetheless, the Phoenicians did not turn their attention away from commerce, the arts and industry, and did not extend their domination further than their mercantile interests demanded. No conquering or warrior people of antiquity, however, had such a persistent importance as this merchant people, a far-reaching civilizing influence. Guided by absolutely egoistic instincts, seeking material gain by ruse and double-dealing, they nevertheless rendered the greatest services to civilization and humanity, especially in Europe. Without the Phoenicians, Europe would not have become what it is today ...
>
> A clever people, the Phoenicians knew how to disappear at the right moment. Endowed with a precise cosmopolitan sentiment, they did not believe that their 'national' civilization was worth universal hatred and hostility. They melted into the peoples among whom they lived; and as a result, they fulfilled more faithfully and exactly, if we can put it like that, the intentions of the natural historical process, than if they had preserved their nationality for long centuries, with untimely and unnatural tenacity ...
>
> These 'Israelite' tribes, as they were subsequently known, were originally pastoral nomadic tribes. After long migrations and vicissitudes, they conquered the land of Palestine, exterminating some of its inhabitants and reducing others to subservience. When civilization began to progress and the population to grow, when this small country was incapable of providing for the needs and demands of its inhabitants, the Jews followed the example of the Phoenicians; they became traders and spread out across the whole

24 Ibid., pp. 180–1.
25 L. Gumplowicz, *La Lutte des races*, p. 212.

> world.
>
> The special organization of their communities in Europe is certainly the reflection of the former Phoenician outposts. There is only one point, the most important perhaps, on which the Jews were unable to follow the example of the Phoenicians; they did not know how to disappear, and they still do not know how.
>
> It is true that this is chiefly the fault of their very developed ancient literature, in the main their theological literature. When Christianity, after emerging victorious from this people, and linked to their ancient tradition, declared that Judaic writings were 'sacred', it seemed – not so much to the blind and ignorant masses, but rather to a proud and deluded caste of scribes – that these were sacred national treasures to be preserved; in an unnatural stubbornness, these scribes preferred to perpetuate, among all peoples and nations, an eternal racial struggle against themselves, rather than sacrifice this outworn and mummified nationality to the newly burgeoning civilization of other countries and new epochs. This persistent attachment to forms of civilization that have long since become decrepit, and which in truth would be more in place in the catacombs of history than in the life of nations, is a grave fault against the great natural law of history, and this fault has been harshly paid for by thousands of generations.
>
> There are already only too many inevitable racial struggles, the result of the natural and necessary development of ethnic and social elements; and it in no way seems necessary, there seems no kind of merit from the point of view of universal human history, in permanently maintaining and perpetually stoking up, by foolish defiance of the eternal laws and all-powerful currents of the natural social process, a struggle of races that, like the struggle of the Phoenicians, could have come to an end long ago.[26]

This is certainly not a philo-Semitic text, but it is not properly speaking anti-Semitic either – even if today it would certainly lead to the prosecution of its author. It is not anti-Semitic because Gumplowicz does not make any kind of value judgement, either on this question or any other. He simply gives what he considers a scientific explanation. There is no place here for moralizing, only for the iron laws of social processes that are assimilated to natural processes, and are thus equally blind and inexorable.

This absolute nihilism of Gumplowicz brought him a rather particular reputation, as if he smelled rather of sulphur. Ward, for example, praises him highly (professing as he does the same kind of theories, simply more humanized or hypocritical), but even he

---

26 Ibid., pp. 326–32.

takes a certain distance, criticizing in particular Gumplowicz's polygenism, with which Gumplowicz had 'so greatly and so unnecessarily weakened his arguments in the eyes of scientific men'.[27] In 1898, R. Worms, in his preface to the French translation of *Sociology and Politics*, recognized Gumplowicz's merits but stressed the disputable boldness of his claims.[28] Novicow, in his 1910 *Critique du darwinisme social*, omits him altogether, perhaps out of decency, as Gumplowicz had just taken his own life. In 1928, Sorokin, in his classification of biological sociologies, clearly does not like him and rejects his theories.[29] In 1953, Gaston Bouthoul wrote in his 'Que sais-je?' volume on war:

> The acme of theoretical warmongering was reached by Gumplowicz, whose doctrine is simply an avid summons to battle ... He presupposes an inherent and undying hatred in the relationship between one group and another: hence an inevitable and mortal struggle between groups. All social forms and institutions are born of war.[30]

Gumplowicz remains nonetheless an eminent representative of the sociology of his time. He was famous; his books were translated into several languages. He had for a long time a great influence both in Europe and the United States, on sociologists such as Lester Ward (pioneer of American sociology) and Vilfredo Pareto (a noted economist), as well as on politicians including Mussolini.[31] Vacher de Lapouge, as we have seen, borrowed many of Gumplowicz's ideas without acknowledging him (though *La Lutte des races* does appear in the bibliography to *Les Sélections sociales*), and he was also an inspiration to Ammon.

This conception of a society based on the struggle of races was certain to have a conspicuous posterity, both in sociology and in politics. At the time, and at least until the Second World War,

27 L. Ward, *Pure Sociology*, p. 32.

28 Gumplowicz, *Sociologie et politique*, with preface by R. Worms (Paris: Giard and Brière, 1898).

29 P. A. Sorokin, *Contemporary Sociological Theories* (New York: Harper, 1928), pp. 194–218.

30 G. Bouthoul, *La Guerre* [1953] (Paris: PUF, 1978), p. 27.

31 D. Gasman, *Haeckel's Monism and the Birth of Fascist Ideology*, pp. 169–70. Chapter 5 of this book ('The monism of Ludwig Gumplowicz and Gustav Ratzenhofer', pp. 165–98) deals in particular with Gumplowicz's relationship to Haeckel.

theories of racial struggle were accepted as an official current in sociology, even if they clearly did not constitute the whole of sociology. They were no more shocking (though no less) than racial hierarchies à la Haeckel.

Here, for example, is a text from 1931 that presents an entertaining version of the struggle of races; the influence of Gumplowicz is very perceptible, as can be seen by comparing it with the quotations given above. The writer here is Sir Arthur Keith (1866–1955), the same to whose authority Ardrey appealed in making the territorial instinct more powerful than sexuality (see pp. 31–2 above). As a good Englishman, Keith was undoubtedly a lover of sport, and for him the struggle of races becomes a kind of sporting competition; each 'racial tribe' is compared to a team (whites, yellows, browns and blacks), participating in a match of 'Nature's League':

> No transfers for her; each member of the team had to be home-born and home-bred. She did not trust her players or their managers farther than she could see them! To make certain they would play the great game of life as she intended it should be played she put them into colours – not of transferable jerseys, but liveries of living flesh, such liveries as the races of the modern world now wear. She made certain that no player could leave her team without being recognized as a deserter. To make doubly certain she did an almost unbelievable thing. She invaded the human heart and organized it so that her tribal teams would play her game – not theirs. She tuned the heart of her teams for her own ends. She not only imbued her opposing teams with an innate love of competition and of 'team-work'; she did much more. What modern football team could face the goal-posts unless it developed as it took the field a spirit of antagonism towards the players wearing opposing colours? Nature endowed her tribal teams with this spirit of antagonism not for her own purposes. It has come down to us and creeps out from our modern life in many shapes, as national rivalries and jealousies and as racial hatreds. The modern name for this spirit of antagonism is race-prejudice.[32]

To read Keith, who was an anatomist and anthropologist, it is perfectly understandable that Hitler should have made Jews wear

32 A. Keith, *The Place of Prejudice in Modern Civilization* (London: Williams & Norgate, 1931), cited inr M. Banton, 'The autonomy of post-Darwinian sociology', in M. Banton (ed.), *Darwinism and the Study of Society: A Centenary Symposium* (London: Tavistock Publications, 1961), pp. 168–9.

a yellow star, and why the Americans in the United States, like the British and Dutch in South Africa, had no need for anything similar when they imposed racist legislation – as against the Jews, the blacks had their 'livery of flesh' that they could not exchange by 'desertion'.

Keith's thesis, that racial prejudice serves the avoidance of interbreeding and the maintenance of distinct races in this competition, is not in itself so original. The same idea can be found in 1918, under the name of 'the colour line', in *Applied Eugenics* by Paul Popenoe and Roswell H. Johnson.[33] (Popenoe, a member of the board of the American Eugenics Society, was one of the California eugenicists who had sympathetic connections with German eugenics.)[34]

Curiously, Gumplowicz is rarely cited in the history of racism, preference always being given to Gobineau and Vacher de Lapouge. In the books we quoted in the Introduction (p. xxii),[35] he is as absent as Haeckel, whose monism Gumplowicz closely followed. Only P.-A. Taguieff mentions him (just once, in a note),[36] to say that his theory 'cannot be summarily dismissed as "racist"', and that 'J. Freund, for his part, recognizes a certain pertinence in Gumplowicz's conception of politics'. It would have seemed better to me if Taguieff had said Gumplowicz's theory was not summarily racist, as its notion of race is somewhat sophisticated, lacking any fixed racial hierarchy or discourse of racial purity. That said, Gumplowicz was a polygenist (unlike Gobineau); he makes races (whether biological or historico-biological – and for him, history is as natural as biology) the basic social units, and their struggle the motor of history. He was certainly not as crude as to utter invectives against black people. But it seems hard to me not to see this kind of theory as a form of racism, even as one of the most dangerous forms. As for the 'pertinence' of his conception of politics, this is dubious to say the least. One may well be fascinated by the radicalism of this character, his

33 P. Popenoe and R. H. Johnson, *Applied Eugenics* (New York: Macmillan, 1918).

34 P. Popenoe, 'The German sterilization law', *Journal of Heredity*, 1934, 25, p. 257.

35 J. Huxley, A. C. Haddon and A. M. Carr-Saunders, *We Europeans* (London: Cape, 1935); UNESCO, *The Race Concept* (Paris: 1952); F. de Fontette, *Le Racisme* (Paris: PUF, 1981).

36 P.-A. Taguieff, *La Force de préjugé, essai sur le racisme et ses doubles* (Paris: La Découverte, 1987) p. 577.

absolute nihilism, or even an intelligence that shines above the grey mediocrity of the majority of racist thinkers (for example, the crass stupidity of a Madison Grant). But this fascination should not lead us to forget the contents of his doctrine. Otherwise one might even find Himmler fascinating, with the same intelligence, the same nihilism, and the same lack of humanity. Besides, Gumplowicz was intelligent enough to realize what his sociological theory implied (remember that he saw politics as an application of sociology, as per the quotation on pp. 17–8). This is what he wrote in the Preface to *La Lutte des races*:

> But I halted for a long while on another consideration: what would happen if the new knowledge contained in this book, as Roscher so well said, were to 'fall on human passion, so prompt to catch fire; if the flame were to flare up and exert by contagion its devastating action'. I weighed this objection for a long time, but ended up by rejecting it along with the others. Certainly, it is possible that human passion might seize on a number of passages in this book to justify odious tendencies; but in this case, the theories presented here would only share the fate of the most sublime doctrines that have ever been taught to humanity, as it is in the name of the sublime doctrines of religion that evil and perversity have spilled torrents of blood.
>
> What is the use, then, of scientific research, given these tendencies of human passion? Passion allied with infamy will continue its course without being stayed. Let science do the same! It has neither the pretension nor the hope of restraining passion, as there is an unbridgeable barrier between its doctrines and everyday thinking. Let science be left with one consolation alone, that of focusing without obstruction on the quest for truth, and proclaiming its findings without prevarication! Let it be spared useless scruples, and its unique dogma be left intact. Far from truth and the royal quest for it being able to damage humanity, this is the only path that can lead to humanity's salvation![37]

Such discourse is nothing new, and science can bear the responsibility. But where is science in Gumplowicz's theory?

Normal and commonplace

What today we find abominably racist in texts of Gobineau, Darwin, Haeckel, Büchner, Vogt, Gumplowicz, and so on, such as those cited here, was in their own day the dominant opinion

37 L. Gumplowicz, *La Lutte des races*, preface, p. ix.

– so commonplace that scarcely anyone thought of criticizing it, no matter what their political views. At most, there was a certain reticence when these theories reached a glacial nihilism, as in the writings of Gumplowicz, or when they took on the form of a caricature, as with Vacher de Lapouge. (Indeed, the latter's main mistake was undoubtedly to apply to European populations the same criteria that no one flinched from applying to people of colour. The differences between dolichocephalic Europeans and brachycephalic Europeans were not always convincing from the scientific point of view, whereas those between Europeans and Africans were universally accepted – and an ill-intentioned critic could say that this tradition is continued by those historians who constantly cite Lapouge but forget Haeckel.) Vacher de Lapouge and Gumplowicz openly spoke the true content of their theories, without bothering overmuch to wrap them up in fine sentiments (the iron laws of nature against which, sadly, nothing can be done; the necessity of progress; etc.). It was this contempt for accepted forms that was disturbing, rather than the content of the discourse itself.

For all these thinkers (aside from a small minority of anti-Darwinists, often idealist and generally considered backward), evolutionary racist theories were scientific, representing modernity and progress. It is precisely this commonplace normality that conceals them today. The flamboyant littérateur Gobineau and the marginal but clairvoyant Vacher de Lapouge are far more visible. That is why it is they who are cited today, and who conveniently serve as cover for the others (Gumplowicz aside, as he is still disturbing today).

At the time, racist legislation, whether explicitly formulated or not, existed everywhere that races came into contact: in the United States and the majority of colonies, in other words, almost the whole of Africa and a good part of Asia (Brazil already had a high degree of interbreeding and aroused a variety of sentiments – see p. 264 above).

If Europe did not have such legislation, it is because it did not have a number of races cohabiting. Persons of colour were too rare to require special laws (we shall examine below the cases of Jews and Gypsies). However, when North Africans and blacks appeared in Germany in the form of French colonial troops after the First World War, this occasioned panic. And when these Africans left behind a few hundred children, born to feckless mothers, the idea immediately arose (well before Hitler came to power) of sterilizing them to prevent the contamination of German blood.

If it was necessary to illustrate the commonplace normality of these racist theories, we can note how in 1919 the Allied Powers rejected the proposal of the Japanese delegation at the Paris Peace Conference that sought to include in the League of Nations charter a declaration proclaiming racial equality.[38] (We had to be able to continue exploiting the 'lower' races in the colonies without undue concern.) This rejection is the equivalent to eugenics and euthanasia being 'forgotten' in 1946 in the definition of crimes against humanity, and in the 1948 Universal Declaration of Human Rights (see pp. 210–1 and 216 above). After the Second World War, those lower races were accepted as fully human and given their rights, but not yet the mentally ill, the handicapped or the deviant; laws providing for compulsory sterilization remained in force more or less everywhere.

It was the rise of Nazism that led certain people to moderate their racial theories, just as in the case of eugenics, though qualification is necessary here. Thus, already in 1935, Nazi anti-Semitic legislation managed to shock world opinion because, quite apart from its echoes of mediaeval pogroms, it targeted white populations that were well integrated into German society. Other racist laws, in a wide range of countries, did not shock many people. In 1937 the sterilization of mixed-race people in the Rhineland and Ruhr took place amid relative indifference – was there a single protest by one of the great anti-Nazi humanists of the time? – and still today, most history books completely fail to mention it. Besides, the Nazis never failed to underline that the Americans and British, who criticized their anti-Semitic laws, had long had racist laws in force against black people (in South Africa as well as in the United States).[39] This point was all the sharper in that Anglo-Saxons were viewed at this time, along with Germans, as pre-eminent representatives of the Nordic race, and Anglo-Saxon eugenics had been the model for German eugenics.

38 J. Comas, 'Les Mythes raciaux', in UNESCO, *Le Racisme devant la science*, p. 56.

39 S. Kühl, *The Nazi Connection. Eugenics, American Racism, and German National Socialism* (Oxford: OUP, 1994), pp. 98–100.

15

Racism, Anti-Semitism and Biology

Racist legislation did not wait for scientific validation to be introduced. But science gave a new justification to existing laws, and actively supported their extension. The model here was that of eugenics: the scientization of racism and racist legislation took its arguments from theories of degeneration and the supposed eugenic remedies for it.

The racial dimension of eugenics was present from the start. Here is what Galton wrote in 1883, in the introduction to the book in which he coined the very word 'eugenics':

> My general object has been to take note of the varied hereditary faculties of different men, and of the great differences in different families and races, to learn how far history may have shown the practicability of supplanting inefficient human stock by better strains, and to consider whether it might not be our duty to do so by such efforts as may be reasonable, thus exerting ourselves to further the ends of evolution more rapidly and with less distress than if events were left to their own course.[1]

As early as 1869, in *Hereditary Genius*, Galton had emphasized the unequal value of different races (one of the chapters in his book being titled 'The Comparative Worth of Different Races').[2] Before him, in the 1862 preface to her translation of Darwin's *The Origin of Species*, Clémence Royer showed, as we see below, that she understood very well the implications of Darwinism, once it was applied to man and society – which Darwin had yet to do. A dimension equally racist and eugenicist is expressed here quite

1 F. Galton, *Inquiries into Human Faculty and its Development* (London: J. M. Dent, 1911), p. 1. (In Vacher de Lapouge's quotation from this text in *Les Sélections sociales*, p. 462, the translation reinforces its racist aspect.)

2 F. Galton, *Hereditary Genius, an Inquiry into its Laws and Consequences*, pp. 325–37.

clearly: the issue at hand was that the higher races should supplant the lower ones (racism), just as superior individuals should supplant inferior ones (eugenics), all in the name of progress. (Note how, here again, Darwin's abolition of the distinction between race and species is utilized in reverse, to make races into quasi-species.)

> Finally, does not Mr Darwin's theory, by giving us some rather clearer ideas on our real origin, by this very fact undermine a great number of philosophical, moral and religious doctrines, and systems of political utopias, which – however generous they may be – are assuredly false in their attempts to realize an equality between all men that is impossible, damaging, and counter to nature? Nothing is more clear than the inequalities between the different human races, nothing more marked than the same inequalities between different individuals of the same race. The findings of the theory of natural selection cannot leave us in any doubt but that higher races are progressively produced; and that, by consequence, according to the law of progress, they are destined to supplant the lower races by progressing further, and not to mix and combine with them in a way that would lower the average level of the species. In a word, human races are not distinct species, but they are very well-defined and highly unequal varieties; and we should think twice before proclaiming political and civic equality among a people composed of an Indo-European minority and a majority of Mongols or Negroes.[3]

This relation between racism and eugenics certainly had its practical limitations. Racist legislation did not demand the sterilization of the 'lower' races, simply segregation and the prohibition of marriage (there were also such prohibitions on the basis of certain illnesses). Racially incorrect individuals could not simply be sterilized en masse the way that ill people, alcoholics, criminals, etc., could be. It was not that the desire was lacking, but this desire was put into practice only in Nazi Germany, and perhaps, more unevenly, in the United States, where it is likely that the application of eugenic laws (depending on the look of the customer) was not exempt from racism. Better to be rich and white than poor and black. In any event, Nazi Germany is the only country in which the process can be followed with continuity, from eugenics to racism, and from sterilization to extermination.

---

3 C. Royer, preface to C. Darwin, *L'Origine de espèces* (Paris: Reinwald, 1882), p. xxxviii.

We have mentioned at several points the case of the racially mixed children that French African troops left behind in Germany. Under the Weimar Republic, some people proposed the necessity of sterilizing them, in particular the racial theorist Hans F. K. Günther.[4] In July 1933, immediately after the adoption of eugenic legislation, Fischer and Verschuer examined a number of these children and concluded that they were mentally deficient, so that this legislation was applicable to them.[5] Sterilization, however, was deferred for the time being. In 1937, a new examination of 385 children was conducted by a number of specialists, again including Fischer.[6] And on 18 April 1937, Hitler himself finally gave the order for their sterilization.[7] Here we can see very plainly the utilization of eugenic legislation for racist ends.

Exactly the same pattern is found in the case of Gypsies. They were considered not only an inferior race but also 'half-breeds', mixed with all the elements counted as society's lowest: antisocial persons, delinquents, degenerates, etc.[8] As a consequence, it was possible to envision sterilizing them, again in the name of the same eugenic laws.

Here is an extract from a letter of 9 January 1938 from Deputy *Gauleiter* Tobias Portschy of the province of Styria to Reich Minister Dr. Lamers:

> For reasons of public health, and particularly because the Gypsies are well-known to have a burdened heredity, as well as being inveterate criminals who are parasitic on our people and can only cause immense harm, endangering the purity of blood of the rural population of the German borderlands, its way of life and its legislation, it is important first of all to prevent them from reproducing and to subject them to the obligation of forced labour in work camps, without however preventing them from choosing voluntary emigration abroad ... The arguments in support of sterilization of the Gypsies can in fact be tacitly developed a priori to the

4 P. Weindling, *L'Hygiène de la race: Hygiène raciale et eugénisme médicale en Allemagne* (Paris: La Découverte, 1998), vol. 1, pp. 213–4.

5 Y. Ternon and S. Helman, Les Médecins allemands et le *National-Socialisme, les métamorphoses du darwinisme* (Tournai: Casterman, 1981), pp. 173–4.

6 B. Müller-Hill, *Science nazi, science de mort, l'extermination des Juifs, des Tziganes et des maladies mentaux de 1933 à 1945* (Paris: Odile Jacob, 1989), p. 26.

7 E. Ben Elissar, *La Diplomatie du IIIe Reich et les Juifs (1933–1939)* (Paris: Bourgois, 1981), pp. 151–5.

8 See, for example, O. von Verschuer, *Manuel d'eugénique et d'hérédité humaine* (Paris: Masson, 1943).

> point at which, simply under the law on the prevention of descendants with damaged heredity [the eugenic law of July 1933], we can wage an effective struggle against the increase of the Gypsy population. We have to make bold and unrestrained use of this law. At least the foreign press will not be able to make a hue and cry about it, given that we can always maintain in good faith that this law on the prevention of descendants with damaged heredity is equally applicable to German citizens.[9]

And here is an extract from a report by Robert Ritter on the same theme, from January 1940. Ritter was a psychologist who had been assigned, already before the war, to do a census and genealogical study of the Gypsies (both while they were still at liberty, and when they had already been herded into concentration camps):

> The Gypsy question can thus not be considered resolved until the great majority of half-breed, anti-social and work-shy Gypsies are collected in large camps for itinerant labour, when they are forced to work and when this half-breed population will have been definitively prevented from reproducing. Only then will future generations of the German people be really freed. The Centre of Eugenic Research is already in a position today to give specialist advice on the degree of interbreeding and the hereditary worth of each of these supposed Gypsies, so that there is no longer any further obstacle to taking eugenic measures.[10]

Only a few Gypsies, however, were sterilized. Instead they were exterminated, in the same fashion as the mentally ill, but later, along with the Jews. We shall examine below the case of the Jews; contrary to the racially mixed and the Gypsies, the Jews were not considered an inferior race (or even a race in the strict sense). In any case, for the racially mixed and the Gypsies, the relationship between eugenics and racism is very clear.

The extension of sterilization to entire racial groups was not only envisaged by the Nazis, but became the object of specific study – particularly in the concentration camps, where experiments with various methods of sterilization were imposed on prisoners. Surgical sterilization turned out to be too complicated and lengthy

---

9 Cited by Y. Ternon and S. Helman, *Les Médecins allemands et le National-Socialisme*, pp. 185–6.

10 Ritter to Breuer (DFG), 20 January 1940, cited in B. Müller-Hill, *Science nazi, science de mort*, p. 60.

a process, especially for women, and so other, more rapid methods were developed that could be applied on a larger scale, either by various intrauterine injections, or by way of X-rays (on men as well as women). Brack even imagined the possibility of sterilizing individuals without their knowledge, using an X-ray machine camouflaged behind a counter, which people were told to stand against for some made-up reason. A single machine of this kind would supposedly have made it possible to carry out 150 to 200 sterilizations per day.[11] In other words, a genocide by sterilization rather than a genocide by extermination.

This is the meaning of the various sterilization experiments carried out in the camps. Though often discussed as an aberration with no further significance, sterilization was in fact clearly integrated into the pursuit of eugenic measures.

Anti-Semitism and false biology

Because this subject involves continually dealing with Nazism, a problem arises – that of the relationship between anti-Semitism and the various forms of racism supported by evolutionary classifications of human races and theories of racial struggle. At first sight, indeed, such a relationship seems to be lacking.

For Haeckel, the highest human species is what he calls the 'Mediterranese' (in fact, simply meaning whites). This species is made up of four races. The first of these is the Indo-Germanic (with High Germans and Anglo-Saxons jointly at their head, these peoples being most advanced in the Industrial Revolution, followed by the Scandinavians – it should be noted that it was these three peoples who would adopt eugenic legislation fifty years later). The second is the Semitic (with Jews and Moors jointly at their head). The third and fourth are the Caucasians and Basques. For Haeckel, therefore, as for Gobineau, Jews are not an inferior race.

Anti-Semitism is either not involved at all in the racism underpinned by this kind of classification, which targets above all people of colour (black, more or less dark brown, and yellow), or at most it is only marginally involved. Whites, even if Haeckel's

11 For these experiments of sterilization on prisoners, see F. Bayle, *Croix gammée contre caducée, les expériences humaines en Allemagne pendant la Deuxième Guerre mondiale* (Neustadt: no imprint, 1950), pp. 665–725, and E. Klee, *La Médecine nazie et ses victims* (Paris: Solin-Actes Sud, 1999), pp. 316–24.

maniacal classification differentiates them into distinct races, are always viewed as the superior race. And Jews, as white Europeans, are part of this (and so they themselves could readily accept this sort of classification). That does not mean that classifying biologists like Haeckel were not anti-Semitic; they were as much so as the rest of the population. But they did not justify their anti-Semitism by biological arguments of a racist kind that targeted people of colour. This is a point that must be stressed: if the officially accepted biological theories gave succour to racism, they did not do the same for anti-Semitism.

Biological anti-Semitism followed a different route from this racism of an evolutionary racial hierarchy. It was based above all in 'traditional' anti-Semitism, which was not specifically racial and referred to social, economic and religious factors. Then came the contact between this anti-Semitism and the Aryan question, and finally the encounter with Darwinism – not acknowledged by biologists.

The Aryan question was originally one of linguistics. It was posed at a public lecture in 1786, when William Jones (1746–94) noted the resemblance among Sanskrit, Persian, Greek, Latin, and several other European languages. These languages were then grouped into a single family, which in 1813 was baptized 'Indo-European' by Thomas Young (1773–1829), simply on the grounds that its speakers were all either Indian or European. The source of all these languages was then called Proto-Indo-European, presumably the language of a people who were then, by extension, also called Proto-Indo-European, and who supposedly lived at one time somewhere between Central Asia and Eastern Europe (the geography varies a good deal). The language was thus made into a people, and the people then into a race.

The slippage from language to race happened 'naturally', but was taken up and accompanied by a Darwinian linguistics. Here is the argument that August Schleicher offered on the subject. He began by remarking that linguistics had imagined well before Darwin the idea of an evolution in which certain related languages derived from a single mother tongue (just as related species arose from the same ancestor); the typical example of this was precisely Indo-European:

> What Darwin maintains for animal and vegetable species also holds good, at least in its essential features, for the organisms of language ... Let us now examine the faculty of transformation over time that Darwin attributes to

> species ... This faculty has long since been generally accepted for linguistic organisms ... For the linguistic stems that we are familiar with, we can compose genealogical trees just as Darwin sought to do for animal and vegetable species. No one today doubts that the entire group of Indo-Germanic – Indian, Iranian (Persian, Armenian, etc.), Greek, Italian and Germanic languages – this whole group with its many species, sub-species and varieties, had its birth in a single mother form, the original Indo-Germanic language. The same holds for the Semitic stem, which includes Hebrew, Syriac, Arab, etc., and in general for all others. The genealogical tree of the Indo-Germanic stem can take its place here, and will form the representation of the imperceptible development that in our view this stem has experienced.[12]

Elsewhere Schleicher proposes that language is closely dependent on the biological – and thus racial – characteristics of individuals:

> The function and activity of the organ, therefore, is no more than a manifestation of the organ itself, even if the scalpel and microscope of the observer do not always succeed in showing the material causes of each manifestation ... Language is the perceptible manifestation to the ear of the activity of a series of conditions that are realized in the conformation of the brain and the organs of speech, together with their nerves, bones and muscles. It is true that the material principle of language and its variations has not yet been shown anatomically, but also no one to my knowledge has as yet undertaken the comparative observation of organs of language in people speaking different tongues. It is possible, and perhaps even likely, that research of this kind would not lead to a satisfactory result: but this would not refute the theory that the form of language depends on certain material conditions.[13]

Finally, on the basis of this principle and taking into consideration that language is the characteristic par excellence of humanity, it is maintained that human races can be classified on the basis of the classification of language – a more certain criterion than the shape of the skull:

> If language is the specific characteristic of humanity, this suggests the idea that language would well serve as a distinctive principle for a scientific and

12 A. Schleicher, *La Théorie de Darwin et la science du langage* [1863] and *De l'importance du langage pour l'histoire naturelle de l'homme* [1863] (Paris: Vrin, 1980), pp. 7–9.

13 A. Schleicher, *De l'importance du langage pour l'histoire naturelle de l'homme*, p. 22.

> systematic classification of humanity, and form the basis for a natural system of the species. Cranial shape and other distinctive racial indications are far too variable. Language, on the other hand, is a perfectly constant character. It is possible for a German to display hair and prognathism that rival that of any Negro head, but he will never perfectly speak a Negro language ... Language does not just seem important to us for the construction of a scientific natural system of humanity, as this is now shown by observation, but also for the history of its development.[14]

No comment is needed. We should note how today the idea of a linkage between the evolution of languages and that of human groups has been resurrected, but this time on a totally different basis: that of a parallelism rather than a relationship of cause and effect.[15]

In any event, this linguistic and racial basis serves as the foundation for an entire mythical history rich in incidents, a further anthropological fable. Gobineau's role here is far from negligible, even if he was not the inventor of Aryanism. These Indo-Europeans soon became Indo-Germanic or Aryan (from the Sanskrit *rya*, meaning 'of good family', 'well-born' in the aristocratic sense, precisely what *eugenes* means in Greek). At the same time, they became very clearly blonde and blue-eyed, and were identified with the barbarian hordes who invaded the Roman Empire – hordes deemed to represent the last Indo-Europeans in a more or less pure state, before their miscegenation on contact with the southern races.

For the aristocratic and Germanic side, the construction of this myth certainly borrowed from the ideas of Henri de Boulainvilliers (1658–1722), according to whom the French aristocracy was descended from the (Germanic) Franks who had invaded Gaul, with the original Gauls and Gallo-Romans furnishing the plebs (Gobineau reproduced the same kind of aristocratic racism). With the Teutonic 'blonde beast', the myth acquires the coloration of German nationalism, pan-Germanism and the proclaimed superiority of the German 'race' – all this on the basis of the war of races already discussed (see the quotation from Boutroux on pp. 32–3).

It is pointless or even impossible to trace the various peripatetics of 'Aryan' history, but it all led to the idea of the superiority of the

14 Ibid., pp. 25–6.

15 See for example L. Cavalli-Sforza, *Genes, Peoples and Languages* (New York: North Point Press, 1999).

Indo-European race – white, blonde, dolichocephalic, blue-eyed, and so on. Anti-Semitism then jumped on the bandwagon. But even here, the question is not that clear, as the opposition between Aryan and Jew was not an original feature of these theories. Ploetz, for example, founder of the German Society for Racial Hygiene, considered Jews to be Aryans.[16] Vacher de Lapouge constructed his own anthropological fable to try to give the Jews an origin in the blonde dolichocephalics, who subsequently interbred to the point of being no more than Phoenicians. Here is his text, which defeats all logic but at least gives an idea of the intellectual confusion that prevailed in this field:

> We do not know anything very precise about the morphological type of the most ancient Hebrews. Perhaps these Terachites were not far removed from the superior Arab type, perhaps they were closer to H. europaeus [i.e. the Aryan], as we can glimpse the moment when they may have linked up with the Oriental branch of the dolicho-blonde race, the first tribes that migrated from Europe to Asia Minor, Armenia and neighbouring regions, then further, and these are the Aryans in the strict sense of the word, in Bactria, Persia and India. They became singularly mixed when they settled in Palestine. At least three races can be found there: the Amorites, a pure dolicho-blonde race, the debris of the great maritime invasions of Western peoples in the Orient; the Canaanites; and in the north certain brachycephalic elements that have not yet been well studied. Despite all claims to the contrary, the people of Israel were a rich mixture of these three elements, the Canaanite being dominant, to the point that they were absorbed. The purest Jews were no more descended from Abraham than we are from the Gauls. At the time of the Babylonian conquest, the people of Israel were profoundly Phoenicianized in their ideas, customs, and particularly their language, as they had completely forgotten their original language, of which we know nothing, and spoke a Phoenician dialect: biblical Hebrew only differs by a hair from the dialect of Tyre and Carthage ... It is the Phoenicians that we should see in the majority of Jews of pure race, and in those who are blonde, the descendants of the Amorites.[17]

Whatever the case with these Aryan Jews, the prevailing notion in theories of this kind was anti-Semitic in one way or another, the

16 L. Poliakov, *Le Mythe aryen* (Paris: Pocket, 1994), p. 384; S. F. Weiss, 'The race hygiene movement in Germany, 1904–1945', in M. B. Adams (ed.), *The Wellborn Science, Eugenics in Germany, France, Brazil, and Russia* (Oxford: OUP, 1990), p. 17.

17 G. Vacher de Lapouge, *Les Sélections sociales*, pp. 136–7.

Semitic races generally being considered as white races bastardized by their contact with southern peoples, including Negroes.[18] Opposed to these inferior or degenerated southerners was then the glorious Nordic, Germanic, Anglo-Saxon or Scandinavian race; the blonde beast, barbarian and full of life, as against the degenerate brown, exhausted by a debilitating civilization. This was an initial biologizing of anti-Semitism, and more generally of racism against the peoples of southern Europe, whose whiteness was less than pure, and who were brachycephalic rather than dolichocephalic.

This Aryan myth, with or without its anti-Semitic component, was widely criticized and never accepted by official science. Despite leaving its traces here (as can be seen with the presence of these Indo-Germanics in Haeckel's classification), Aryanism – and still more so, anti-Semitism – remained above all characteristic of ideologically and politically motivated individuals and movements, whereas evolutionist racism, which considered people of colour as biologically inferior, was universally accepted and supported by science, with a few exceptions such as Quatrefages.

The second step along the road was the recuperation of Darwinian theses. This rapprochement between Darwinism and anti-Semitism seems to have had its pioneers early on. According to Léon Poliakov,[19] one of its first manifestations was a book by Ottmar Beta, *Darwin, Deutschland und die Juden, oder der Juda-Jesuitism*, published in Berlin in 1876.

There was nothing so original in itself about the Darwinization of anti-Semitism (or, if you prefer, the anti-Semitization of Darwinism), as it simply consisted in applying to the Jews the same arguments that were deployed vis-à-vis people of colour (officially recognized as inferior) and 'degenerate' or 'genetically damaged' whites. It therefore led to demanding identical measures against them, such as segregation, a ban on interracial marriage, even sterilization.[20]

18 For Vacher de Lapouge: 'The index of negrescence, as evinced by the varying darkness of colour of the subjects, is the best means of ethnic analysis . . . to measure Aryan blood, light colour being the best characteristic of H[omo] europaeus' (ibid., p. 94).

19 L. Poliakov, *Le Mythe aryen*, p. 83.

20 This is certainly the point at which Gobineau's book played its part, only then really entering into the history of racism. His racial conception of history was used to link the Aryan myth with another racial conception that was then prevalent: social Darwinism, with its evolutionary hierarchy and struggle of races. This link was an awkward one (we have seen how different the two racisms were), but these ideologists were not that demanding. It is undoubtedly at this point as

Here again, this type of theory is characteristic of individuals and movements that were ideologically motivated. It was not accepted by established science. It is found in an ideologist like Chamberlain (Wagner's son-in-law), writing for a broad public, as well as a marginal figure such as Vacher de Lapouge – but not in Darwin, Wallace, Vogt, Büchner, Haeckel, etc., even though none of these had any scruple in saying the most atrocious things about blacks. (At that time there were as yet no black biologists, but there were a number of Jewish ones, and even if a biologist detested a colleague, he could not classify him scientifically as a member of an inferior race.)

No more does it seem that official anthropology accepted such a thesis, particularly in Germany, where this discipline was long headed by the anti-Darwinian Rudolf Virchow (1821–1902). Thus, among the thousands of articles published between 1890 and 1914 in the four major German anthropological periodicals, there were only six on the subject of Jews, five of which were written by Jews, with the sixth being a critique of anti-Semitism.[21]

We should not, however, believe that this linkage of anti-Semitism to Darwinism (and its racist component) was achieved in a clear manner. First of all, because it was difficult to characterize a Jewish race, even if Haeckel in the 1860s did accept the existence of such a race. There were indeed enclosed communities of Jews in Eastern Europe who, by their enforced practice of endogamy, ended up constituting characteristic genetic groups. (Certain hereditary diseases that appeared in these populations so rarely spread beyond them that they are, still today, particularly characteristic of individuals of Ashkenazi descent.)[22] But leaving

---

well that Gobineau's doctrine became associated with anti-Semitism, which it was not originally, and that it began to have a degree of success. In a certain sense, it was necessary to wait until social Darwinism had scientifically 'biologized' and 'racialized' society for Gobineau to be recognized and used as a competing variant of this racialization, a variant that did not fit at all well with the evolutionist biology then in force, and was of scarcely any use other than in an Aryanist and anti-Semitic perspective.

21 B. Massin, 'From Virchow to Fischer: physical anthropology and "modern race theories" in Wilhelmine Germany', in G. W. Stocking Jr. (ed.), '*Volksgeist*' *as Method and Ethic: Essays on Boasian Ethnography and the German Anthropological Tradition* (Madison: University of Wisconsin Press, 1996), pp. 79–154 (quotation from p. 90).

22 On the life of these Eastern European Jewish communities, see the reportage of Albert Londres, 'Le Juif errant est arrivé' (1930), in *Oeuvres complètes* (Paris: Arléa, 1992), pp. 613–709.

this case aside, the Jews had long since intermixed with indigenous European populations. According to Juan Comas, 49 per cent of Polish Jews were blonde, as were 32 per cent of German Jews. In Germany between 1921 and 1925, 42 of every 100 marriages in which at least one partner was Jewish was with a non-Jew. In Berlin in 1926, there were 554 mixed marriages of this kind as against 861 marriages between two Jews.[23] Contrary to what is sometimes made out, therefore, Jews seem to have been quite well integrated into German society.

Vacher de Lapouge himself did not believe in the existence of a Jewish race. He saw Jews as a nationality (compare the quotation on p. 290 above, where he explains what a nation is):

> To sum up, the Jews are not a race, but a nationality, whose chief common characteristic lies in a particular religion and psychology, due to a small quantity of Canaanite infiltration. Israel is in reality well spread across all nations, even making up a substantial part, but there is no Israelite race in the anthropological sense of the term. A marvellous aptitude for proselytism enabled small groups of immigrants to attract in all countries of the world elements of like mind to their own, and the persecution that resulted almost everywhere gradually eliminated from these groups all those elements that were not sufficiently homogeneous, so that all that is left of the supposed ubiquity of the Jews is simply a further example of social selection, and a truly curious one at that.[24]

A Nazi such as Verschuer professed the same kind of theory. He begins by indicating a number of physical characteristics (the nose, the protruding lower lip, etc.) and diseases that are more frequently met with among Jews. He accepts that Jews have been considerably mixed by marriage with non-Jews. But he claims that these non-Jews who married Jews must necessarily have had a certain mental affinity with them (for want of having the physical characteristics), so much so that, even if it is impossible to define a Jewish race in the taxonomic sense, there certainly is a Jewish group that is characteristic in terms of mentality. (Exactly the same argument was put forward almost half a century earlier by Vacher de Lapouge, in *L'Aryen, son rôle social*, pp. 465–6.) Individuals who fall under this rubric do not have an identical 'genealogy', says

23 J. Comas, 'Les mythes raciaux', in UNESCO, *Le Racisme devant la science*, pp. 37–8.

24 G. Vacher de Lapouge, *Les Sélections sociales*, pp. 139–40.

Verschuer; they do not belong to the same 'taxonomic lineage', but they present for all that a 'psychic convergence', i.e., the different lineages converge as far as this characteristic is concerned: 'These processes of selection have maintained and constantly re-formed a Jewish type that is above all mental, whilst the somatic racial type has remained less uniform. The concept of a systematic race is not applicable to the Jews.'[25]

We see here therefore that it is the vagueness of the definition of race (and the integration into this definition, in the human case, of social, historical and other factors) that makes it possible to differentiate a Jewish group as a quasi-race, and thus include it in biological racism. An essentialist definition would not have allowed this.

Here are Hitler's own assertions, collected shortly before his death by Martin Bormann:

> The Jew is a foreigner to the root of his being ... The Jewish race is above all else a community of spirit. Added to this is a kind of connection of destinies, the consequence of persecution experienced over the centuries ... And what is determinant for the race, what is supposed to serve as the sorry proof of the superiority of the 'spirit' over the flesh, is precisely this incapacity for adaptation.[26]

This is more or less the same thesis as in the texts quoted above. Müller-Hill, in citing this passage, comments on it as follows: 'If this was Hitler's intimate conviction [concerning Jewish nature], those specialists in the human sciences who had biological conceptions would have been for him no more than accomplices in murder: idiots, no doubt, but useful ones.'

Müller-Hill is mistaken here: these racial theorists – at least those whom I have read – held exactly the same views as Hitler, who indeed took them over from these sources, quite simply because the notion of race (Jewish or otherwise) never had the essentialist character that historians wrongly ascribe to it. It is precisely because this notion of race did not have a very strict definition that it was possible to make Jews into a race, or a quasi-race, despite the lack of well-defined biological criteria in their case.

---

25 O. von Verschuer, *Manuel d'eugénique et d'hérédité humaine*, pp. 125–6.

26 *Hitlers politisches Testament: Die Bormann-Diktate vom Februar und April 1945* (Hamburg 1981), pp. 54ff., cited in B. Müller-Hill, *Science nazi, science de mort*, p. 95.

Was this Jewish quasi-race considered inferior? Not really. It was rather just 'undesirable'. Vacher de Lapouge, an anti-Semite himself, not only does not classify the Jews as an inferior race, but actually offers them as a kind of model for the establishment of a new aristocracy, a natural and eugenic one. He sometimes refers to the Jews as a 'factitious race', and this artificial character shows how it would be possible to fabricate, through racist eugenic measures, a new and Aryan racial aristocracy:

> The constitution of a natural aristocracy is relatively easy. It can be achieved without the direct involvement of the public powers, as long as it is not obstructed by laws designed to prevent it. Individual will, fructified by association, can carry the undertaking to its conclusion. The example of the Jews shows how simple it is for a particular category of individuals to isolate themselves, while still playing a most active part in social life ... The creation of a ubiquitous dominant race is not significantly more difficult than the constitution of a natural aristocracy in a single country. This problem has today been more or less resolved by the Jews. The people of Israel are far from constituting a pure race, and despite their merit, they are far from representing the ideal natural aristocracy. Nonetheless, their existence demonstrates the relative ease with which a programme of this kind can be realized.[27]

The Aryan version of this natural and eugenic new aristocracy was presented in a book by Richard Walther Darré (the Nazi minister for agriculture and supplies, and Reich peasant leader), *Neuadel aus Blut und Boden*.[28] This is a curious mélange of genetics, agronomy, pseudo-medieval Teutonics and esoteric mysticism. In any case, this new nobility owes very little to that of the ancien régime. It emerges straight from racial genetics, onto which is pasted a bastardized version of Gobineau's notions.

Let us return to Vacher de Lapouge. His considerations on the new aristocracy do not prevent him from describing the Jewish spirit, which according to him takes the place of a quasi-racial characteristic, in the following terms:

> The Jewish nation of today is the best example of psychic convergence: its cephalic index varies from 77 in Algeria to 83 in Poland; Jews are blonde

---

27 G. Vacher de Lapouge, *Les Sélections sociales*, pp. 480–2.

28 R. W. Darré, *La Race, nouvelle noblesse du sang et du sol* (Paris: Sorlot, 1939).

> and brown-haired, but they are everywhere the same, arrogant in success, servile in setbacks, sly, swindlers as far as possible, great hoarders of money, with a remarkable intelligence and yet no creative ability. They have thus been odious at all times, and beset by persecution that they have always put down to their religion, but which they seem to have deserved by their bad faith, covetousness, and spirit of domination. If we reflect on the fact that anti-Semitism goes back well before Christianity, at least to the fifteenth century bc, it is hard to see the suffering of Christ as the sole cause of the hatred with which they have been persecuted by Christians.[29]

We find here the habitual themes of anti-Semitism, but no biological criterion comparable to what was current at the time in the case of the black and other coloured races that were officially and 'scientifically' inferior (see quotations on pp. 257–60). On the contrary, biological characteristics are completely missing.

Concerning the last point made in the above quotation,[30] it is notable that the argument of the 'people who killed God' is totally absent from Vacher de Lapouge, who was at least as anti-Christian as he was anti-Semitic. Along with the majority of thinkers of this genre (Haeckel in particular, whose monism he is happy to follow), he was in fact anti-religious altogether:

> We are marching through monism towards the complete elimination of the idea of religion. We are marching, by way of new formulas based on social hygiene, towards the elimination of the idea of morality. This is an evolution that has its advantages and its disadvantages, but one that the advance of human knowledge makes inevitable.[31]

> It is science that will give us the new religion, the new morality, and the new politics – how different from those of another age! ... The debate today has shifted from dogmas that have long been questionable onto the terrain of ethics, always regarded as intangible, and the twentieth century will see a still more formidable battle between scientific and religious moralities, selectionist and other policies, than the battles of the Reformation and the French Revolution ... Politics itself is affected, since to the celebrated

29 G. Vacher de Lapouge, *L'Aryen, son rôle social*, p. 466.

30 According to Jules Isaac, it was only in the first century BC that any kind of anti-Semitism was detectable (J. Isaac, *Genèse de l'anti-sémitisme* [1956] (Paris: Calmann-Lévy, 1985), pp. 72–84.

31 G. Vacher de Lapouge, *L'Aryen, son rôle social*, p. 509.

> formula that sums up the secular Christianity of the Revolution: liberty, equality, fraternity – we reply: determinism, inequality, selection![32]

Nor do we find with Vacher de Lapouge recourse to such productions of his day as the *Protocols of the Elders of Zion*, which may have enjoyed a certain credit among the more backward social milieux of Eastern Europe and Russia, but which in Western Europe had little more success than it has today. Vacher de Lapouge does not seem to have believed in a 'Jewish plot', rather that Jews were not sufficiently aware of their social role. It is socio-economic criteria that are introduced, sometimes specifically. (The 1890s, which saw the flourishing of the Aryan theses of Chamberlain and Vacher de Lapouge, were a period of economic and social turmoil, like the 1930s that saw the rise of Nazism.)

According to S.F. Weiss, the German geneticist Lenz, though an anti-Semite and a Nazi, recognized the same qualities in the Jews as in the Aryans. What he criticized them for had more to do with such characteristics as their taste for money and their political liberalism.[33]

Verschuer, who was a genuine scientist despite being a Nazi sympathizer, also did not justify his anti-Semitism in terms of the supposed inferior biology of the Jews. In his manual of eugenics, Gypsies were broadly seen as an inferior race – even, what is more, an inferior race that had intermixed with all that society saw as lowest: delinquents, criminals, and other anti-social types.[34] But this kind of critique is not found with regard to Jews. In order to give a biological justification to the anti-Semitic legislation that banned any sexual relationship between Jews and Aryans, Verschuer was obliged to bring up the idea that it was interbreeding in itself that was harmful, a thesis of Gobineau's that was completely at odds with the genetic theories of the day (see above, pp. 261–6). Verschuer thus takes his stand here on a psychological rather than a biological argument, which could be interpreted any way you like: 'The ban on racially mixed marriages is based on experience obtained with their progeny: a social and economic position "between two races", a disharmonic

32 G. Vacher de Lapouge, preface to his translation of E. Haeckel, *Le Monisme, lien entre la religion et la science* (Paris: Schleicher, n.d.), pp. 1–2.

33 S. F. Weiss, 'The race hygiene movement in Germany, 1904–1945', in M. B. Adams (ed.), *The Wellborn Science*, p. 31.

34 O. von Verschuer, *Manuel d'eugénique et d'hérédité humaine*, p. 130.

combination of very different aptitudes, often giving rise to a spiritual unbalance.'[35]

In other words, Jews are a race without really being a race – a piece of nonsense made possible by the vague definition of race itself. And without being really a low race, they are not really a high race either. They are undesirable for political, social, economic and, for a section of the populace, religious reasons. These various reasons are then vaguely transformed into characteristics that are racial without really being racial, and the whole package is hooked up to the biological racism discussed above.

The question of the opposition between Jews and Aryans is equally vague within this schema. I shall try to sum this up as it is presented in Chapter 8 (pp. 463–514) of Vacher de Lapouge's *L'Aryen, son rôle social.*

The problem is this: How is Aryan superiority to be combined with anti-Semitism to establish an opposition between Aryans and Jews – given that there were some ideas according to which Jews were Aryans, and others that proclaimed Aryan superiority over all races but without the duality of Aryan and Jew? We should bear in mind that alongside these two groups there was an entire human population, neither Aryan nor Jewish, that remained outside this opposition and whose role remained uncertain, likewise the reason for its exclusion from history. In Haeckel's classification, for example, Indo-Germanics and Jews are two of the thirty-six races composing twelve human species, two races that represent only a small section of humanity. The question thus remains how they come to be opposed in this way, and how the other thirty-four races, representing 98 or 99 per cent of the world's population, fit in.

For Vacher de Lapouge, obsessed with cephalic index, the primary opposition is that between dolichocephalics and brachycephalics. In other words, between *Homo europaeus*, the Aryan or Nordic race – blonde, Protestant and liberal – and *Homo alpinus*, the peoples of southern Europe – dark, Catholic and statist. The other human races[36] – African, Asiatic and American – are out of the game, being already colonized and dominated by these two white races. Jews do not make an appearance, as they are not a biological

35 Ibid., p. 227.

36 Vacher de Lapouge claimed to base himself on Linnaeus in dividing humanity into five races: Homo europaeus, H. alpinus, H. americanus, H. asiaticus and H. africanus. In fact, he adapted Linnaeus to his purpose (see p. 247 above for the human races in Linnaeus's classification).

race in this pseudo-Linnaean picture. Besides, their cephalic index is variable (as per the quotation on pp. 316–7 above), and they are thus neither dolichocephalic nor brachycephalic.

Upon this primary opposition between dolichocephalic and brachycephalic is superimposed 'the struggle for world domination' [*sic*], a struggle that is political in the broad sense (economic, diplomatic, military, cultural, etc.) but that clearly has a racial component. In this struggle, continental Europe is eliminated, as it has definitively become brachycephalic – in other words, statist, ridden with functionaries, and Catholic – with only northern Germany preserving some Aryan residues. Vacher de Lapouge does indeed imagine the possibility of a unification of this Europe under Jewish control, by virtue of their cosmopolitan character, and because 'the plutocratic regime – misnamed democratic' can only favour the development of a 'Jewish feudal power, in control of the land, the factories, and capital'. But he does not believe in the success of such an undertaking, as 'the Jew disorganizes everything he touches, possessing neither the spirit of government, nor the military instinct needed for the preservation of empires ... The Jew has never had a political sense.'[37]

Vacher de Lapouge moves on to review the various non-European countries that are possible winners in the struggle for world domination: Russia, the United States, China, and so on. Though he does not pronounce a definitive verdict (the first sentence of this chapter is 'The trade of the prophet is filled with blighted hopes'), his preferred candidate is clearly a combination of England and the United States (the main reserves of Aryans, according to him). The Aryan thus seems the designated victor in the struggle for world domination, with England in control of the oceans and possessing a number of colonies, and the United States a virgin land full of future promise. It is in this Anglo-Saxon form that the Aryans will come up against the Jews, who are not a biological race but a 'factitious' one, though contending in the same fashion for world domination:

> The only dangerous competitor of the Aryan at present is the Jew. For my part, however, I do not see the Jewish question in the same fashion as do the anti-Jews, and Drumont in particular, if we consider only the French aspect. For Drumont and his friends, the Aryan is the indigenous, the original French stock, but in actual fact this means the more or less mixed brachycephalic europaeus, result of long selections in the past. Through

37 G. Vacher de Lapouge, *L'Aryen, son rôle social*, p. 475.

> the Revolution, the brachycephalic conquered power, and by a democratic evolution this power tended to concentrate, in theory, in the lower and most brachycephalic classes. Political anti-Semitism aims at preserving the work of the Revolution, and preventing power from passing into the hands of the Jews, and of foreigners more generally ... The Aryan as I have defined him is something else, Homo europaeus, a race that made France great, but today has become rare and almost extinct. This is a race and not a people, and those peoples that it comprises, English, Dutch, Americans, are foreigners to us, and more or less enemies. The question as I understand it is therefore to know who, out of English, Americans and Jews, has the greatest chance in the struggle for existence.[38]

We need only go back a few pages here (Chapter 6, 'Psychology of the Aryan') to understand the significance of this opposition between Anglo-Americans and Jews. For Vacher de Lapouge, as we have seen, racial criteria are transparent and only poorly conceal what they cover. Here are his 'Aryan heroes', some of whom we have already met:

> It is scarcely possible to accumulate great wealth without speculation. Americans thus often make colossal fortunes. The Carnegies, Blairs, Rockefellers and Armours are as rich as the Rothschilds, if not more so, but the American millionaire has a grandeur that is lacking in his Jewish counterpart. While the latter piles up millions by an atavistic instinct, just as his ancestor raked in ducats, the other operates like a gentleman placing his stakes on the green baize table. Nothing is more common than to see an American philosophically lose tens of millions, and regain the same fortune with equanimity. Nothing is more common than to see the colossal sums an American has made by good luck and business genius pass entirely or in part into great works of public utility. The fortune that the Californian Stanford made in railways served to create Stanford University. This was endowed with more than a hundred million dollars, a university founded not to manufacture graduates, but – to your astonishment, Frenchmen! – instructed men able to earn their living. I do not know the current situation with Rockefeller and the University of Chicago, but I think that this has already passed the thirty million mark. The use that American speculators make of their fortunes often absolves them of their sins. That which they have taken from their neighbour largely returns to their neighbour.[39]

38 Ibid., pp. 464–5.
39 Ibid., pp. 390–1.

I do not believe it would force the sense of this text to maintain that, for our deputy librarian from Montpellier and ancestor of all racisms, the opposition between Aryan and Jew for world domination could be summed up as an opposition between WASP capitalism and Jewish finance; in other words, Rockefeller versus Rothschild. I leave it to economic historians to decide whether there is any truth in this opposition, but that at least is what Vacher de Lapouge believed at this time, and there is not a trace of biology about it. The Jewish race is explicitly described as an artifact, and the Aryan character of Rockefeller or Carnegie never discussed.

Vacher de Lapouge gave his lectures in 1889–90 and published them in 1899, when Hitler was still in his childhood. A few years later, however, his ideas underwent a change. Perhaps this was due to the rise of German power, perhaps to the fact that racial hygiene began to be fashionable (after the essay competition in 1900 organized by Krupp, and the establishment of the German Society for Racial Hygiene in 1905 by Ploetz); perhaps again it was because Aryan theories had proved popular – Chamberlain was already a successful author. At all events, Vacher de Lapouge amended his ideas. In 1909 he would write:

> It is impossible not to take into account, by pretending to ignore it, an idea as powerful as that of the mission of the Aryans; and if we do not make use of it, we can be sure that someone else will. I foresaw this development a long time back, and indicated the remedy for it, but it was not Germany that I had in mind. What I admire most about the Germans, however, is their having created a navy in a country lacking in ports, and having staked a claim for hegemony, in the name of Aryanism, against the Anglo-Saxons of Europe and America, who are the most Aryan peoples in the world.[40]

This time, his Aryan is very clearly the Teuton who 'stakes his claim', even if, for Vacher de Lapouge, the English and Americans are still the true Aryans. This is the form that this kind of theory would now adopt, to be taken over by Nazism. Certainly these ideas have different genealogies in the case of other writers, but this is how we find them in Vacher de Lapouge, generally viewed along with Gobineau as one of the major sources of Aryanist theories.

40 G. Vacher de Lapouge, *Race et milieu social*, p. xxv.

I have tried to understand what Chamberlain's position on this subject was, but have to admit a certain discouragement in the face of his *Foundations of the Nineteenth Century*, a work close to two thousand pages long, written in the expected style of a salon ideologist who married the daughter of a great composer. Very likely I am not alone in this discouragement, and if Vacher de Lapouge is more commonly cited today than Chamberlain, it is not because he had a bigger readership or was a more important thinker, but simply because he is clear to the point of caricature, and very easy to read.

Chamberlain's work is exactly contemporary with Vacher de Lapouge's *L'Aryen* (1899). It is a kind of epic, comparable with Gobineau in this respect, though twice as long and twice as tedious. Chamberlain was undoubtedly inspired by this predecessor (far more so than by Darwin, contrary to his claim – see quotation on p. 269), but he totally transformed Gobineau's ideas. In many essentials, what is attributed to Gobineau by people who have not read him is taken in fact from Chamberlain.

He begins by describing the great creations of antiquity and the collapse of civilization in the 'ethnic chaos' of the Roman Empire in its decadent period. Then comes his presentation of the 'heirs' of antiquity, of whom there are only two: Jews and Teutons, the rest of humanity presumably lost for good in the 'ethnic chaos', or at any rate dispatched from history still more rapidly than they are in Vacher de Lapouge. (Gobineau, for his part, had a more realistic view of the dimensions and importance of different civilizations than either of his epigones. He does not dismiss 98 per cent of humanity with a snap of his fingers, even if it is true that in 1900 the world was almost entirely dominated by a few Western European countries, soon joined by the United States, whereas in the 1850s colonization was less advanced and the bourgeois economy not yet everywhere dominant.)

As far as the Jews are concerned, Chamberlain's line of argument is comparable with Vacher de Lapouge's in seeing them as a model for a new aristocracy: from the fact of their endogamy, they are alone in having preserved the purity of their race (here the Jews are a race), and thus alone in having traversed the ethnic chaos without degenerating:

> Out of the midst of the chaos towers, like a sharply defined rock amid the formless ocean, one single people, a numerically insignificant people – the

> Jews. This one race has established as its guiding principle the purity of the blood; it alone possesses, therefore, physiognomy and character.[41]

If we move on to a later point in his book, one which is far less flattering to the Jews, we immediately come to the other heirs, the Teutons. Here, Chamberlain is much less stinting of praise (in his fashion), but he finds it hard to justify their claim to antique heritage. It seems in fact that the Teutons, like Asterix's Gauls, originally formed a small band of unconquerable warriors lost in the depths of the Prussian forests, and it was thus that they preserved the Indo-European race in its pure state:

> The entrance of the Jew into European history had, as Herder said, signified the entrance of an alien element – alien to that which Europe had already achieved, alien to all it was still to accomplish; but it was the very reverse with the Germanic peoples. This barbarian, who would rush naked to battle, this savage, who suddenly sprang out of woods and marshes to inspire into a civilized and cultivated world the terrors of a violent conquest won by the strong hand alone, was nevertheless the lawful heir of the Hellene and the Roman, blood of their blood and spirit of their spirit ... But for him the sun of the Indo-European must have set.[42]

The rest is not hard to imagine, even if it takes a further thousand pages: the rivalry between the two heirs, Jews and Teutons. This is a new anthropological fable, which borrows from Gobineau (among others) the themes of Aryan superiority and the rejection of interbreeding, but makes a quite different use of them, as in Gobineau there is not this reduction of history to an Aryan-Jewish dualism, nor the craziness about the Teutons as creators of a new culture.

If for the Frenchman Vacher de Lapouge the ideal Aryan was the WASP capitalist, for the Englishman Chamberlain it was the Teuton in his blonde-beast costume, hurling himself naked into the midst of battle and singing Siegfried at Bayreuth in his father-in-law's theatre. There is not the least sign of any socio-economic underpinning, and the racial discourse is as opaque in his case as it was transparent in that of Vacher de Lapouge. The Germans preferred this version, which was far more romantic. In this

41 H. S. Chamberlain, *The Foundations of the Nineteenth Century* [1899] (London: Lane, 1910), vol. 1, p. 253.
42 Ibid., p. 494.

ideological-decorative domain, Chamberlain prefigured Nazism far more than did Vacher de Lapouge, who was much more down to earth.

As we have said, this current was the work of writers who were marginal and ideologically driven. It was not present in established and recognized science, except for German science after 1933. This is undoubtedly why Jewish biologists continued to support 'scientific' eugenics even when this was drifting on a racist path. From the standpoint of science, as white Europeans they belonged to the superior race and so did not feel targeted.

The problem arises from several facts: that these marginal thinkers came to power in Germany, where they put into practice their eugenic programme against the sick, the disabled and the abnormal, in a more systematic and intensive fashion than in any other nation. That they extended it to representatives of peoples in their country who were publicly classified as inferior (the sterilization of the 'Rhineland bastards' in 1937; the internment of Gypsies in concentration camps in 1938, and their projected sterilization), moving on to convert their sterilization eugenics into an extermination of those sterilized. And finally that they applied to the Jews a programme that was not originally designed for them (at least not in democratic countries, or in Germany before 1933).

As in the case of the Gypsies, some traces of a project for the sterilization of Jews can be found, but in neither case was this really put into effect, overtaken as it was by the shift to extermination.

According to Ternon and Helman, at the Wannsee conference of 20 January 1942, when the extermination of Jews was decided, there was a proposition to sterilize *Mischlinge* (German 'half-Jews', with one Aryan parent), but this was not followed up, as in wartime the authorities did not want to encumber hospitals with this kind of surgery. Also according to these authors, 'at Westerbork, a transit camp for Auschwitz, it was proposed on 15 May 1943 to allow Jews to avoid deportation on the next train ... if they agreed to be sterilized'. Finally, in Holland in March 1943, the 'Jewish section' of the Gestapo in The Hague supposedly planned to sterilize 8,000 Jews married to non-Jews, and a number of sterilizations were performed.[43]

As we know, it was extermination that carried the day, but these facts trace the red thread that links all Nazi exterminations.

---

43 Y. Ternon and S. Helman, *Histoire de la médecine SS, ou le mythe du racisme biologique* (Paris: Casterman, 1969), pp. 146–9.

The origin of the extermination of the Jews thus lies in the joining of two paths, one being 'traditional' anti-Semitism (with its vague biologization in the Aryan myth), the other the social Darwinist current (with its eugenic and racist component). It is the latter current that provided the theories and methods, initially conceived for those assessed as genetically incorrect and for coloured races officially deemed inferior. It is likely that, without this, traditional anti-Semitism, however violent this often was, would never have led to the massive and systematic exterminations imposed by Nazism.

Does this mean, as is sometimes claimed, that science, and in particular biology, should be held responsible for Nazism? This is not totally false, but qualifications are needed. First of all, it is necessary to distinguish two levels, that of practice and that of theory. On the practical level, the Nazis deployed industrial means of extermination, and thus made use of science, or at least technology. But as Simone Weil wrote in 1939, the sword was as effective for massacres as the airplane:

> Many people today, moved by the horrors of all kinds that our era brings in a profusion that is paralyzing for any sensitive temperament, believe we are entering into a period of greater barbarism than the centuries that humanity has traversed in the course of its history – whether this is through the effect of too great a technological power, or a kind of moral decadence, or some other reason. Not at all. To convince yourself to the contrary, you need only open any text of antiquity – the Bible, Homer, Caesar, Plutarch ... None of the past centuries that history records is lacking in atrocious events. The power of weapons, in this respect, is unimportant. The simple sword, even a bronze one, is as effective an instrument of massacre as the aeroplane.[44]

Recent events in Rwanda have proved her right: up to a million people were killed by machete in just a few weeks, a rate far higher than occurred in the Nazi gas chambers.

It is not at the practical level that we need to look, but rather at that of theory, and the place that this theory accords to man. As Ternon and Helman wrote in 1973,

> Thus the prodigious development of the human sciences in the nineteenth century was not foreign to the advent of National Socialism. Anthropology

44 S. Weil, 'Refléxions sur la barbarie' (1939), in *Oeuvres* (Paris: Gallimard, 1999), pp. 503–7.

> led to racism, Malthusianism to the will to elimination, evolutionism to Social Darwinism, and the discovery of the laws of heredity to genetic delirium. As Wilhelm Röpke reminds us, even if all these precursors of National Socialist racism had nothing in common with the horrifically demented character of this doctrine, the fact remains that it is the abyss into which one finally plunges if one takes the false path of this biologism, whose origin lies in Darwin and those who followed him. The racial doctrine of National Socialism is the final product ... of a mental operation that, in the course of the nineteenth century, with the zeal of a misunderstood science, reduced man by making him the object of zoology and natural selection.[45]

The questions remains whether the Darwinian and genetic hysteria that characterized the late nineteenth century and the first half of the twentieth (and which still exists today) deserves the name of science.

Whatever the case, the Nazi exterminations display an order and a logic that are perfectly clear. Hitler did not invent anything new; he simply put into practice, and took to their logical conclusion, processes that had already been envisaged long before him. And he extended them to the Jews, for whom they had not originally been intended.

The logic of these exterminations was befogged after the Second World War by the focus on Nazi anti-Semitism, and the obscuring of eugenics as well as what has become known as euthanasia, or at least their relegation to the background, where they were more or less ignored in relation to the extermination of the Jews – as was also the extermination of the Gypsies.

If this refocusing is not correct, it is at least understandable. On the one hand, anti-Semitism incontestably occupied an important place in Nazi ideology. On the other hand, the number of Jewish victims was far greater – even if in Germany itself euthanasia was responsible for more deaths. And finally, Jewish organizations guaranteed both the memory of the victims and the defence of the survivors, whereas Gypsy groups did not possess comparable financial or media resources, and there were no organizations to defend the victims of eugenics and euthanasia.

This refocusing of the history of Nazism on anti-Semitism, however, had at least two disturbing consequences. The first of these is that non-Jewish victims were rather forgotten; the

45 Y. Ternon and S. Helman, *Les Médecins allemands et le National-Socialisme*, p. 9.

extermination of Gypsies is generally mentioned only for the sake of completeness, as a minor complement to that of the Jews; while that of diseased and handicapped people is almost always passed over in silence, or at least extremely minimized. So much so that, once the immediate aftermath of the war had passed, there were hardly any further prosecutions and condemnations apart from those pursued for the extermination of Jews.

Ritter, the scientist in charge of a census of Gypsies with a view to their extermination, was tried in 1950 and committed suicide before a verdict was passed.[46] Those responsible for the 'euthanasia' programme largely slipped through the net. In 1949 Pfannmueller, whose speciality was the murder of handicapped children by starvation, was sentenced to half a dozen years in prison for homicide, a sentence then cut in half (see above, pp. 209–10). The leading eugenicists were never disturbed: Verschuer taught at German universities until his retirement in the 1960s; Fischer and Lenz likewise escaped justice.

Even today, these 'other' groups remain second-class victims. For example, there was a mobilization of international press and diplomacy against the establishment of a Catholic convent at Auschwitz, whereas the protests of Gypsy associations against the building of a piggery in the Gypsy concentration camp at Lety in Czechoslovakia generated a few lines on an inside page of *Le Monde*, in an article basically devoted to the educational segregation of the Roma.[47] As for the extermination centres of ill and handicapped people, only a handful of specialists know where they were located, and to my knowledge no one has bothered to find out what has become of them since.

This all gives the impression that what the Nazis are attacked for is less their programme of political biology, which they took over from the various sources discussed in this book, than for having extended it to Jews for whom it was not originally intended. And that, without this extension, no one would have had any complaint about the programme, which before the war was broadly accepted, at least in its 'moderate' form of the sterilization of 'genetically incorrect' individuals, the segregation or sterilization of races categorized by science as 'inferior', and

46 E. Klee, *La Médecine nazie et ses victims*, pp. 59–61. B. Massin, 'La science nazie et l'extermination des marginaux', *L'Histoire*, 217, January 1998, pp. 52–9.

47 M. Plichta, 'Les Roms tchèques dénoncent la "ségrégation raciale systématique à l'école"', *Le Monde*, 26 June 1999.

sometimes even the 'euthanasia' of individuals whose life 'was not worth living'.

Apart from this categorization of victims, the other result of this refocusing of the history of Nazism on anti-Semitism has been a complete obfuscation of the logic of Nazi exterminations.

This has gone along with what we could call a Hollywoodizing of the extermination of the Jews. This extermination, in other words, has been framed and lit in such a way as to appear to be an isolated object, with well-defined, sharp contours against an obscure and vague background. The consequence of this has been to wipe out the historical context, left out of the frame or placed in shadow – a context that does not of course excuse the crime, but that at least partly explains its origins. Once dehistoricized in this way, the extermination is projected against a general and a-historical background of anti-Semitism (itself made up of elements that are largely dehistoricized, from the ban on marriage between Christians and Jews decreed by the Synod of Elvira in 306, via the expulsion of Jews from Spain in 1492, through to the Russian pogroms of 1905).

This, for example, is the procedure of Raul Hilberg, a completely respectable historian, who devotes a whole chapter to these anti-Semitic precedents,[48] but scarcely more than a page to the gassing of mental patients (historically and logically rather more closely connected), and has not a word to say on the racist legislation in force almost everywhere (for example, the Mines and Works Act, also known as the Colour Bar Act, of 1926 that banned blacks in South Africa from certain jobs, or the ban on sexual relations between whites and blacks, the Immorality Act, that was imposed the following year), even though it was these racist laws that served as a model for anti-Semitic legislation in Germany, rather than the exclusion of Jews from public functions decreed by the Synod of Clermont in 535, which it is unlikely that Hitler had ever heard of.[49]

When the extermination of Jews is cut off from its context and rendered a-historical in this way, it becomes something singular and inexplicable, a kind of diseased symptom of a quasi-religious

48 R. Hilberg, *The Destruction of the European Jews* (New York: Harper & Row, 1979), Chapter 1: 'Precedents', pp. 5–28.

49 Eugenic legislation in the United States directly served as a model for the German law of July 1933, and the relationship between American eugenic and racist movements and their German counterparts is well known (cf. S. Kühl, *The Nazi Connection*.)

nature, the expression of 'absolute evil'. It is then given a proper name (the Shoah or the Holocaust – both popularized by cinema and television), unlike the extermination of Gypsies or of ill or handicapped people (the latter ambiguously referred to as 'euthanasia'), and this proper name reinforces its singularity. From then on, any attempt to explain it in rational terms is forbidden, seen as a profanation that questions, not the reality of the event, but its consecrated character as the ultimate infamy.

Nazi policy then itself becomes inexplicable without resort to myth and irrationality, and the responsibility for it falls entirely on Hitler – mentally deranged, no doubt, but diabolically (devilishly) fascinating. Joseph Rovan can thus write:

> The singularity of Nazi crimes, that has been so much discussed in recent years, constitutes in my view a particular case of the general singularity of an effort through which a hallucinating leader has sought to lead a great modern nation to the modes of being, thinking and feeling of the Germanic tribes and their myths ... In the whole history of the West, from the end of antiquity, there has been no individual who can be compared with Hitler. The diabolical side mixed with the incongruous side, the charismatic leader trapped in his hallucinating dreams, and who – despite his clown-like air – is endowed with a prodigious skill and readiness for action ... Perhaps the Bible can better explain what happened to the German people between 1933 and 1945 than the teachings of Marx or Freud.[50]

Rovan is in fact a good historian, but one may well believe that the crimes of the Nazis borrowed rather less from Germanic mythology and the Devil than from the social Darwinism that was fashionable at this time. And that these borrowings give a better explanation than the Bible or the esoteric claptrap in which Nazism was dressed up. Haeckel, too, drifted into an esoteric mysticism (see note 20 on p. 81 above), but what interested Krupp was racial hygiene, not Haeckel's musings on the souls of crystals.

At all events, the Shoah has become in this way the symbol of Nazism, the paradigm of its horrors. By this token, the extermination of Gypsies and mental patients, handicapped and abnormal people, can only appear as a pale reflection, 'collateral damage' that cannot be put on the same plane, the Shoah being the

---

50 J. Rovan, *Histoire de l'Allemagne des origins à nos jours* (Paris: Points-Seuil, 1998), pp. 705–6.

'radical evil' that gives a measure without itself being measurable.[51] Whereas in actual fact, the Shoah was the extension to the Jews – a 'factitious race' – of theories and practices that were designed for ill, handicapped and abnormal people, and for races classified as inferior by official science.

The Shoah, as singular object and manifestation of radical evil, thus serves to screen off everything in Nazism that did not arise from anti-Semitism. By making the one the essence of the other (with a few anecdotal and marginal remarks on the Gypsies, the racially mixed, mental patients, etc.), the entire social Darwinist trajectory is hidden, and the logic of the historical process completely obscured. The upshot is that, today, ideas on eugenics and even euthanasia have resurfaced, with the claim being made that these practices are tolerable so long as they are not racist (or anti-Semitic), and that they have nothing in common with Nazism.

I shall not go so far as to say that the extermination of the Jews is actually used to mask the responsibilities of the authors and champions of social Darwinism, but there can be no doubt that if certain events (such as eugenics, the sterilization of the racially mixed, 'euthanasia', etc.) are generally relegated to the shadows, this is because they give a rather embarrassing image of the European society of the first half of the twentieth century that gave birth to Nazism.

51 A. Badiou, *L'Éthique* (Paris: Hatier, 1993), pp. 55ff.

16

The Present Perspective

After the Second World War, although biological theories supportive of racism did not disappear, they became far more uncommon and were now defended almost wholly by scientists with a strong ideological bias. Contrary to the case with eugenics, in which silence was the rule and the decline steadily pursued its own course, racism was explicitly combated by political organizations, which demanded that biologists put their house in order. To this extent, the lesson of the war had been learned.

The most characteristic example of this struggle against racism is that of UNESCO. An initial declaration condemning racism and asserting there was no scientific proof of racial inequality was published in 1950. However, certain anthropologists and geneticists (disciplines under-represented on the drafting committee, which was more biased towards sociologists) protested, maintaining that neither was there any proof of the intellectual equality of races; at that moment, not all biologists were ready to abandon cherished beliefs. Accordingly, in 1951, a second declaration was drawn up.[1] UNESCO – in other words, the United Nations – thus carried out what the Allies had refused to do in 1919 in the Charter of the League of Nations (see p. 302 above). The controversy over these successive declarations, and the commentaries they aroused, were soon presented in a short book.[2] It showed that Lenz and Fischer, pillars of Nazi biology, were back at work and had no hesitation in opposing the equality of races. Muller, for his part, followed them in this view, with a few rather sophistic qualifications. We also discover a British anthropologist claiming that, if there is no

1 The two declarations were republished in UNESCO, *The Race Question in Modern Science* (London: Allen & Unwin, 1975).

2 UNESCO, *The Race Concept: Results of an Inquiry* (Paris: UNESCO, 1952).

such thing as a French or a German race, there is nevertheless an English one. As a whole, the publication is not especially brilliant, and even bears the mark of a certain scientistic mediocrity, despite the good intentions behind it.

The two declarations, that of 1950 and the more reserved one of 1951, are also very general and full of good intentions. The most interesting thing about them is undoubtedly the individuals involved in helping to draft them. That of 1950 bears the mark of the inevitable compères Julian Huxley and Hermann J. Muller – whose views on the questions of race and eugenics we have already explored. Huxley was also a co-author of the 1951 declaration, but he was now joined by Hans Nachtsheim, director of the institute of genetics at the Freie Universität Berlin. This is the same Nachtsheim we met with in Part Two, when he worked on experimental genetic pathology and enjoyed financial support from the Rockefeller Foundation (above, p. 182).

Nachtsheim also crops up in contexts very different from the UNESCO declaration against racism. Here is what the racial theorist H.F.K. Günther wrote about him in his preface to the reissue of *Platon, eugéniste et vitaliste*, a book that Günther published in 1928 and that served as an apologia for Nazi eugenics:

> In December 1962, the eminent geneticist, Dr Hans Nachtsheim, well-known for his fierce hostility to National Socialism, published in Informations Médicales, no. 48, an article titled 'La Loi en faveur de la lutte contre les maladies héréditaires, datant de 1933, considérée dans l'optique actuelle'. This scientist had been charged with examining the decisions of public health tribunals on questions of heredity, decisions bearing on sterilization under the law of 1933 in support of the struggle against hereditary diseases, which was abrogated as a Nazi law in 1945. The result of these examinations was that Nachtsheim found no case of abuse and declared that this law was in no way a 'Nazi law' but had been promulgated after the example of North American laws on eugenics, benefiting from subsequent reflection on these. The terrifying increase in physical and mental degeneration of a population defenceless against the dangers of hereditary diseases, and in particular the increase in imbecility, led Nachtsheim to advise the reintroduction of a 'law providing for struggle against hereditary diseases'.[3]

3 H. F. K. Günther, *Platon, eugéniste et vitaliste* (Puiseaux: Pardès, 1987), pp. 10–11.

We thus know what Nachtsheim's view on Nazi eugenics was in the 1960s – the same as it had been in the 1930s and 1940s. According to S. Kühl, immediately after the war Nachtsheim re-established his links with American colleagues and played a major role in the 'laundering' of German biologists at the time of de-Nazification. It was he in particular who enabled Verschuer to shake off the accusation of having worked with human 'material' that Mengele had sent him from Auschwitz.[4] And it was Nachtsheim who enabled Lenz and Fischer to slip in among the commentators on the UNESCO declarations, in the booklet on *The Race Concept*.

Let us now look at what Ernst Klee says of Nachtsheim. To understand properly the quotation that follows, it is necessary to explain that the low-pressure chamber of Professor Strughold that is mentioned here was initially located at Dachau and used to study the behaviour and death of human guinea-pigs who were placed in it before its air was removed. These experiments, undertaken on behalf of the Luftwaffe, were designed to improve flying conditions at high altitude and the pilots' chances of survival. For the most part, the doctors involved escaped punishment after the war. Strughold, the major culprit, along with a few others, was taken to the United States by American secret agents (the US Air Force wanting to profit from his experience, which came in quite handy in the conquest of space). Strughold's name does not even appear in the report that Bayle gives of that part of the Nuremberg proceedings dealing with these matters. Here, however, is Klee's text:

> We do not know what experiments the low-pressure chamber was used for after Dachau. Details of only one human experiment eventually filtered out. Its instigator was Professor Hans Nachtsheim, who from 1942 to 1945 was head of the department of experimental genetic pathology at the Kaiser-Wilhelm Institute for Anthropology. With the financial backing of the Deutsche Forschungsgemeinschaft, he conducted research on epilepsy, or more precisely, on the distinction between hereditary and non-hereditary epilepsy. According to him, hereditary epileptics should react more quickly than the non-hereditary to convulsions that were triggered artificially.
>
> Nachtsheim began by conducting experiments on Viennese white rabbits ... In 1943, Nachtsheim was working with Dr Gerhard Ruhenstroth-

4 S. Kühl, *The Nazi Connection*, p. 102. Verschuer's son, on the other hand, charged Nachtsheim with having falsely accused his father (B. Müller-Hill, *Science nazi, science de mort*, p. 132).

Bauer, of the Kaiser-Wilhelm Institut for Biology (director Professor Adolf Butenandt). Instead of white rabbits he used children as guinea-pigs. Ruhenstroth-Bauer obtained his samples from the psychiatric service of Brandenburg-Görden.

[Letter from Nachtsheim to Dr Gerhard Koch, 20 September 1943]:
'Thanks to the good offices of Obermedizinalrat Dr Brockhausen, we received six epileptic children (four severe epileptics, two with symptoms of epilepsy), with whom we conducted experiments in Professor Strughold's low-pressure chamber last Friday ... But in the initial phase it was not possible to say that rabbits and human beings behave differently in a low-pressure situation, as the children on whom we experimented were aged between 11 and 13, which corresponds to an age of 5 or 6 months in rabbits. Epileptic rabbits of 5 or 6 months, however, no longer have the same reactive response as animals of 2 or 3 months, who have fits almost constantly. We need to experiment on epileptic children aged 5 or 6 years, but this is not possible at the moment, as this age category is not present at Görden ...'

Hans Nachtsheim, who was never a member of the Nazi party, remained in Berlin after 1945 ... He held first of all the position of department head at the Kaiser-Wilhelm Institute for Anthropology, subsequently being appointed professor of genetics at the Humboldt University in the eastern sector, then moving in 1949 to the Free University ... In 1955, Nachtsheim received the Federal Grand Cross of Merit, and in 1958 became a member of the Federal Health Council.

Hans Nachtsheim, who used handicapped children as guinea-pigs in his laboratory, was appointed in 1961 to the committee for the restitution of victims of Nazism.[5]

Klee overlooks here one more of Nachtsheim's titles to fame: he was one of the drafters of the UNESCO declaration on racism. We should also note that, as against Verschuer, his name did not appear in the *Brown Book* in which the GDR denounced former Nazis still in service in the Federal Republic.

History does not say what happened to Nachtsheim's human guinea-pigs, but we saw in Part Two the usual fate of these kinds of patients. We can at least conclude that UNESCO did not select its collaborators with a great deal of care.

In any event, the content of the two UNESCO declarations on racism did not do the organization a disservice. The box below

5 E. Klee, *La Médecine nazie et ses victims*, pp. 167–9.

gives the definition of race proposed in 1951, which is still quite acceptable today. (That of 1950 noted that all men belonged to the same species, *Homo sapiens*.)

UNESCO Statement on the Nature of Race and Race Differences, June 1951

1 ... The concept of race is unanimously regarded by anthropologists as a classificatory device providing a zoological frame within which the various groups of mankind may be arranged and by means of which studies of evolutionary processes can be facilitated. In its anthropological sense, the word 'race' should be reserved for groups of mankind possessing well-developed and primarily heritable physical differences from other groups. Many populations can be so classified but, because of the complexity of human history, there are also many populations which cannot easily be fitted into a racial classification.

2. Some of the physical differences between human groups are due to differences in hereditary constitution and some to differences in the environments in which they have been brought up. In most cases, both influences have been at work. The science of genetics suggests that the hereditary differences among populations of a single species are the results of the action of two sets of processes. On the one hand, the genetic composition of isolated populations is constantly but gradually being altered by natural selection and by occasional changes (mutations) in the material particles (genes) which control heredity. Populations are also affected by fortuitous changes in gene frequency and by marriage customs. On the other hand, crossing is constantly breaking down the differentiations so set up. The new mixed populations, in so far as they, in turn, become isolated, are subject to the same processes, and these may lead to further changes. Existing races are merely the result, considered at a particular moment in time, of the total effect of such processes on the human species. The hereditary characters to be used in the classification of human groups, the limits of their variation within these groups, and thus the extent of the classificatory subdivisions adopted may legitimately differ according to the scientific purpose in view.

3. National, religious, geographical, linguistic and cultural groups do not necessarily coincide with racial groups; and the cultural traits of such groups have no demonstrated connection with racial traits ...

Immediately following the two declarations, UNESCO undertook the publication of a series of studies (*Race and Society*, *Race and Culture*, *Race and Biology*, *Race and Psychology*, etc.), authored by competent figures of the time.[6] The only one of these to have stood the test of time is *Race and History* by Claude Lévi-Strauss. It cannot be said that the relative oblivion into which these studies have fallen is undeserved and unjust, as they contain much in the way of generalizations and good intentions, but little analysis or original ideas.

If one wanted to caricature the UNESCO procedure (the two declarations and the above-mentioned studies), it could be said that it consisted in bringing together eminent specialists and asking them to declare that, regarding the question of race, what was scientific before the war no longer was so. This explains the reactions provoked by the first declaration, among certain scientists who found themselves deprived of their favourite toy – hence the second, vaguer declaration, and the various specific studies that decisively 'drowned the fish' in well-intentioned generalities.

I am not going to criticize the UNESCO initiative here. It was completely justified in declaring that the pre-war notions about race were not scientific – though the interesting question is why they should have seemed scientific at that time. The problem is that this shift was not made on the basis of any new study: from the scientific point of view, race after the war was exactly the same as it had been before the war. What forced the change of discourse, in other words, was not science but Auschwitz; it was impossible to silently overlook the extermination of Jews as it was that of ill and handicapped people.

It was not science that imposed the pre-war racist discourse, no more than it was science that justified the post-war anti-racist discourse. Besides, all that the assembled specialists could say to UNESCO was that the pre-war racist theories had no foundation in science. They were not asked about who had developed them, or why. When, in the 1950 declaration, they wanted to go a bit further in maintaining the intellectual equality of races, some people – those who still maintained the pre-war theories – objected that this was going beyond the limits of the assertions made possible by science. What is perhaps yet more embarrassing is that none of these specialists raised the real question, i.e., whether science had anything to say on this subject, and whether biological categories were in any way applicable to politics.

6 These were gathered together and reissued in the collective work *The Race Question in Modern Science* (London: Allen & Unwin, 1975).

Thus, however justified the UNESCO initiative may have been, it was still far from adequate. Certainly the 'scientific' racism of the pre-war period had to be criticized. But this was done in an underhand way, so as not to discuss it too openly, and above all to avoid any serious analysis of the question, which would have displayed only too clearly where the responsibilities lay – instead of simply attributing them to Hitlerism, or to Gobineau and Vacher de Lapouge. Hence the generalities and fine sentiments of these declarations and studies, which smothered the question without really lancing the abscess.

Yet this did have a certain effect, and as we have said, after the war only a few biologists, above all Anglo-Saxon, continued to represent a more or less overtly racist current. Anti-racist discourse, however, became entangled in scientistic stereotypes. Before the war, biology was seen as supporting racism; after the war, the appeal to biology was meant to support anti-racism. In both cases, biology had nothing to say on the subject. But in both cases, it did what was asked of it, adapting itself to the dominant ideology and providing all the arguments desired.

Not in our genes

As with sociobiology and eugenics, the racist current has recently been somewhat reactivated, in this case by the way in which molecular geneticists have reacted to the theoretical difficulties that their discipline has encountered. They have extended themselves into untimely assertions about the importance of their work, multiplying in a ridiculous fashion 'genes' for diseases, predispositions, and all manner of things. Some biologists, and intellectual and political figures in their wake, concluded that if everything was so hereditary, there was no reason why racial characteristics should be unimportant, in particular as far as intellectual capacities are concerned. (This was already one of the main racist themes in the late nineteenth century; see the various quotations from Büchner, Vogt, and others on pp. 257–60.) Hence, for example, Le Pen's declarations on the inequality of races, or a book such as *The Bell Curve*, which claims to prove the intellectual inferiority of certain races, with the more or less explicit project of using this as a basis for political inequality.[7]

---

7 R. J. Herrnstein and C. Murray, *The Bell Curve: Intelligence and Class Structure in American Life* (New York: The Free Press, 1994).

I do not believe it is necessary here to criticize this work on the unequal intelligence of different races, as it is such a threadbare subject. More than a century after this line of research was initiated, it has still failed to come up with anything serious. It rambles round in circles, in a kind of mania with which certain biologists, psychologists and anthropologists are afflicted.

We can simply recall that the notion of IQ, to which this research often refers, has only a relative value. Certain psychological tests may well make it possible to assess the aptitude of an individual to fulfil a determinate task, but certainly not to 'measure his or her intelligence' (assuming that this has any meaning). In general tests, moreover, questions and answers represent two different things: the questions express the intelligence of the conceivers of the test (or their idea of intelligence), while the answers quantify the capacity of the subject to bend to the fantasies of these conceivers.

Because it is inherently impossible for these tests, in themselves, to distinguish between the innate and the acquired components of what they measure (whether it is 'intelligence' or simply some kind of skill), they are supplemented by studies that resort to the phenomenalist and mathematical methods already discussed, methods inspired by those of pre-war genetics, which are devoid of value in the conditions under which they are used in these studies.

In any event, independent of their lack of scientific foundation, these works are of no interest here, as in a democracy equality is not based on IQ: the countless imbeciles of all races have the same rights as do geniuses (also countless), and it matters little whether imbecility and genius are due to heredity, education, or alcohol. Those who are unhappy about this should meditate on the lessons of history.

Let us finally return to the definition of race and the supposed applications of this definition, in order to sum up the above remarks.

Faced with the resurgence of racist theories, some geneticists have thought it proper to declare that human races do not exist (see Introduction, p. xii above). Julian Huxley already said the same thing back in 1935, referring to comparable principles; only the form of the argument has been modernized in the meantime. These declarations are not new, and they have never prevented the existence of racism.

The box on p. 336 above presents the definition of race agreed by the experts convened by UNESCO. Though now half a century old, and counting a former Nazi doctor among its authors, the definition remains valid today. Certain points in it are certainly debatable, but the essence still holds. It is obviously quite general, and the taxonomic category of race that it defines has a quite relative value, but it is not thereby to be rejected, and its interest is clearly demonstrated by its bearing on the history of human populations. Such classifications have always been subject to criticism (Buffon already criticized that of Linnaeus), and biologists have always known the limits of their validity. But it does not follow that the classifications are devoid of interest, and must be rejected on the grounds that a dodgy politician might misuse them.

Besides, looking more closely, we may note that, for Huxley in 1935 as for geneticists today, the point is less to reject the taxonomic category of race than to change its name: 'ethnic group' or 'population' is now preferred to 'race', with its rather charged past. This is a petty procedure: it does not matter whether one says 'race', 'subspecies', 'variety', or a still vaguer expression, as long as the same thing is denoted. The problem is not one of words. Moreover, vocabulary cannot be changed by waving a magic wand and cautioning people to watch their language.

When the UNESCO declaration was drawn up, the structure of DNA was still unknown; it was not yet even universally accepted that this was the support of heredity. The development of molecular genetics, however, did not change a great deal in terms of the definition of taxonomic categories. Analysis of the genome can certainly provide information on the relationships between populations, and can thus make classifications more precise, but there is nothing it can add – or subtract – from the definition of race.

The emphasis placed today on the genome is even rather harmful, as it ends up by reducing humans – and living beings in general – to their genes. The very slight difference between the genomes of individuals of various races can certainly be used as an argument for the singularity of the human species, this difference being no more pronounced than that between individuals of the same race. But one might equally maintain that the difference between the human genome and that of the chimpanzee is also very slight, being less than 1 per cent. And this opens the way for racist theories which say that if a difference of 1 per cent is sufficient to divide human and chimpanzee, then a difference of 0.1 per cent between whites and blacks is sufficient to make the latter into semi-chimpanzees. This kind of argument turns around all too easily, and can be made to say whatever one likes.

If the genomes of human and chimpanzee differ by 1 per cent, though the respective phenotypes differ far more, this simply shows how absurd it is to reduce living beings to their genomes, whether in the interest of classification or for any other purpose. Taken to its logical conclusion, this resort to genes leads to arguments such as those of Crick, which would make the right of the newborn to humanity subject to a genetic examination (see quotation on p. 219).

In fact, using the argument of the quasi-identity of genomes to combat racism implicitly means basing equality of rights on the relative genetic singularity of the human species. It simply switches the gears of the racist thesis that seeks to base inequality on perceptible anatomical differences such as skin colour or skull shape, and instead justifies inequality by a reductionist prejudice that would ascribe to certain genes a higher dignity – which may be true in genetics, but not in anatomy, let alone in social relations, where skin colour is a matter of more than just the gene for melanin. The fragility of this process is all the more glaring in that the same biologists who have argued that the uniformity of human genomes undermines racist assumptions have also often praised human genetic diversity. In both cases, genetic unity and genetic diversity, this is equally absurd. Equality of rights is no more based on genes than on IQ. In fact, with the best intentions in the world, these geneticists fall back into the excesses of the biology of the late nineteenth century, and for the same reasons.

Against these temptations and attempts, it has to be reasserted that the universality of human rights is not based on the genetic identity of the human species. Such a notion leads straight to the differentiation of social and political rights as a function of variations in the genome – whether these are racial variations or not. It is not up to biology to lay down the law, to make decisions of a political and social order, whether on matters of race or of 'genetic correctness'.

As we have explained, there are two quite distinct social uses of biology (see Part One, p. 6). On the one hand, there are uses, such as Pasteurianism, that are essentially technical, and these are perfectly acceptable and even desirable. On the other hand, there are uses, such as those made of genetics and Darwinism, that claim the right (or even the obligation) to intervene in the political-social order and modify this to make it correspond to a supposedly natural order – which in reality is more like an order of profitability. This second category of social uses is totally unacceptable.

In these matters of society and politics, geneticists have nothing to say; it is up to political philosophers to make comments and recommendations. As these latter keep silent and abandon the field to biologists, which they certainly should not do, I shall attempt, for better or worse, to step into their place and maintain that, although the objective physical and intellectual qualities of individuals may be different – whether this difference is hereditary or acquired – this does not affect these individuals in their essential being, because they cannot be reduced to a set of objective qualities. Persons are not objects, 'human resources' whose profitability or contribution to progress is to be measured. In this respect, they are neither unequal nor different; they are in fact incomparable. And it is because they are incomparable that they are equal, in an equality that is based neither on measurement nor on comparison, but on an equality of dignity and right. Biological criteria have no place here.

BIBLIOGRAPHY

Abbé Le Noir, *Dictionnaire des harmonies de la raison et de la foi, ou Exposition des rapports de concorde et de mutuel secours entre le développement catholique, doctrinal et pratique, du christianisme et toutes les manifestations rationelles, philosophiques, scientifiques, littéraires, artistiques et industrielles, de la nature humaine individuelle et sociale*, vol. 19, col. 1339, Paris: Encyclopédie Migne, 1856.
Adams, Mark B., ed., *The Wellborn Science. Eugenics in Germany, France, Brazil and Russia*, Oxford: Oxford University Press, 1990.
Allen, Garland E., 'The Eugenics Record Office at Cold Spring Harbor, 1990–1940, An Essay in Institutional History,' *Osiris*, 2.1 (1986), 225–64.
Ambroselli, Claire, *L'Éthique médicale,* Paris: Presses Universitaires de France, 1994.
Ammon, Otto, *L'Ordre social et ses bases naturelles, esquisse d'une anthroposociologie*, trans. H. Muffang, Paris: Fontemoing, 1900.
———, 'Des biologists et des races', *La Recherche*, 295, February 1997.
Andrieu, Bernard, 'Le capital génétique,' *La Pensée*, 306 (1996), 75–89.
———, '*Médecin de son corps,* Paris: Presses Universitaires de France,1999.
Anonymous, '*Die Aristocratie des Geistes als Lösung der soxialen Frage*,' Leipzig: Friedrich, undated (1894 or 1895).
Anonymous, 'L'extermination des malades mentaux sous le régime national-socialiste,' *La Raison*, 2 (1951), 14–45.
Anonymous, 'La stérilisation des fous, des infirmes et des pauvres,' *La Raison*, 3 (1951), 33–61.
Apert, Eugéne, *Eugénique et sélection*, Paris: Alcan, 1922.
Ardrey, Robert, *African Genesis*, London: Collins, 1961.
Badiou, Alain, *L'Éthique*, Paris: Hatier, 1993.
———, *Ethics*, trans. by Peter Hallward, London: Verso, 2001.
Bagehot, Walter, *Physics and Politics*, London: H. S. King, 1872.
Baldwin, James Mark, *Darwin in the Humanities*, Baltimore: Review, 1909.
———, *Le Darwinisme dans les sciences morales*, trans. G.-L. Duprat, Paris: Alcan, 1911.
Banton, Michael, ed., *Darwinism and the Study of Society. A Centenary Symposium*, London: Tavistock Publications, 1961.
Baur, Erwin, Eugen Fischer and Fritz Lenz, *Menschliche Erblehre und Rassenhygiene*, Munich: Lehmans, 1921.
Bayle, François, *Croix gammée contre caducée, les expériences humaines en Allemagne pendant la Deuxième Guerre mondiale*, Neustadt (Palatinat): Centre de l'Imprimerie nationale, 1950.
Béjin, André, 'Le sang, le sens et le travail: Georges vacher de Lapouge, darwiniste social fondateur de l'anthropolosociologie,' *Cahiers internationaux de sociologie*, 73 (1982), 323–43.
———, 'Théories socio-politiques de la lutte pour la vie,' in *Nouvelle Histoire des idées politiques*, ed. Pascal Ory, Paris: Hachette-Pluriel, 1996.

Bernardini, Jean-Marx, *Le Darwinisme social en France (1859–1918): fascination et rejet d'une idéologie*, Paris: Centre National de la Recherche Scientifique, 1997.
Beta, Ottomar, *Darwin, Deutschland und die Juden, oder der Juda-Jesuitismus*, Berlin: der Verfasser, 1876.
Biddiss, Michael D., 'Gobineau and the Origins of European Racism,' *Race* 7 (1966), 255–70.
Blumenbach, Johann Friedrich, *De l'unité du genre humain et de ses variétés*, trans. F. Chardel, Paris: Alut, An XIII (1804).
Bouglé, Célestin, *La Démocratie devant la science*, 3rd edn, Paris: Alcan, 1923.
Bouthoul, Gaston, *La Guerre*, 6th edn, Paris: Presses Universitaires de France, 1978.
Boutroux, Émile, Preface to *Le Darwinisme et la Guerre*, by Peter Chalmers-Mitchell, trans. M. Solovine, Paris: Alcan, 1916.
Bowler, Peter J., *Darwin, The Man and His Influence*, Oxford: Blackwell, 1990.
Brix, Emil, ed., *Ludwig Gumplowicz oder die Gesellschaft als Natur*, Vienna: Hermann Bohlaus, 1986.
Büchner, Ludwig, *L'Homme selon la science*, trans. C. Létourneau, Paris: Reinwald, 1870.
Burnet, Frank Macfarlane, *Credo and Comment*, Melbourne: Melbourne University Press, 1979.
Carlson, E. A., 'H. Müller,' in *Dictionary of Scientific Biography*, ed. C. Gillispie, New York: Scribener and Sons,1974.
Carrel, Alexis, *L'Homme, cet inconnu*, Paris: Plon, 1941.
Carol, Anne, *Histoire de l'eugénisme en France*, Paris: Le Seuil, 1995.
Cavalli-Sforza, Luca, *Genes, Peoples and Languages*, New York: North Point Press, 1999.
Chalmers-Mitchell, Peter, *Le Darwinisme et la Guerre*, trans. M. Solovine, Paris: Alcan, 1916.
Chamberlain, Houston Stewart, *The Foundations of the Nineteenth Century*, vol. 1, London: Lane, 1910.
———, *La Genèse du XIXe siècle*, 2 vols., 3rd edn, Paris: Payot, 1913.
Chambon, Philippe and Dodet Betty, 'Le marché des gènes,' *Sciences et Avenir*, 565 (1994): 22–35.
Changeux, Jean-Pierre, *L'Homme neuronal*, Paris: Fayard, 1983.
Claparède, Éduoard, *Comment diagnostiquer les aptitudes chez les écoliers*, Paris: Flammarion, 1929.
Cohen, Francis J. et al., *Science bourgeoise et science prolétarienne*, Paris: Éditions de la Nouvelle Critique, 1950.
Cohen, Daniel, *Les Gènes de l'espoir*, Paris: Lafont, 1993.
Comas, Juan, 'Les mythes raciaux', in UNESCO, *Le Racisme devant la science*, Paris: Unesco-Gallimard, 1960.
Combris, Andrée, *La Philosophie des races du comte de Gobineau et sa portée actuelle*, Paris: Alcan, 1937.
Conry, Yvette, *L'Introduction du darwinisme en France*, Paris: Vrin, 1974.
Corner, G. W., 'Carrel,' in *Dictionary of Scientific Biography*, ed. C. Gillipsie, New York: Scribner and Sons, 1981.
Cowdrey, Edmund V., ed., *Human Biology and Racial Welfare*, New York: Hoeber, 1930.
Cuvier, Georges, *A Discourse on the Revolutions of the Surface of the Globe*, London: Whitaker, Treacher & Arnot, 1829.
———, *Le règne animal distribué d'après son organisation, pour servir de base à l'histoire naturelle des animaux et d'introduction à l'anatomie comparée*, 4 vols., Paris: Déterville, 1817.
Le Dantec, Félix, *La Lutte universelle*, Paris: Flammarion, 1906.
———, *L'Égoisme, base de toute société*, Paris: Flammarion, 1911.
Darré, Richard Alther, *La Race, nouvelle noblesse du sang et du sol*, trans. P. Mélon and A. Pfanstiel, Paris: Sorlot, 1939.

Darwin, Charles, *The Origin of Species*, Harmondsworth: Penguin, 1985.
———, *The Descent of Man, and Selection in Relation to Sex*, Harmondsworth: Penguin, 2004.
Darwin, Charles and Alfred R. Wallace, 'On the Tendency of Species to Form Varieties; and on the Perpetuation of Varieties and Species by Natural Means of Selection,' *Journal of the Proceedings of the Linnean Society (Zoology)*, 3 (1858): 45–62.
Darwin, Leonard, 'Practical Eugenics,' trans. from E. Apert et al., *Eugénique et selection*, Paris: Alcan, 1922.
Davenport, Charles B., *The Feebly Inhibited: Nomadism, or the Wandering Impulse, with Special Reference to Heredity; Inheritance of Temperament, with Special Reference to Twins and Suicides*, Washington: Carnegie Institution, 1915.
———, 'The effects of race intermingling,' *Proceedings of the American Philosophical Society*, 56 (1917): 364–68.
———, *The Genetical Factor in Endemic Goiter*, Washington: Carnegie Institution, 1932.
———, *Heredity in Relation to Eugenics*, New York: Holt, 1911.
Davenport, Charles B., and Morris Steggerda, *Race Crossing in Jamaica*, Washington: Carnegie Institution, 1929.
Dawkins, Richard, *The Selfish Gene*, Oxford: Oxford University Press, 1976.
———, *The Extended Phenotype*, Oxford: Oxford University Press, 1989.
———, *River Out of Eden*, London: Weidenfeld & Nicolson 1995.
Delage, Yves, *L'Hérédité et les grands problèmes de la biologie générale*, Paris: Reinwald-Schleicher, 1903.
Diehl, Karl and Otmar von Verschuer, *Zwillingstuberkulose. Zwillingsforschung und erbliche Tuberkulosedisposition*, Jena: Fischer, 1933.
Dressen, Wili, 'Euthanasia,' in E. Kogon et al., *Nazi Mass Murder*, New Haven: Yale University Press, 1993.
Dreyfus, François-Georges, *Histoire de Vichy*, Paris: Perrin, 1990.
Dumont, Léon, *Haeckel et la théorie de l'évolution en Allemagne*, Paris: Germer-Baillière, 1873.
Edelman, Bernard, and Marie-Angéle Hermitte, *L'Homme, la nature et le droit*, Paris: Bourgois, 1988.
Elissar, Eliahu Ben, *La Diplomatie du IIIe Reich et les Juifs (1933–1939)*, Paris: Bourgois, 1981.
Engels, Friedrich, *The Origin of the Family, Private Property, and The State*, New York: Pathfinder Press, 1972.
Executive Council of the National Front of Democratic Germany (DDR state publication), *The Brown Book, Nazi and War Criminals in the Federal Republic and West Berlin*, Dresden: Zeitim Bild, 1968.
Ferrière, Regis, 'Help and You Shall Be Helped', *Nature*, 393 (1998), 517–19.
Ferro, Marc, ed., *Chronologie universelle du monde contemporain*, Paris: Nathan, 1993.
Fisher, Ronald A., *The Social Selection of Human Fertility*, Oxford: Clarendon Press, 1932.
———, *The Genetical Theory of Natural Selection*, New York: Dover, 1958.
de Fontette, François, *Le Racisme*, Paris: Presse Universitaire de France, 1981.
Friedlander, Henry, *The Origins of Nazi Genocide, from Euthanasia to the Final Solution*, Chapel Hill: The University of North Carolina Press, 1995.
Galton, Francis, 'A Theory of Heredity', *Contemporary Review*, 27 (1875), 80–95.
———, *Inquiries into Human Faculty and its Development*, London: Dent, 1911.
———, *Hereditary Genius, an Inquiry into its Laws and Consequences*, London: Macmillan, 1914.
Gantz, Kenneth Franklin, *The Beginnings of Darwinian Ethics: 1859-1871*, PhD diss., The University of Chicago Libraries, 1937, reprinted from The University of Texas Studies in English, 1939.
Garrod, Archibald Edward, *Inborn Errors of Metabolism*, London: Frowde and Holder and Stoughton, 1909.

Gasman, Daniel, *Haeckel's Monism and the Birth of Fascist Ideology*, New York: Lang, 1998.
———, *The Scientific Origins of National Socialism: Social Darwinism in Ernst Haeckel and the German Monist League*, New York: Lang, 1998.
Gerard, R. W., 'A Biological Basis for Ethics,' *Philosophy of Science*, 1942.
Gillispie, C. C., ed., *Dictionary of Scientific Biography*, New York: Scribner & Sons, 1981.
Ginsberg, Morris, 'Social Evolution,' in M. Banton, ed., *Darwinism and the Study of Society, A Centenary Symposium*, London: Tavistock, 1961.
Girard, Jean, *Considérations sur le loi eugénique allemande de 14 juillet 1933*, Thesis in medicine, Strasbourg, 1934.
Gobineau, Arthur de, *Essai sur l'inégalité des races humaines*, Paris: Belfond, 1966.
Godron, D. A., *De l'espèce et des races dans les êtres organisés de la période géologique actuelle,* Nancy: Grimblot-Raybois, 1848.
Goldschmidt, R., *Im Wandel des Bleibenden. Mein Lebensweg*, Hamburg, 1963.
Gould, Stephen Jay, *The Mismeasure of Man*, New York: W.W. Norton & Company, 1981.
Graham, Loren R., 'Science and Values: The Eugenics movement in Germany and Russia in the 1920s,' *The American Historical Review*, 84:4 (1977), 1133–64.
Grant, Madison, *Le Déclin de la grande race*, Paris. Payot, 1926.
Gros, François, *Les Secrets du gène*, Paris: Odile Jacob/Points-Seuil, 1991.
Gruenberg, H., 'Men and Mice at Edinburgh: Reports From the Genetics Congress,' *Journal of Hereditary*, 30:9 (1939), 371–73.
Gumplowicz, Ludwig, *La Lutte des races*, Paris: Guillaumin, 1893.
———, *Sociologie et politique*, Paris: Giard, 1898.
———, *Outlines of Sociology,* New York: Arno Press, 1975.
Günther, Hans F. K., *Platon, eugéniste et vitaliste*, trans. Elfrida Popelier, Puiseaux: Pardès, 1987.
Guyénot, Émile, *L'Hérédité*, 4th edn, Paris: Doin, 1948.
Haddon, Alfred C., *The Races of Man and their Distribution*, Halifax: Milner, 1912.
———, *Soviet Genetics and World Science*, London: Chatto & Windus, 1949.
Haeckel, Ernst, *The History of Creation*, London: H. S. King, 1876.
———, *Kristalseelen. Studien über das anorganische Leben*, Leipzig: Kröner,1917.
———, *Le Monisme, lien entre la religion et la science*, trans. Georges Vacher de Lapouge, Paris: Schleicher, 1897.
———, *Le Monisme, profession de foi d'un naturaliste*, trans. Georges Vacher de Lapouge, Paris: Schleicher, undated.
———, *Origine de l'homme*, trans. L. Laloy, Paris: Schleicher, 1900.
———, *The Riddle of the Universe*, New York: Prometheus Books, 1992.
Haldane, John Burden S., *The Causes of Evolution*, Ithaca: Cornell University Press, 1966.
———, *Heredity and Politics*, London: Allen & Unwin, 1938.
Hamilton, William D., 'The Genetical Theory of Social Behaviour, I, II', *Journal of Theoretical Biology*, 7:1 (1964), 1–52.
———, 'Selfish and spiteful behaviour in an evolutionary model,' *Nature*, 228 (1970), 1218–20
———, 'Geometry for the selfish herd,' *Journal of Theoretical Biology*, 31:2 (1971), 295–311.
———, 'Selection of Selfish and Altruistic Behaviour in Some Extreme Models,' in J. F. Eisenberg and W. S. Dillon, eds., *Man and Beast: Comparative Social Behavior*, Washington: Smithsonian Institution Press, 1971.
———, 'Altruism and Related Phenomena, Mainly in Social Insects,' *Annual Review of Ecology and Systematics*, 3 (1972), 193–232.
Haycraft, John B., *Darwinism and Race Progress*, London: Sonnenschein, 1894; also New York: Scribner, 1894.
Herrnstein, Richard J. and Charles Murray, *The Bell Curve Intelligence and Class Structure in American Life*, New York: The Free Press, 1994.

Hersch, Jeanne, 'Sur la notion de race,' *Diogéne*, 59 (1967), 125–42.
Hilberg, Raul, *The Destruction of the European Jews*, New Haven: Yale University Press, 2003.
Hillel, Marc, *Au nom de la race*, Paris: Fayard, 1975.
Hofstadter, Richard. *Social Darwinism in American Thought*, Boston: Beacon Press, 1955.
Holmes, Samuel Jackson, *A Bibliography of Eugenics*, Berkeley: University of California Press, 1924.
Huxley, Julian, *Essais d'un biologiste*, trans. J. Castier, Paris: Stock, 1946.
———, *Evolution, the Modern Synthesis*, London: Allen and Unwin, 1942.
———, *La Génétique soviétique et la Science mondiale*, trans. J. Castier, Paris: Stock, 1950.
———, *La Revanche du darwinisme*, Paris: Conférences du Palais de la Découverte, 1945–1948.
———, *The Uniqueness of Man*, London: Chatto & Windus, 1941.
Huxley, Julian., Alfred Haddon, and A.M. Carr-Saunders, *We Europeans*, London: Jonathan Cape, 1935.
———, *Nous Européens,* trans. J. Castier, Paris: Minuit, 1947.
Huxley, Thomas Henry, *Collected Essays*, II (*Darwiniana*), London: Macmillan, 1893.
———, *Evolution and Ethics, and Other* Essays, London: Macmillan, 1893.
Huxley, Thomas Henry, and Julian Huxley, *Evolution and Ethics (1893–1943)*, London: The Pilot Press Ltd, 1947.
Isaac, Jules, *Genèse de l'anti-sémitisme*, Paris: Calmann-Lévy, 1985.
Jacob, François, *The Logic of Living Systems*, trans. Betty E. Spillman, London: Allen Lane, 1974.
———, *The Logic of Life and the Possible and the Actual*, Harmondsworth: Penguin, 1989.
Jacquard, Albert, *Éloge de la différence*, Paris: Seuil, 1978.
Kanner, Leo, 'Exoneration of the Feeble-Minded,' *American Journal of Psychiatry*, 99 (1942), 17–22.
Kaplan, J.-C. and M. Delpech, *Biologie Moléculaire et médecine*, Paris: Flammarion, 1993.
Keith, Arthur, *The Place of Prejudice in Modern Civilization*, London: Williams & Norgate, 1931.
Kennedy, Foster, 'The Problem of Social Control of the Congenital Defective. Education, Sterilization, Euthanasia,' *American Journal of Psychiatry*, 99 (1942), 13–6.
Kevles, Daniel J., *In the Name of Eugenics,* London: Harvard University Press, 1995.
Klee, Ernst, *Euthanasie im NS-Staat, Die Vernichtung lebensunwerten Lebens*, Frankfurt: Fischer, 1983.
———, *La Médecine nazie et ses victims*, trans. O. Manoni, Paris: Solin-Actes Sud, 1999.
Kogon, Eugen, Herman Langbein and Adalbert Ruckerl, *Les Chambres á gaz, secret d'État*, trans. H. Rollet, Paris: Minuit, 1984.
Kohler, Robert E., 'The Management of Science: the Experience of Warren Weaver and the Rockefeller Foundation Programme in Molecular Biology,' *Minerva,* 14 (1976), 279–306.
Kropotkin, Peter, *La Morale anarchiste*, Paris: Groupe libertaire Kropotkine, 1969.
Kropotkin, Peter, *Mutual Aid: A Factor of Evolution*, Montreal: Black Rose Books, 1989.
Kühl, Stefan, *The Nazi Connection. Eugenics, American Racism, and German National Socialism,* Oxford: Oxford University Press, 1994.
Lafont, Max, *L'Extermination douce, La mort de 40 000 malades mentaux dans les hôpitaux psychiatriques en France sous le régime de Vichy*, Ligné: Éditions de l'AREFPPI, 1987.

de Lamarck, Jean-Baptiste Pierre Antoine de Monet, *Philosophie zoologique*, Paris: GF-Flammarion, 1994.
Lange, Maurice, *Le comte Arthur de Gobineau, étude biographique et critique*, Strasbourg: Publications de la Faculté des Lettres de l'Université de Strasbourg, 1924.
Lanessan, Jean-Louis de, *La Lutte pour l'existence et l'Association pour la lutte*, Paris: Doin, 1881.
Laughlin, Harry H., 'Die Entwicklung der gesetzlichen rassenhygienischen Sterilisierung in den Vereinigten Staaten,' *Archiv für Rassen- und Gesellschaftsbiologie*, F. Lehmanns Verlag, 1929, 253–62.
Laurent, Goulven, *Paléontologie et évolution en France, 1800*–1860, Paris: Éditions du Comité des Travaux Historiques et Scientifiques, 1987.
Lecourt, Dominique, *Proletarian Science?: The Case of Lysenko*, London: New Left Books, 1977.
Lemaine, Gérard, and Benjamin Matalon, *Hommes supérieurs, hommes inférieurs? La controverse sur l'hérédité de* l'intelligence, Paris: Armand Colin, 1985.
Lémonon, Michel, *Le Rayonnement du gobinisme en Allemagne*, 2 vols., typescript thesis, University of Strasbourg II, 1971.
Lenoir, Noëlle, *Le Monde*, 9 December 1998.
Lester, Paul and Jacques Millot, *Les Races humaines*, Paris: Armand Colin, 1936.
Lettich, André, *Trente-quatre mois dans les camps de concentration*, Tours: Imprimerie, 1946.
Lévi-Strauss, Claude, *Race et Historie*, Paris: Gontheir, 1970.
Lifton, Robert Jay, *Les Médecins nazis: le meurte médical*, Paris: Laffont, 1989.
Lindqvist, Sven, *Exterminez toutes ces brutes*, trans. A. Gnaedig, Paris: Le Serpent á Plumes, 1998.
Linné, Charles, *Systema naturae per regna tria naturae secundum classes, ordines, genera, species cum characteristibus, differentiis, synonymis, locis*, 2 vol. Holm: Laurentii Salvii, 1758–9.
Londres, Albert, *Oeuvres completes*, Paris: Arléa, 1992.
Lucas, Prosper, *Traité philosophique et physiologique de l'hérédité naturelle dans les états de santé et de maladie du système nerveux, avec l'application méthodique des lois de la procréation au traitement général des affections dont elle est le principe*, Paris: Éditions Baillière, 1847–50.
Lukács, Georg, *The Destruction of Reason*, London: Merlin Press, 1980.
Luçat, François, *La Science suicidaire, Athéns sans Jérusalem*, Paris: Guibert, 1999.
Lysenko, Trofim Dennissovitch, *Agrobiologie*, Moscow: Éditions en Langues Étrangères, 1953.
MacBride, E. W., 'Cultivation of the Unfit,' *Nature* 137:3454 (1936), 44–5.
MacKenzie, Donald, 'Eugenics in Britain,' *Social Studies of Science*, 6 (1976), 499–532.
Malthus, Thomas R., *An Essay on the Principle of Population*, Cambridge: Cambridge University Press, 1992.
Marchand, Henri-Jean, *L'Évolution de* l'idée *eugénique*, PhD. diss., Bordeaux, 1933.
Martial, René, *La Race Française*, 3rd edn, Paris: Mercure de France, 1934.
———, *Race, hérédité, folie, Étude d'anthroposiologie appliquée à l'immigration*, 3rd edn, Paris: Mercure de France, 1938.
Marx, Karl, and Frederick Engels, *Selected Correspondence*, London: Lawrence & Wishart, 1941.
Massin, Benoit, 'Anthropologice raciale et national-socialism: heurs et malheurs du paradigme de la "race",' in Olff-Nathan, J., *La Science sous le Troisième Reich*, Paris: Seuil, 1993.
———, De l'eugénisme à l'opération euthanasia: 1890–1945,' *La Recherche* 227 (December 1990): 1562–8.
———, 'La science nazie et l'extermination des marginaux,' *L'Histoire*, 217 (January 1998): 52–9.

———, 'L'Euthanasie psychiatrique sous le IIIe Reich, La question de l'eugénisme,' *L'Information psychiatrique*, 8 (October 1966): 811–22.
———, 'From Virchow to Fischer: Physical Anthropology and "Modern Race Theories" in Wilhelmine Germany', in George W. Stocking, Jr., ed., *"Volksgeist" as Method and Ethic, Essays on Boasian Ethnography and the German Anthropological Tradition*, Madison: University of Wisconsin Press, 1996.
Matthew, Patrick, *Naval Timber and Arboriculture*, Edinburgh: Black, 1831.
Maupertius, Pierre Louis, *Vénus physique*, Paris: Aubier-Montagne, 1980.
———, *Système de la nature*, Paris: Vrin, 1984.
Medvedev, Zhores A., *The Rise and Fall of T. D. Lysenko*, New York: Columbia University Press, 1969.
Milliez, Jacques, *L'Euthanasie du foetus*, Paris: Odile Jacob, 1999.
Morel, Benedict-Augustin, *Traité des dégénérescences physiques, intellectuelles et morales de l'espèce humaine et des causes qui produisent ces variété maladives*, Paris: Baillière, 1857.
Morgan, Thomas H. et al., *The Mechanism of Mendelian Heredity*, New York: Johnson Reprint Corporation, 1972.
Mourre, Michel, *Dictionnaire d'histoire universelle*, Paris: Éditions universitaires, 1968.
Muller, Hermann J., *Out of the Night: A Biologist's View of the Future*, London: Gollancz, 1936.
Müller-Hill, Benno, *Science nazie, science de mort, L'extermination des Juifs, Tziganes et des maladies mentaux de 1933 à 1945*, Paris: Odile Jacob, 1989.
Naccache, Bernard, *Marx critique de Darwin*, Paris: Vrin, 1980.
Nature, 'Announcements', 1936, vol. 137, no. 3475.
Nisot, Marie-Thérèse, 'La sterilisation des anormaux,' *Mercure de France*, 1929.
Noble, Elizabeth Shor, 'C. B. Davenport,' in C. Gillispie, ed. *Dictionary of Scientific Biography*, New York: Scribner & Sons, 1874.
Novicow, Jacques, *La Critique du darwinisme social*, Paris: Alcan, 1910.
Nowak, Martin A., Karl Sigmund, 'Evolution of Indirect Reciprocity by Image Scoring,' *Nature*, 393 (1998), 573–77.
Olff-Nathan, Josiane, ed., *La Science sous le Troisième Reich*, Paris: Le Seuil, 1993.
Ory, Pascal, *Nouvelle histoire des idées politiques*, Hachette-Pluriel, 1996.
Pasteur, Louis, *Écrits scientifiques et médicaux*, André Pichot, ed., Paris: Flammarion, 1994.
Petzoldt, Joseph, *Sonderschulen für hervorragend Befähigte*, Leipzig: Drud und Derlag von B.G. Teubner, 1905.
Pichot, André, 'Des biologistes et des races,' *La Recherche*, 295 (1997), 9–10.
———, *L'Eugénisme, ou les Généticiens saisis par la philanthropie*, Paris: Hatier, 1995.
———, *Histoire de la notion de vie*, Paris: Gallimard, 1993.
———, *Histoire de la notion de gène*, Paris: Champs-Flammarion, 1999.
———, 'Racisme et biologie,' *Le Monde*, 4 October 1996.
Pius IX, *Encyclique Quanta cura et Syllabus (1864)*, documents collected by Jean-Robert Amogathe, Paris: Pauvert, 1967.
Piux IX, *Encyclique Casti connybii sur le mariage chrétien (31 December 1930)*, Fr. trans., Paris: Spes, 1964.
Plichta, Martin, 'Les Roms tchèques dénoncent la "ségrégation raciale systématique à l'école",' *Le Monde*, 26 June 1999.
Poliakov, Léon, *Le Bréviaire de la haine*, Paris: Calmann-Lévy, 1951.
———, *Le Mythe aryen*, Paris: Pocket, 1994.
Popenoe, Paul, 'The German Sterilization Law,' *Journal of Heredity*, 25 (1934), 257.
Popenoe, Paul and Roswell H. Johnson, *Applied Eugenics*, New York: Macmillan, 1918.
Portmann, Heinrich, *Cardinal von Galen*, Norwich: Jarrold, 1957.
Provine, William B., 'Geneticists and the Biology of Race Crossing,' *Science*, 182 (1973), 790–6.

Quatrefages, Armand de, *L'Espèce humaine*, 6th edn, Paris: Germer-Baillière, 1880.
Quételet, Adolphe, *A Treatise on Man and the Development of His Faculties*, Gainesville, FL: Scholars' Facsimiles and Reprints, 1969.
Ribot, Théodule, *L'Hérédité, étude psychologique sur ses phénomènes, ses lois, ses causes, ses consequences*, Paris: Librairie Philosophique de Ladrange, 1873.
Richard, Gaston, *L'Idée d'evolution dans la nature et dans l'histoire*, Paris: Alcan, 1903.
Richter, Claire, *Nietzsche et les theéories biologiques contemporaines,* Paris: Mercure de France, 1911.
Rifkin, J., *The Biotech Century*, New York: Putnam, 1998.
Robert, Jean-Louis, '1841, la première loi sociale en France,' *Le Monde*, 20 April 1999.
———, 'Enfermer les démunis,' *Le Monde*, 16 November 1999.
Roberts, Windsor H., *The Reaction of American Protestant Churches to the Darwinian Philosophy, 1860–1900*, PhD diss., University of Chicago, distributed by The University of Chicago Libraries, 1938.
Roger, Jacques, *Pour une histoire des sciences à part entière*, Paris: Albin Michel, 1995.
Rostand, Jean, *L'Atomisme en biologie*, Paris: Gallimard, 1956.
———, *Biologie et medicine*, Paris: Gallimard, 1939.
———, *L'Homme, introduction à l'étude de la biologie humaine*, Paris: Gallimard, 1942.
———, *De la mouche à l'homme*, Paris: Fasquelle, 1930.
———, *La Nouvelle Biologie*, Paris: Fasquelle, 1937.
———, *Peut-on modifier l'homme?*, Paris: Gallimard, 1956.
Rouche, Max, *Herder précurseur de Darwin, Histoire d'un mythe*, Paris: Les Belles Lettres, 1940.
Roux, Wilhelm, *Der Kampf der Theile im Organismus, Gesammelte Abhandlungen über Entwicklungsmechanik der Organismen*, 2 vols., Leipzig: Engelmann, 1895.
Rovan, Joseph, *Histoire de l'Allemagne des origines à nos* jours, Paris: Points-Seuil, 1998.
Royer, Clémence, 'Preface' in Charles Darwin, *L'Origine de espèces*, Paris: Reinwald, 1882.
Ruffié, Jacques, *Naissance de la médecine predictive*, Paris: Odile Jacob, 1993.
Ruffié, Jacques, and J.-C. Sournia, *Les Épidémies dans l'histoire de l'homme*, Paris: Flammarion, 1984.
Rupke, Nicolaas A., ed., *Science, Politics and the Public Good: Essays in Honour of Margaret Gowing*, London: MacMillan Press, 1988.
Sahlins, Marshall, *The Use and Abuse of Biology*, Ann Arbor, MI: University of Michigan Press, 1976.
Schleicher, August, *La Théorie de Darwin et la science du langage* (1863) – *De l'importance du langage pour l'histoire naturelle de l'homme* (1865), trans. M. de Pommayrol, Paris: Vrin, 1980.
Shirer, William L., '"Mercy Deaths" in Germany', *Reader's Digest*, 1941, 55–8.
———, *Berlin Diary*, New York: Knopf, 1941.
Simpson, G. G., *L'Évolution et sa signification*, trans. A Ungar-Levillain and F. Bourliére, Paris: Payot, 1951.
Smith, Adam, *The Wealth of Nations*, New York: Penguin, 1982.
Smith, Annette, *Gobineau et l'histoire naturelle*, Paris: Droz, 1984.
Sorokin, Pitirim A., *Contemporary Sociological Theories*, New York: Harper, 1928.
Soupault, Robert, *Alexis Carrel*, Paris: Plon, 1952.
Stark, M. B., 'An Hereditary Tumour in the Fruit Fly Drosophilia,' *Journal of Cancer Research*, III (1918), 279–300.
Stocking, George W. Jr., ed., *'Volksgeist' as Method and Ethic, Essays on Boasian Ethnography And the German Anthropological Tradition*, Madison: University of Wisconsin Press, 1996.

Sutter, Jean, *L'Eugénique, problème, méthodes, résultats*, Cahier no. 11 of the Institut National d'Études Démographiques, Paris: Presses Universitaires de France, 1950.
Taguieff, Pierre-André, 'L'eugénisme, objet de phobie idéologique,' *Esprit*, 156 (1989), 99–115.
———, *Force du préjugé, essai sur le racisme et ses doubles*, Paris: La Découverte, 1987.
———, 'Retour sur l'eugénisme. Question de définition (réponse à J. Testart),' *Esprit*, 200 (1994), 198–214.
Taine, H., *De l'intelligence*, 2 vols., Paris: Hachette, 1928.
Ternon, Yves and Socrate Helman, *Histoire de la médecine SS, ou le mythe du racisme biologique*, Tournai: Casterman, 1969.
———, *Le Massacre des aliénés, des théoriciens Nazis aux practiciens SS*, Tournai: Casterman, 1971.
———, *Les Médecins allemands et le National-socialisme, les métaphorphoses du darwinisme*, Tournai: Casterman, 1973.
Ternon, Yves., *L'État criminel, Les génocides au XXe siècle*, Paris: Seuil, 1995.
Testart, Jacques, *Des hommes probables, de la procréation aléatoire à la reproduction normative*, Paris: Seuil, 1999.
———, 'Les risques de purification génique: questions á Pierre-André Taguieff,' *Esprit*, 199 (1994), 178–84.
———, 'Sur l'eugénisme,' *Esprit*, 205 (1994), 175–82.
Thomas, Jean-Paul, *Les Fondements de l'eugénisme*, Paris: Presses Universitaire de France, 1995.
Thuillier, Pierre, *Les biologistes vont-ils prendre le pouvoir?*, Brussels: Complexe, 1981.
———, 'La génétique et le pouvoir, ou les rêves fous d'un prix Nobel,' *La Recherche*, 119 (1981), 231–3.
———, 'Les biologistes vont-ils prendre le pouvoir?,' *La Recherche*, 98 (1979), 302–6.
———, 'Les scientifiques et le racisme,' *La Recherche*, 45 (1974), 456–66.
———, 'La tentation de l'eugénisme,' *La Recherche*, 155 (1984), 743–8.
UNESCO, *The Race Concept*, Paris: 1953.
———, *Le Racisme devant la science*, Paris, Unesco-Gallimard: 1960.
———, *The Race Question and Modern Science*, London: Allen & Unwin, 1975.
Uschmann, George, 'Haeckel,' in C. Gillispie, ed., *Dictionary of Scientific Biography*, New York: Scribner & Sons, 1981.
Vacher de Lapouge, Georges, *L'Aryen, son rôle social*, Paris: Fontemoing, 1899.
———, *Les Sélections sociales*, Paris: Fontemoing, 1896.
———, *Race et milieu social*, Paris: Rivière, 1909.
Vapereau, G., *Dictionnaire universel des contemporains, contenant toutes les personnes notables de la France et des pays étrangers*, 6th edn., Paris: Hachette, 1893.
Veuille, Michel, *La Sociobiologie*, Paris: Presses Universitaire de France, 1986.
Villermé, Louis-René, *Tableau de l'état physique et moral des ouvriers employés dans les manufactures de coton, de laine et de soie* (1840), texts selected and presented by Yves Tyl, Paris: Union Générale d'Édition (10–18), 1971.
Verschuer, Otmar von, *Manuel d'eugénique et d'hérédité humaine*, trans. G. Montandon, Paris: Masson, 1943.
Vogt, Karl, *Leçons sur l'homme, sa place dans la création et dans l'histoire de la terre*, trans. J. J. Moulinié, Paris: Reinwald, 1878.
Vogt, William, *The Road to Survival*, London: Gollancz, 1948.
Vries, Hugo De, *The Mutation Theory*, vol. 1, New York: Kraus Reprint Co., 1969.
Wallace, Alfred Russel, 'On the Tendency of Species to Form Varieties; and On the Perpetuation of Varieties and Species by Natural Means of Selection', *Journal of the Proceedings of the Linnaean Society (Zoology)*, 1858: 3.
———, *Contributions to the Theory of Natural Selection*, London: Macmillan, 1871.
Ward, Lester F., *Pure Sociology*, London: Macmillan, 1903.
Weil, Simone, 'Refléxions sur la barbarie,' in *Oeuvres*, Paris: Gallimard, 1999.

Weindling, Paul and Benoit Massin, eds., *L'Hygiène de la race, hygiène raciale et eugénisme médical en Allemagne*, vol. 1, Paris: La Découverte, 1998.
Weingart, Peter et al., *Rasse, Blut und Gene, Geschichte der Eugenik und Rassenhygiene in Deutschland*, Frankfurt: Suhrkamp, 1988.
Weismann, August, *The Germ-Plasm, a Theory of Heredity*, trans. W. Newton Parker and H. Rönnfeldt, London: Walter Scott, 1893.
———, *Essays on Heredity and Kindred Biological Problems*, Oxford: Clarendon Press, 1899.
Weiss, Sheila Faith, 'The Race Hygiene Movement in Germany, 1904–1945,' in M. B. Adams, ed., *The Wellborn Science, Eugenics in Germany, France, Brazil and Russia*, Oxford: Oxford University Press, 1990.
Wiesenthal, Simon, *The Murderers Among Us*, London: Heinemann, 1967.
Wilson, Edward O., *Sociobiology, the New Synthesis*, Cambridge, MA: Harvard University Press, 1972.
Zaloszyc, Armand, *Le Sacrifice au dieu obscur, ténèbre et pureté dans la communauté*, Nice: Z'éditions, 1994.

INDEX